RESEARCH TECHNIQUES in NONDESTRUCTIVE TESTING

Edited by

R. S. SHARPE

Head of the Nondestructive Testing Centre,
Atomic Energy Research Establishment,
Harwell, Berkshire, England

 1970

ACADEMIC PRESS
London and New York

ACADEMIC PRESS INC. (LONDON) LTD.
Berkeley Square House
Berkeley Square
London, W1X 6BA

U.S. Edition published by

ACADEMIC PRESS INC.
111 Fifth Avenue
New York, New York 10003

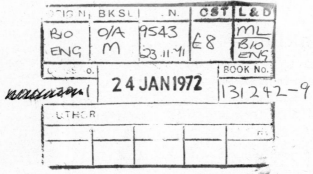
Copyright © 1970 By ACADEMIC PRESS INC. (LONDON) LTD.

Library of Congress Catalog Card Number: 72-109038
SBN: 12-639050-9

Printed in Great Britain by
HARRISON AND SONS LIMITED
BY APPOINTMENT TO HER MAJESTY THE QUEEN,
PRINTERS, LONDON, HAYES (MIDDX.) AND HIGH WYCOMBE

List of Contributors

E. E. ALDRIDGE, *Electronics and Applied Physics Division, Atomic Energy Research Establishment, Harwell, Berkshire, England*

F. L. BECKER, *Battelle Memorial Institute, Pacific Northwest Laboratory, Richland, Washington, U.S.A.*

H. BERGER, *Argonne National Laboratory, Argonne, Illinois, U.S.A.*

W. H. B. COOPER, *Electronics and Applied Physics Division, Atomic Energy Research Establishment, Harwell, Berkshire, England*

C. W. COX, *Department of Electrical Engineering, South Dakota School of Mines and Technology, Rapid City, South Dakota, U.S.A.*

D. S. DEAN, *Rocket Propulsion Establishment, Westcott, Aylesbury, Buckinghamshire, England*

A. E. ENNOS, *Division of Optical Metrology, National Physical Laboratory, Teddington, Middlesex, England*

O. R. GERICKE, *Department of the Army, Army Materials and Mechanics Research Center, Watertown, Massachusetts, U.S.A.*

R. HALMSHAW, *Royal Armament Research and Development Establishment, Fort Halstead, Sevenoaks, Kent, England*

P. H. HUTTON, *Battelle Memorial Institute, Pacific Northwest Laboratory, Richland, Washington, U.S.A.*

J. E. JACOBS, *BioMedical Engineering Center, Technological Institute, Northwestern University, Evanston, Illinois, U.S.A.*

L. A. KERRIDGE, *Rocket Propulsion Establishment, Westcott, Aylesbury, Buckinghamshire, England*

W. D. LAWSON, *Royal Radar Establishment, Malvern, Worcestershire, England*

H. L. LIBBY, *Battelle Memorial Institute, Pacific Northwest Laboratory, Richland, Washington, U.S.A.*

R. N. ORD, *Battelle Memorial Institute, Pacific Northwest Laboratory, Richland, Washington, U.S.A.*

R. L. RICHARDSON, *Battelle Memorial Institute, Pacific Northwest Laboratory, Richland, Washington, U.S.A.*

J. W. SABEY, *Royal Radar Establishment, Malvern, Worcestershire, England*

D. L. WAIDELICH, *Department of Electrical Engineering, University of Missouri, Columbia, Missouri, U.S.A.*

Preface

The broad objective I set for this book was to include between single covers a collection of informative papers reviewing the current scientific understanding of those techniques and principles on which forward thinking and future practice in materials inspection will be based. To achieve this objective, I have invited chapters from authors who can review, from their own first-hand experience, topics in which there is significant research activity and which—by a suitably liberal definition of the scope of the subject—can be considered within the general framework of nondestructive testing.

Nondestructive testing is the technology by which standards of materials quality, components reliability and systems safety are monitored and maintained. The economic and commercial significance of these factors is becoming increasingly realized and standards and specifications are being continually improved. This has revealed inadequacies in some of the nondestructive testing techniques conventionally being used in current industrial practice and shortcomings in the understanding of the principles on which they are based.

I am convinced that nondestructive testing technology will improve its effectiveness and increase its recognition, if more scientific effort is directed towards developing techniques which are less subjective, less dependent on the operator, more quantitative and more suited to automated production control; removing, in fact, the aura of suspicion and scepticism that still surrounds the subject in many management circles. As one of the authors so rightly says, " Tests . . . yield signals that are rich in information, but much of it comes as by a foreign language for which a translating key is not yet available ". I hope that within these chapters some keys will be found to the language barriers that exist between the scientists and technologists caught up in the subject, and the deeper barriers of communication and understanding that so often exist between the technologists and their managements.

I have been privileged to act as Head of the Harwell Nondestructive Testing Centre since it was set up in 1967 to provide a centralized research and development unit for British Industry, and hence know only too well how big and how real the gulf is between the enthusiasm and ingenuity of the laboratory and the practical requirements and harsh economic realities of the industrial inspection department. Because of this problem I have persuaded the authors wherever possible to volunteer pertinent information justifying their development programmes, and factual information indicating present and potential applications. Indeed, I have tried myself to emphasize an essentially practical purpose by indexing the text only in relation to applications.

I trust that the book will help to arouse a wider scientific interest in many quarters in the subject of nondestructive testing and be a source of ideas and contacts to forward-thinking research staffs, design engineers, and quality control managers in industry, a source of inspiration to those responsible for setting up University courses in materials science and a source of relief to the mechanical and constructional engineer that some of his future inspection problems are being anticipated before they arise.

Fortunately there is a strong international friendship amongst those who have an interest in, and concern for, the scientific development of non-destructive testing techniques and my task of editing this book has been made easy by being able to prevail upon this close circle of friends to obtain authors for the various chapters. I would like to thank them all for their co-operation and I hope the sense of urgency which I have tried to inject into my correspondence with them has not irrevocably strained the bonds between us!

January, 1970   R. S. SHARPE

Contents

Chapter 1. Acoustic Emission

P. H. HUTTON AND R. N. ORD

Chapter 2. Ultrasonic Spectroscopy

O. R. GERICKE

Chapter 3. Ultrasound Imaging Systems

J. E. JACOBS

Chapter 4. Ultrasonic Critical Angle Reflectivity

F. L. Becker and R. L. Richardson

Chapter 5. Ultrasonic Holography

E. E. Aldridge

Chapter 6. Optical Holography and Coherent Light Techniques

A. E. Ennos

Chapter 7. Data Handling Techniques

C. W. Cox

Chapter 8. Direct-View Radiological Systems

R. HALMSHAW

Chapter 9. Neutron Radiography

H. BERGER

Chapter 10. A Communication Network Approach to Sub-millimetric Wave Techniques in Nondestructive Testing

W. H. B. COOPER

Chapter 11.   Multiparameter Eddy Current Concepts

H. L. Libby

Chapter 12.   Pulsed Eddy Currents

D. L. Waidelich

Chapter 13.   Microwave Techniques

D. S. Dean and L. A. Kerridge

Chapter 14.   Infrared Techniques

W. D. Lawson and J. H. Sabey

CHAPTER 1

Acoustic Emission

P. H. HUTTON AND R. N. ORD

*Battelle Memorial Institute, Pacific Northwest Laboratory,
Richland, Washington, U.S.A.*

I. INTRODUCTION AND BACKGROUND

Industry in general has a continuing interest in and need for new and improved methods of assessing the integrity of structures. "Structures" in this sense include physical assemblies of solid materials ranging from items such as a small capsule to contain a radioactive isotope to assemblies such as a bridge or an aircraft frame or a nuclear reactor coolant system. A new technique with exciting potential in this area consists of detection and analysis of "acoustic emission" in solids.

Acoustic emission is a phenomenon arising from energy released as a solid material undergoes plastic deformation and fracture. Part of this energy is converted to elastic waves which propagate through the material and can be detected at the material surface using high sensitivity sensors. Two outstanding features of acoustic emission as an integrity monitoring device are the ability to detect crack formation or movement at the time it occurs and to accomplish this remotely. Also, simultaneous interrogation of multiple sensors (three or more) provides the necessary information to locate the source of the signals by triangulation.

Some of the more significant of the many potential applications of acoustic emission are:

(a) Continuous surveillance of nuclear reactor primary pressure boundaries to detect and locate active flaws;

(b) Detection of incipient fatigue failure in aircraft structures;

(c) Monitor weldments for cracking during weld cooling;

(d) To detect penetration of space vehicles by micrometeorites;

(e) To determine the onset of stress corrosion cracking in susceptible structures;

(f) To serve as a study tool for investigation of fracture mechanisms and material behavior

The first basic study of the acoustic emission phenomenon appears to have been done in the late 1940's and early 1950's in the United States and Germany (Mason et al., 1948; Kaiser, 1953). Efforts to develop the potential of acoustic emission for high resolution, nondestructive surveillance of structures and pressure systems to detect growing flaws, however, are of more recent origin. To the author's knowledge, the first application of acoustic emission as a true surveillance device was made by Aerojet General Corporation in the United States in 1964 (Green et al., 1964). In this application it was used as a means of detecting flaw growth during hydrostatic testing of Polaris missile chambers. When a growing flaw was detected in this manner, the chamber could be depressurized in time to avoid failure and permit repair of the faulty area, thus salvaging a chamber which otherwise might have been destroyed in test.

The Pacific Northwest Laboratories of Battelle Memorial Institute (Battelle-Northwest) at Richland, Washington, has carried on an active research and development program in the area of acoustic emission since early 1966. The major effort has been one sponsored by the United States Atomic Energy Commission's (USAEC's) Division of Reactor Development and Technology. This program, which has continued since February 1966, is aimed at development of technology and instrumentation required to apply acoustic emission as a reliable continuous integrity monitor for the primary pressure boundaries of nuclear power reactors. Three levels of refinement are visualized in such a system. The first level would detect and locate growing flaws under hydrostatic test conditions; the second would detect and locate growing flaws under on-line reactor operating conditions;

the third would detect, locate, and describe growing flaws under on-line reactor operation conditions. Extending the application of acoustic emission surveillance from monitoring hydrostatic testing of vessels to continuous monitoring of an operating system such as a nuclear reactor primary coolant system imposes significantly more demanding requirements. These requirements include such things as the necessity for sensors to function for periods of at least $1\frac{1}{2}$ years to 2 years in an environment of nuclear radiation and temperature in the 600-700 °F range, and the ability to distinguish acoustic emission in the presence of intense background noise.

A companion program, also sponsored by the USAEC, is in progress at the Phillips Petroleum Co.'s Idaho Falls, Idaho Division. Its objective is to develop plant application techniques to utilize the technology developed under the Battelle-Northwest program. Some of the other locations conducting active work in the investigation of acoustic emission are: Lawrence Radiation Laboratory at Livermore, California; Aerojet General Corporation, Sacramento, California; Teledyne Materials Research, Waltham, Massachusetts; and the University of Michigan, Ann Arbor, Michigan.

A. State-of-the-Art Summary

Acoustic emission technology and instrumentation has been developed to the stage where it can be reliably used in applications such as detection and location of flaw growth in pressure containers undergoing hydrostatic testing, and detection of flaw formation in welds during welding and cooling phases (Day, 1968; Jolly, 1968). It can also be applied to surveillance of localized areas of known high potential for flaw development in operating systems such as the primary coolant systems of a nuclear reactor. Additional development work, however, is required to apply the technique to general surveillance of a complete reactor pressure system during operation with a high degree of confidence that all active flaws will be detected without generating false indications that cause unnecessary interruption of reactor operation. Improved detection methods must be adapted to field use, data analysis methods need to be optimized, and maximum detection range with various geometries must be established. Also, a "catalog" of emission characteristics is required to establish a set of standards for juding the significance of a given quantity of acoustic emission data.

Evidence indicates that the acoustic emission signal, at least in steels, is a very short period transient—less than a 0.03 μsec period. Wave propagation studies show that use of a sensing technique which is sensitive primarily to one propagation mode offers a much shorter, better defined, detected signal.

High sensitivity transducers using PZT-5A piezoelectric crystals have been fabricated which are sensitive primarily to Rayleigh and shear waves. These are suitable for long term service at temperatures up to about 625°F. Lithium niobate piezoelectric material offers a possibility of extending the temperature range of these transducers to at least 1500°F.

Solid state signal conditioning and processing instrumentation has been developed. This equipment is rugged, small and lightweight to facilitate easy transportation to field test locations. The system consists of a tuned, 400 kHz bandwidth, low noise preamplifier with a voltage gain of 100 to 150. This operates into a filter-amplifier combination which provides additional voltage gain ranging from 10 to 100. At this point, the signal is split going both to a magnetic tape recorder for raw data storage and to a detector-count rate section which generates data on emission rate and total number of emissions. Emission rate and total number of emissions are read out on a strip chart recorder. Source location is accomplished by interrogation of the data stored on magnetic tape. Development of an instrument to automatically determine the difference in time of signal arrival at various sensors for use in source location triangulation is nearing completion. This information can then be used to either manually determine source location or perform the determination with a computer. This equipment will eliminate the requirement to store data on magnetic tape.

Analysis of nuclear reactor background noise during operation has shown that acoustic emission detection on an operating plant involving turbulent fluid flow and cavitation must be accomplished in the low megahertz frequency range to avoid noise interference. Laboratory tests have demonstrated that acoustic emission detection in the frequency range of 1–3 MHz is relatively immune to interference from hydraulic noise including moderate cavitation. At the same time, this frequency range is low enough to avoid the extreme signal attenuation with distance in the material which is experienced at higher frequencies.

Acoustic emission detection has been successfully applied to field tests such as pipe rupture studies performed with 24 in. diam., 1.7 in. wall pipe specimens, surveillance of known flaws in a test pressure vessel, and monitoring suspected slow fatigue crack growth in reactor process piping. The technique has also been demonstrated to be applicable to detection of flaw formation in welds, stress corrosion cracking, and fatigue failures. The resolution attainable appears to significantly exceed that of any of the conventional nondestructive testing techniques such as ultrasonics, eddy current, and radiography.

II. THEORETICAL ASPECTS AND PHYSICAL PRINCIPLES

Considerable evidence has been accumulated by a number of researchers to indicate that most of the acoustic emission observed from metals can be related to dislocation movement and fracture when the material is undergoing strain. This evidence has led us to assume that the initial elastic–plastic wave is a discrete phenomenon and can be treated as a nearly point source for detection frequencies in the low megahertz region. However, propagation of these disturbances in a finite media produces some alteration to the signal and it is this modified signal sensed by a transducer that we term acoustic emission.

The signals are detected by appropriately positioned sensors which, due to their finite size, are simultaneously sensitive to a range of angles. Because of this angular aperture, signals having ray paths other than those on a straight line between the source of the emission and the sensor are detected. For a finite thickness material, the total emission signal can be a combination of many ray path reflections between boundaries of the sample.

For a point receiver, signals are detected having time delays ranging from a minimum corresponding to the direct rays, to a maximum corresponding to rays which reflect between boundaries an infinite number of times. The signal time length could actually then be infinite if attenuation and spherical wave spreading did not cause the eventual loss of energy. The divergence angle for a sensor of diameter D (or other controlling dimension) is given by the relation

$$\sin \alpha \approx \frac{1 \cdot 22 V'}{fD}$$

where α is the half angle of the beam, V' is the couplant velocity, and f is the frequency. This relationship shows that as the frequency is decreased, finite size sensors appear increasingly more like point receivers, thereby being more susceptible to signal time spreading. Conversely, as the frequency is increased or the sensor size is increased (or both) the angular aperture decreases. For very high frequencies and very large receiver sizes angle selectivity could be limited to only direct ray paths. This, of course, would reduce the signal duration from one emission event to a minimum value.

It is important in certain applications to accurately count acoustic emission events. It is obvious that this becomes more difficult when signals overlap in time due to increased duration. Also, in other applications it is essential to locate the emission sites as closely as possible. Since signals from successive reflections tend to obscure first arrival information it also becomes more difficult to locate emission sites by triangulation.

From the above discussion it is evident that sensors should be made as large, and detection frequencies as high as possible. However, as the sensor size is increased, practical limits and problems with coupling energy out of the specimens are eventually encountered. Also, as the test frequency is increased, attenuation becomes so great that emission signals do not propagate substantial distances.

In analyzing acoustic emission wave behavior, the signal pulse shapes need to be determined as well as their durations. Knowledge of pulse shapes as a function sample thickness, receiver distance from the source, etc., are important for obtaining source location by triangulation and to enable counting systems to be properly devised for determining crack growth.

In order to determine actual signal pulse shapes from acoustic emission events, it would also be necessary to have knowledge of the failure mechanism so as to be able to describe the initial disturbance. Then, since it is likely that the initial disturbance will have a wide band of frequencies, attenuation coefficients must be known over the range of frequencies in the pulse. This would be necessary in order to determine pulse shape changes due to excessive high frequency scattering which removes higher frequencies from the pulse in greater amounts than the lower frequencies. It would be necessary finally to have accurate knowledge of receiving transducer transfer functions and other acousto-electrical quantities to obtain the eventual pulse shape estimates.

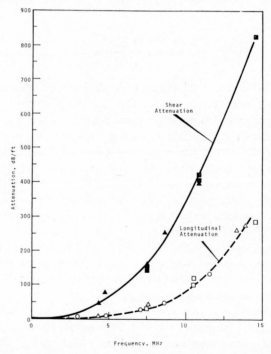

FIG. 1. Attenuation versus frequency for ultrasonic signals in type 304 stainless steel

These quantities are not entirely available at the present time and some would be difficult to obtain.

Another approach can be taken which only requires a limited knowledge of these effects. In most practical applications it is desirable to propagate acoustic emission signals over substantial distances. Although the initial disturbance may have many higher frequency components, most of these are quickly removed due to scattering. For example, in Fig. 1 a typical plot of attenuation coefficient as a function of frequency for shear waves in steel

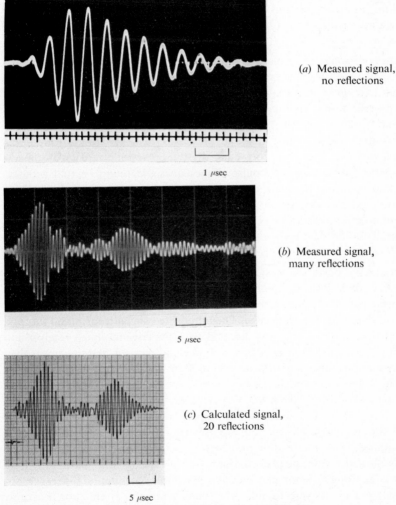

(a) Measured signal, no reflections

1 μsec

(b) Measured signal, many reflections

5 μsec

(c) Calculated signal, 20 reflections

5 μsec

FIG. 2. Measured and calculated signals.

(304 stainless steel) is shown. Scattering losses increase beyond the knee in the curve in approximate proportion to the fourth power of frequency. It is evident, then, that frequencies beyond a few MHz do not propagate more than a few inches before they are much smaller in amplitude than the lower frequencies. Therefore, emission events initially having a wide frequency band become filtered so that they essentially have dominant frequencies less than a few MHz after traveling a few inches. Signal frequencies below 1 MHz are often undesirable in practical applications because of hydraulic noises, etc. Also, these lower frequencies have the effect of excessive received signal durations. For these reasons it appeared that relatively narrow band receivers having a center frequency of 1–2 MHz should be used in field tests.

To test the feasibility of signal synthesis methods, signals were measured near the emission site in a plate so that interference from reflections would not be present. Then using the same receiver and electronic system, the signal could be measured again at a greater distance from the source to include the effects of many reflections. The number of reflections, their reflection factor amplitudes and phases, and their differences in attenuation (because of different path lengths) and the pulse shape at the closer distance are known. It was thought possible, therefore, to predict the pulse shapes at greater distances. This method of determining pulse shapes does not require actual acoustic disturbances within the plate or transducer transfer functions to be known. Provided that the emission signals for a particular test application can be measured before reflections occur and assuming the test sample geometry is known, it should then be possible to predict pulse shapes in most of the field test applications encountered in practice.

Preliminary pulse shape synthesis experiments were conducted on $\frac{1}{2}$ in. thick A212-B carbon steel plate. A small point source shear wave transducer was used as the signal source. The receiver was also a small shear wave sensor mounted at the plate edge. The unreflected signal transmitted through a small distance (~ 1 in.) was detected with the edge-mounted receiver and recorded. Using the same transmitter source and receiver, the signal was then detected at a distance from the source of 30 in. so that many reflections were present. These two signals are shown in Fig. 2(a) and (b). The unreflected pulse was also recorded on an X–Y recorder at the output of a sampling oscilloscope. This X–Y recording then permitted accurate determination of amplitudes incrementally along the pulse. The pulse was then added to itself in amplitude with phase shift information determined for the different ray paths reflected within the plate. These reflection factors and phase shifts were determined using an available computer program. Attenuation differences for different ray paths were found to be small since the attenuation was only about 1·3 dB/ft at the frequency used (about 1·6 MHz).

TABLE I. Total Time Delays

Ray Path Reflections $(n-1)$	θ	$\Delta\phi$	$(n-1)$ $\Delta\phi$	$(n-1)$ $\Delta\phi P/360°$ (μsec)	Δt (μsec)	Total Delay (μsec)
Direct Ray from Top	90°	0	0	0	0	0
Direct Ray from Bottom	89·05°	173·6°	173·6°	·298	·0325	·330
1	88·1°	167·4°	167·4°	·287	·13	·417
2	87·15°	161°	322°	·552	·293	·845
3	86·2°	154·7°	464°	·796	·52	1·32
4	85·2°	148·2°	594°	1·02	·812	1·83
5	84·3°	142·4°	712°	1·22	1·11	2·33
6	83·35°	136·4°	819°	1·40	1·51	2·91
7	82·4°	130·5°	914°	1·565	2	3·57
8	81·5°	125°	1000°	1·71	2·49	4·20
9	80·5°	119°	1071°	1·84	3·23	5·07
10	79·6°	113·7°	1137°	1·95	3·7	5·65
11	78·7°	108·5°	1193°	2·045	4·4	6·45
12	77·8°	103·5°	1242°	2·13	5·2	7·33
13	76·85°	98·2°	1277°	2·10	6·05	8·24
14	75·98°	93·2°	1305°	2·24	7	9·24
15	75·05°	88·8°	1331°	2·285	8	10·29
16	74·2°	84·52°	1350°	2·315	9	11·32
17	73·35°	80·34°	1365°	2·34	10	12·34
18	72·45°	76·03°	1370°	2·35	11·2	13·55
19	71·55°	71·83°	1365°	2·34	12·7	15·04
20	70·7°	67·98°	1360°	2·33	13·6	15·93

$f = 1.62$ MHz; $P = 0.617$ μsec

In Table I the total time delays (in μsec) of the first twenty reflected ray paths are given in Column 7. In addition, Columns 1–6 give the ray path reflection $n-1$, the reflection angle θ, the phase delay per reflection $\Delta\phi$ (from "Snell Detail"), the total reflection phase delay $(n-1)\,\Delta\phi$, the total reflection phase delay $(n-1)\,\Delta\phi P/360°$ (in μsec), and the time difference in μsec due to geometric differences in ray paths. The frequency f, and the period P, for the pulse are given also. The total delays were then used to displace in time each of the first twenty reflected ray paths. All the rays used had θ values such that the reflected ray amplitudes were within less than 0·1 per cent of being totally reflected as calculated by a computer program called "Snell Detail". In Fig. 2(c) the resulting synthesized signal is shown and its similarity to the measured signal at 30 inches is apparent.

Following the above experiments, photographs or other signals were taken with the transmitting source located at different elevation positions. In Fig. 3(a), (b) and (c) the source was located first at the same side as the re-

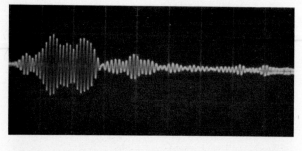

(a) Source near same side of plate edge as receiver

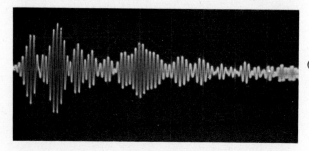

(b) Source at center and receiver at one side of plate edge

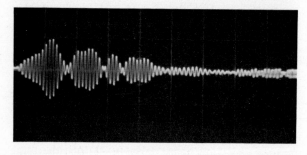

(c) Source and receiver located near opposite surfaces of plate edge

Fig. 3.   Signals as received from various source elevations.

ceiver, then at the center of the plate, and then at the side opposite the receiver respectively. This illustrates an effect important to practical test applications. Since the three signals are substantially different it proves the feasibility of not only locating the range of an acoustic emission source, but also its distance below the surface. In order to make possible this capability it will probably be necessary to generalize the ray reflection studies and perform more elaborate calculations and demonstration experiments.

III. SYSTEM STUDY

Acoustic emission monitor system requirements and equipment development are discussed in functional blocks—the received signal, sensor, signal conditioning, and data analysis and readout.

A. Signal Analysis

One of the major areas of investigation in acoustic emission has been signal analysis and characterization. This, together with development of appropriate detection techniques, is felt to be the key to realizing the full potential of acoustic emission as a reliable integrity surveillance tool.

A tentative frequency categorization of signals produced by plastic deformation, ductile fracture, and brittle fracture was proposed on the basis of early test results (Hutton, 1968a). This categorization is now considered to be very questionable. The data are valid, but the interpretation appears to be erroneous. What was thought to be a failure mode-dependent frequency appears instead to represent acoustic emission event rate for a given circumstance. One of the observations which supports this latter line of reasoning is illustrated by Fig. 4. The signal shown is the detected response from acoustic emission generated by a flaw increment during a test to study fracture characteristics of 24 in. diameter by $1\frac{3}{4}$ in. wall carbon steel pipe. Applying the frequency category concept, this would represent failure of only one grain. A flaw increment restricted to one grain is not logical in this circumstance. The response becomes more rational, however, if one considers that the high frequency packet at the front of the signal represents a number of individual emission events.

Another problem which became evident was the duration of the transducer response generated by a single emission event. The response signal lasted at

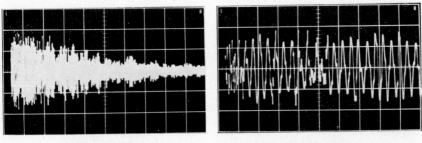

25K H.P. Filter
1·0 msec/cm
2 V/cm

25K H.P. Filter
0·1 msec/cm
2 V/cm

FIG. 4. Acoustic emission signal observed from flaw growth during pipe burst test.

least 6–8 milliseconds as shown in the photo on the left in Fig. 4. Depending on the type of transducer used, the duration could be much greater. Since these times were in excess of the transducer ring down time, it became apparent that reflected signals may be extending the response duration. The signal duration becomes a problem when several emission events occur in close succession. Under this circumstance it can become impossible to accurately resolve individual events and the accuracy of emission rate and total emission analysis is compromised.

The above considerations lead to the conclusion that an analysis of acoustic emission wave propagation and development of improved sensing techniques was a necessary next step before flaw mechanisms and materials could be reliably characterized in terms of acoustic emission. Such work is now in progress and results to date were discussed under "Theoretical Aspects and Physical Principles".

Reactor noise measurements (Hutton, 1968b) and a test of emission detection in the presence of hydraulic noise (Hutton, 1968c) both substantiate the need to work in the low megahertz frequency range. A brief study of the acoustic emission signal *per se* shows that it is not the limiting factor on high frequency detection. Sensors up to 30 MHz were used to study emission and

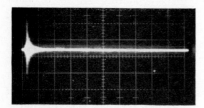

SHEAR SENSOR
1.7 MHz
EDGE MOUNTED
GAIN OF 100
HI-PASS FILTER-1.2 MHz
0.05 VOLTS/CM
0.001 SEC/CM

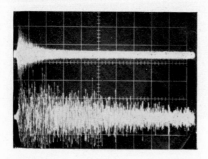

ULTRASONIC SENSOR
1 MHz CUT, SURFACE MOUNT
GAIN OF 100
HI-PASS FILTER-600 kHz
0.01 VOLTS/CM
0.001 SEC/CM

ACCELEROMETER
ENDEVCO #2213E, SURFACE MOUNT
GAIN OF 100
HI-PASS FILTER-30 kHz
0.05 VOLTS/CM
0.001 SEC/CM

FIG. 5. Response of different sensors to the same acoustic emission signal.

the signal transmitted in the metal (steel) was sufficiently short to cause the sensor to respond at its fundamental frequency of 30 MHz. This indicates that the transmitted emission signal is a pulse transient with a period of less than 0·03 μsec. Thirty MHz is well above any practical level for field surveillance application because of frequency-dependent attenuation in the material.

The different modes of signal transmission—longitudinal, shear, and Rayleigh—were studied for merits of one over the other as a vehicle for emission detection. It appears that detection of shear waves would involve the least interference from other modes and produce the sharpest signal. Figure 5 shows an example of the variation in detected signal produced by the same acoustic emission event using different sensors and detection techniques. Results of the wave propagation study and direction obtained therefrom for improving detection techniques is reported in a separate topical report (Fitch, 1969).

To date, the primary acoustic emission characterization parameters have been emission rate and a running total of emission count. Pulse height and pulse duration as characterization parameters are being investigated. The type of correlation attainable between acoustic emission and fracture parameters is illustrated in Fig. 6. In this test a double cantilever beam specimen of D2 tool steel with a preformed starter crack was loaded until the crack started to grow. At this point, the load was held and allowed to decay by slow crack growth. This process is shown in the upper curve. The lower

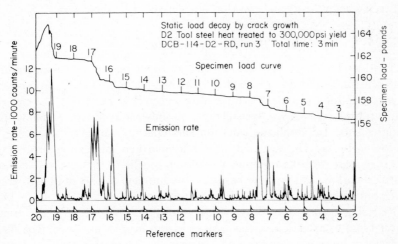

FIG. 6. Acoustic emission rate as a function of load decay by crack growth in a fracture specimen.

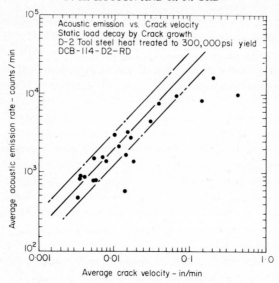

FIG. 7.   Acoustic emission versus crack velocity.

curve shows the accompanying emission rate curve. Each inflection in the load decay curve denotes a change in crack velocity. Examination of the emission rate curve shows that a change in crack velocity is accompanied by a proportionate change in emission rate. Figure 7 shows the relationship between crack velocity and emission rate derived from these data.

B. Sensor Development

Considerable effort has been devoted to developing a high temperature, high sensitivity transducer capable of withstanding extended operation in an environment of about 800°F temperature and nuclear radiation. The electrostatic transducer showed considerable promise initially for fulfilling the requirements. Development work on this concept, however, showed two major problems. The sensitivity was marginal and fabrication procedures were complex with the resulting assembly too delicate for field application. Further work on this concept has been discontinued as part of this program in favour of high temperature piezoelectric materials.

Referring again to Fig. 5, the response of three currently available transducer types to excitation by acoustic emission in steel is shown. The three transducer response signals were generated simultaneously by a single acoustic emission event. This approach is necessary to obtain a valid comparison of response from different transducers because of the randomness of acoustic emission. The shear sensor shown in the top trace produces the most desirable signal. The front is quite sharp and the duration of the main signal is less than 10 μsec which provides good resolution of other closely following signals. Transducer material is PZT/5A piezoelectric cut to a 1·7 MHz primary frequency and poled in the shear direction. This transducer is sensitive almost exclusively to shear waves. With a Curie temperature of $\approx 685°$F this material can be used for extended periods in a temperature environment up to 600–625°F. It also has good resistance to nuclear radiation damage. In order to obtain optimum results, the sensor must be mounted on the edge or thickness plane of the specimen and at least $\frac{1}{16}$ in. below the surface. Adaptation of this technique to field application is still in progress; however, it can be applied currently for laboratory specimen study. Several techniques for field application are being investigated for feasibility.

The middle trace shows the response of a surface mounted, longitudinally poled, PTZ/5A piezoelectric sensor which is being used routinely in both laboratory and field tests to study acoustic emission. The response is primarily to shear and Rayleigh waves. The wave front is quite well defined; however, the undesirable duration of the signal ($1\frac{1}{2}$–2 milliseconds) is evident. Also, although the transducer is being operated in a high frequency mode (≈ 700 kHz) by filtering, it has, in addition, strong modes at frequencies as low as 100 kHz. In a strong noise field such as hydraulic cavitation, these lower frequency responses can become a problem by saturating pre-amplifiers and/or passing through a filter with sufficient strength to show up as erroneous data. Circuit tuning is being used to minimize this problem. With a PZT/5A sensing element, this tranducer has the same environmental capabilities described above.

Acoustic emission generated response from a conventional, commercial accelerometer is shown in the bottom trace. The problems of low frequency response and extremely long signal duration are apparent. The sensing element (basically PZT-5) does have a higher frequency response component; however, it is quite weak compared to that at 30–50 kHz. This produces the same system problem described under the surface mounted ultrasonic sensor discussed above only in a much more aggravated form.

Lithium niobate (LiNbO$_3$) is being studied for application as a high temperature (1500–2000°F) sensing element. The material is a piezoelectric which has shown marginal sensitivity as a longitudinally poled sensor. Recent work, however, using it in a shear poled, edge mounted form shows sensitivity comparable to PZT/5A.

Work is in progress to test LiNbO$_3$ crystals for their ability to withstand exposure to gamma and fast neutron (>1 MeV) nuclear radiation at temperatures up to 1000°F. This information is vital to ultimate application for reactor pressure system surveillance.

Transducer coupling to the specimen surface is an important consideration to detection of acoustic emission. General Electric Company RTV-116 silicone rubber has been found to perform well to couple surface and longitudinal waves to the transducer face. Although it is rated for a maximum service temperature of 600°F, it has been used as a couplant in applications approaching 700°F quite successfully. One of the advantages of this material is the fact that it remains sufficiently pliable that the transducer can be removed for reuse without destroying the front electrode. It also acts as a good insulator to prevent grounding the system at the transducer and thus incurring the problem of electrical ground loops.

A "hard" mount has proven to be much better than silicone rubber for coupling shear waves. An adhesive such as Eastman 910 within its 200°F temperature limit or one of the several ceramic strain gauge bonding materials have shown good shear wave coupling properties. They do have the disadvantage that once a transducer is mounted, it usually cannot be removed without destroying the crystal.

C. Signal Conditioning

Although the transducer is obviously the primary key to successful detection of acoustic emission, the supporting electronics system is vital to beneficial use of the detected information. One currently used monitor system is shown in schematic form in Fig. 8. The signal from the transducer is fed into a low noise preamplifier located as close as possible to the transducer where it is amplified by a factor of 100. From here it goes through a variable, 24 dB/octave, electronic filter to screen out unwanted low frequency information.

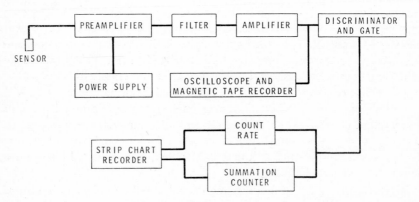

FIG. 8. Diagram of acoustic emission monitor system.

The remaining data goes to a ×100 amplifier. At this point the system divides with one branch going to a magnetic tape recorder for raw data storage and an oscilloscope for continuous data surveillance. The other branch goes to a discriminator and gate circuit wherein an incoming signal above a set threshold voltage value will produce a shaped output signal (square wave) to be fed into a count rate circuit and a count totalized circuit. These two parameters are then read out on a two-pen strip chart recorder $v.$ time or some other test parameter.

Since many of the acoustic emission signals are very low level, it is important to minimize electronic noise. With the preamplifier being the component which primarily establishes the final system noise level, considerable attention has been given to producing a low noise, high frequency preamplifier. Figure 9 shows the resulting product. This is an all transistorized device with a noise level of $7·6\mu V$ r.m.s. referred to input, a bandwidth of 30 kHz to 2·5 MHz, and a gain of 100. The coin (a quarter) shown in the lower right corner provides a measure of size of the assembly. In a very recent modification, this preamplifier was tuned to operate over a 400 kHz band about a selected center frequency. Center frequencies of 700 kHz and 1·5 MHz have been used to date based on characteristics of the transducers being used. Advantages of the tuned version are that the noise level is reduced to $3·5\ \mu V$ while increasing the gain to 150, the inherent roll off on either side of

FIG. 9. Low noise, broadband preamplifier developed for acoustic emission detection.

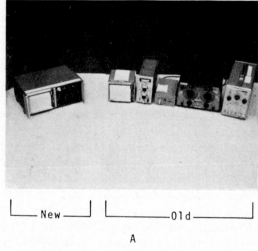

|___ New ___| |_____ Old _____|

A

Emission Rate Analyzer

|___ New ___| |_____ Old _____|

A

Six Channel Signal Conditioner System

FIG. 10. Solid state acoustic emission monitor systems and laboratory equipment counterparts.

the pass band (effectively 14 dB/octave) permits elimination of a separate filter, it enhances selected transducer response characteristics, and by blocking low frequency signals at the preamplifier, it alleviates the problem of pre-amplifier saturation by high level, low frequency extraneous signals. It is stable and unaffected by long output leads (500 ft long leads have been used satisfactorily).

The necessity to transport the monitor system considerable distances to monitor special tests has prompted efforts to minimize the size and weight of the system. The solid-state version of the system currently in use is illustrated in Fig. 10. For comparison, the functional counterpart of each instrument in general purpose laboratory instruments is shown in each photograph. The lower photograph shows a six channel filter–amplifier instrument and the upper photograph shows an instrument incorporating a single channel filter, amplifier, discriminator, gate, and count rate circuit, with integral strip chart recorder.

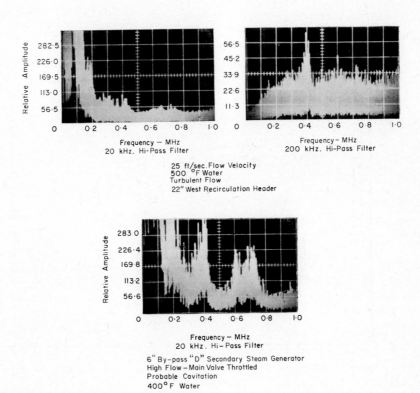

FIG. 11. Hydraulic noise spectra—Dresden I nuclear power reactor.

D. Data Analysis and Readout

Three levels of refinement in a monitor system are currently available. The first is that shown in Fig. 8. With this system, emission rate and total emission only are printed out on a real time basis. The data are stored on magnetic tape for subsequent analysis to determine difference in time of signal arrival at various transducers which permits identification of source location. This system is relatively inexpensive—approximately $30,000 for a seven channel system, including a magnetic tape recorder and a monitor oscilloscope. Such a system is suitable for laboratory study work and limited field application where immediate resolution of results is not vital.

The next level of refinement incorporates an analyzer to automatically determine the relative time of signal arrival at various transducers on a real time basis. This information is then used to manually determine source location. A magnetic tape recorder is not required with this system for data storage. The approximate cost for the second level system is $50,000. The

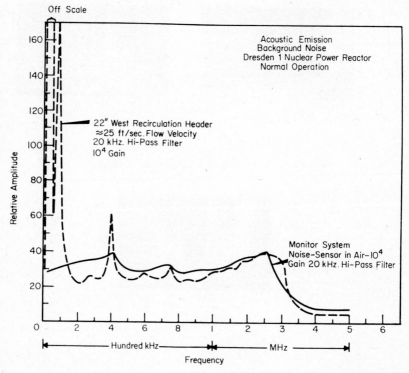

FIG. 12. Noise spectrum—22 in. west recirculation header—Dresden I.

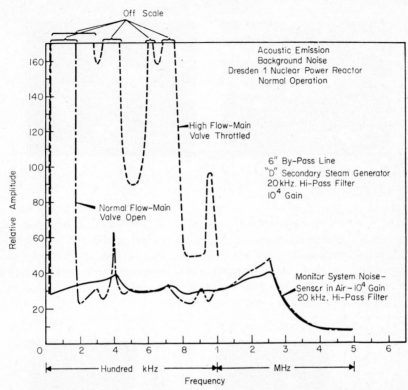

FIG. 13. Noise spectrum—6 in. by-pass line, "D" secondary steam generator—Dresden I.

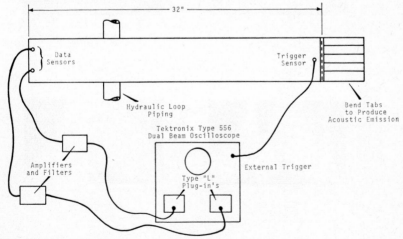

FIG. 14. Diagram of test system for detection of acoustic emission in the presence of hydraulic noise.

FIG. 15. Test arrangement for detection of acoustic emission in the presence of hydraulic noise.

system is suited to field applications such as hydrostatic testing pressure vessels and limited surveillance of operating systems.

The third level of refinement adds a special purpose computer to the second system which gives an immediate indication of source location. The cost of this system will fall in the $100,000–$150,000 range. It is suited to on-line surveillance of complex systems during operation.

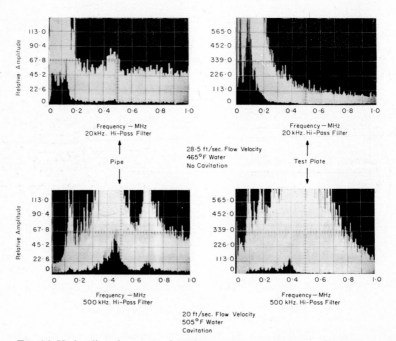

FIG. 16. Hydraulic noise spectra for turbulent flow and cavitation conditions.

System descriptions given above are intended to outline typical equipment for the three levels of capability. There are several variations of each system.

IV. APPLICATIONS

With this being an application oriented research effort, it is appropriate to discuss this aspect at this point. One of the basic questions related to the intended application of acoustic emission for continuous surveillance of nuclear reactor pressure boundaries concerns, "Can acoustic emission be

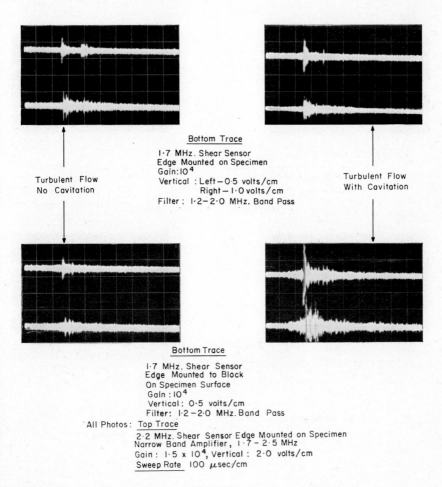

Turbulent Flow
No Cavitation

Bottom Trace
1·7 MHz. Shear Sensor
Edge Mounted on Specimen
Gain:10⁴
Vertical : Left — 0·5 volts/cm
Right — 1·0 volts/cm
Filter : 1·2 — 2·0 MHz. Band Pass

Turbulent Flow
With Cavitation

Bottom Trace
1·7 MHz. Shear Sensor
Edge Mounted to Block
On Specimen Surface
Gain : 10⁴
Vertical : 0·5 volts/cm
Filter: 1·2 — 2·0 MHz. Band Pass

All Photos : Top Trace
2·2 MHz. Shear Sensor Edge Mounted on Specimen
Narrow Band Amplifier, 1·7 — 2·5 MHz
Gain : 1·5 x 10⁴, Vertical : 2·0 volts/cm
Sweep Rate 100 μsec/cm

FIG. 17. Acoustic emission signals detected in the presence of hydraulic noise—A.

detected and identified in the presence of the inevitable hydraulic noise background?" Two major tests have been conducted in an effort to answer this question. The first consisted of monitoring the background noise spectrum of the Dresden I nuclear power reactor during start-up to full power operation. The results of this test are reported in detail elsewhere (Hutton, 1968b). However, briefly, they showed that, for this reactor (General Electric Co. boiling water design), the noise field above 500 kHz is quite low. Figure 11 shows typical spectrum analyzer response for turbulent flow and for turbulent flow with probable cavitation. Figures 12 and 13 provide typical compilations of analyzed data. There is a frequency region nominally from 750 kHz to

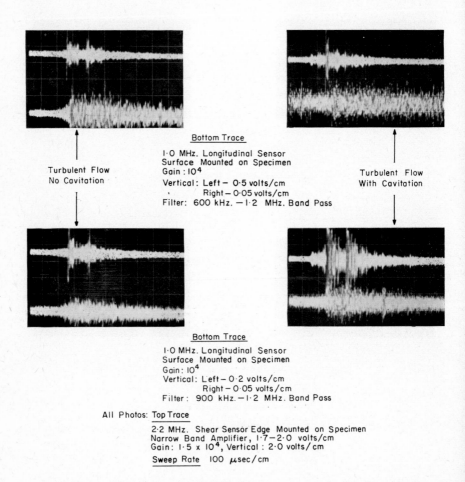

Turbulent Flow
No Cavitation

Bottom Trace

1·0 MHz. Longitudinal Sensor
Surface Mounted on Specimen
Gain : 10⁴
Vertical : Left — 0·5 volts/cm
Right — 0·05 volts/cm
Filter: 600 kHz. —1·2 MHz. Band Pass

Turbulent Flow
With Cavitation

Bottom Trace

1·0 MHz. Longitudinal Sensor
Surface Mounted on Specimen
Gain: 10⁴
Vertical: Left — 0·2 volts/cm
Right — 0·05 volts/cm
Filter: 900 kHz. —1·2 MHz. Band Pass

All Photos: Top Trace

2·2 MHz. Shear Sensor Edge Mounted on Specimen
Narrow Band Amplifier, 1·7 — 2·0 volts/cm
Gain: 1·5 x 10⁴, Vertical : 2·0 volts/cm
Sweep Rate 100 μsec/cm

Fig. 18. Acoustic emission signals detected in the presence of hydraulic noise—B.

about 3 MHz (where signal attenuation with increasing frequency rises sharply for ferrous materials) which is relatively free of noise and conducive to reliable detection of acoustic emission signals.

The above work was followed up with a test wherein a hydraulic test loop was used to generate noise spectra similar to those observed at Dresden I and detection of actual acoustic emission signals in such an environment was evaluated. Results, reported in a separate document (Hutton, 1968c), showed that an acoustic emission sensing system operating in the 1·5 to 2·5 MHz frequency range with shear sensors can detect emission signals with no noticeable interference from the hydraulic noise—either turbulent flow or cavitation noise. A schematic diagram of the test system and photos of the test arrangement are given in Figs. 14 and 15 respectively. Comparison of Fig. 16 with Fig. 11 shows the similarity of the noise environment produced for this test with that measured on the Dresden I reactor. Acoustic emission signals were produced by bending the "fingers" on the test plate shown in the lower right corner of the right-hand photograph in Fig. 15. Responses of various sensors to the acoustic emission signals thus produced are presented in Figs. 17 and 18. A sensor near the emission source was used to trigger the oscilloscope in order that the response of the data sensors three feet from the source would appear in the middle of the screen providing a view of the full response form, and also by position on the screen, verify that the detected signal originated at the emission source. The top trace in all photographs was produced by the same sensing system to provide a basis for comparing the various other sensor arrangements. As the photographs show, good response was obtained from all of the shear sensors for all test conditions and a usable response was

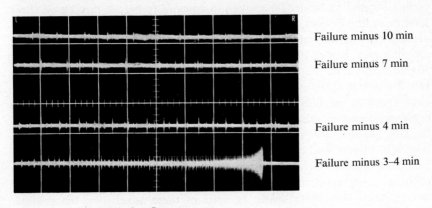

Failure minus 10 min

Failure minus 7 min

Failure minus 4 min

Failure minus 3–4 min

5·0 sec/cm Sweep rate
5·0 V/cm

FIG. 19. Data trace samples from the final 10 min–burst test of 24 in. dia. × 1·6 in. wall steel pipe.

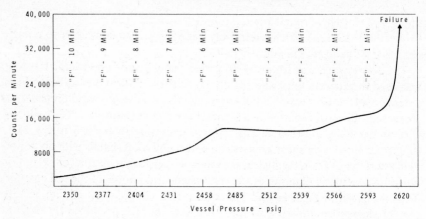

Fig. 20. Emission rate curve from final 10 min–burst test of 24 in. dia. × 1·6 in. wall steel pipe

obtained from the surface mounted longitudinal sensor for turbulent flow noise only. The response of the surface mounted longitudinal sensor was seriously obscured when cavitation noise was introduced.

Acoustic emission was included as part of the monitor network for a program to study rupture of ductile reactor pressure piping materials. Four of these tests have been monitored for acoustic emission to date. An example of the results is shown in Figs. 19 and 20. The test specimen in this case was a 12 ft long section of 24 in. diameter by 1·64 in. wall A106-B carbon steel pipe with a milled flaw at one point in the pipe wall. One of the functions of acoustic emission monitoring was to determine the onset of crack growth and to predict impending failure far enough in advance to permit turning on short term recording devices, such as a movie camera, to monitor the final stages of crack growth. Real time data interpretation was accomplished by measuring emission rate. Figure 19 shows acoustic emission data samples from the last 10 min of the test. Figure 20 shows the corresponding emission rate curve. Initial crack movement was indicated about 9 to 10 min prior to failure. High speed movie film of the flaw showed this to be well before a visible crack developed on the specimen surface. Figure 21 shows the magnitude of the final rupture. This is an example of detecting flaw growth in steel under relatively adverse field conditions. The specimen was pressurized by heating contained water which resulted in boiling water noise throughout most of the test. The specimen temperature reached about 700°F. It was necessary to use 300 ft long signal leads. In spite of these adversities, the acoustic emission was detected rather easily.

Preliminary study of acoustic emission from Type 304L stainless steel specimens that failed in tension-tension high cycle fatigue showed significant

FIG. 21. Ruptured 24 in. dia. × 1.6 in. wall steel pipe test vessel.

acoustic emission activity well before formation of a visible surface crack. The specimen configuration is shown in Fig. 22A. This specimen was pre-loaded, as shown in the loading diagram (Fig. 22B), and then cycled at 1800 cycles/min until failure. Representative acoustic emission produced by these samples is shown in Fig. 23. Practically no emission was observed until after approximately 6000 cycles. From this point, emission rate increased gradually until about 22,000 cycles was reached, after which the emission rate increased sharply. A visual surface crack was detected at approximately 34,000 cycles and it grew to ⅛ in. in length by 55,000 cycles. The large fluctuations in the acoustic emission rate after 34,000 cycles are apparently due to the crack growing in steps. It is significant that acoustic emission activity was first evident at 20 % of the cycles necessary to form a visible surface crack. Also, the most intense activity occurred in the region from shortly before to shortly after initial formation of a visible crack. Further study is necessary to answer questions such as, "What is the material condition that

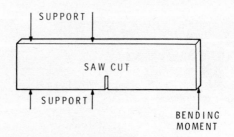

A. FATIGUE SPECIMEN

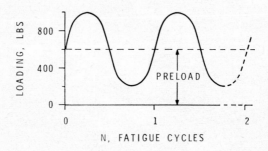

B. SPECIMEN LOADING DIAGRAM

Fig. 22. Cantilever beam specimen fatigued by cyclic loading.

gives rise to the initial emission activity?", and "Can the emission rate fluctuations be identified with step increments in the fatigue crack?"

An acoustic emission system is currently in use on the reactor simulator vessel used in reactor containment studies to monitor the status of known laminar flaws in the vessel heads. During proof testing, only one small indication of flaw action was observed, but as an indication of system response, initial flexing of new support bracing and plastic deformation of flange blank plates were detected and the source located by means of acoustic emission. The system is being used on a continuing basis to monitor the status of the flaws during routine program tests which are conducted at a temperature of 550°F.

V. AREAS FOR FURTHER DEVELOPMENT

One of the problems in applying acoustic emission for integrity surveillance at this point is the degree of resolution inherent. With limited calibration information available, it is difficult to judge the significance of limited indications of flaw movement. In many cases, the resolution of conventional nondestructive test methods is not sufficient to confirm the acoustic emission indication and measure the amount of flaw movement represented by the acoustic emission. Work must be done to develop technology which will permit a valid estimate of crack area increment represented by a given quantity of emission data, and to develop standards or criteria for judging the significance of a given amount of emission.

Other areas of future work include adaptation of the improved shear sensing technique to practical field application and work on characterization of different fracture modes (brittle, ductile, etc.) in terms of emission parameters using a high frequency (1–3 MHz) detection system. The maximum distance at which emission signals can be reliably detected needs, also, to be more firmly resolved.

In summary, there is little question that acoustic emission constitutes a powerful tool for both structural integrity surveillance and for basic study of material failure mechanisms. Full utilization of the technique will require considerable additional development work; however, in the meantime, the current state-of-the-art permits reliable and beneficial application to many selected industrial and laboratory problems. Some examples of these are: (1) monitoring for flaw growth during hydrostatic proof test of pressure

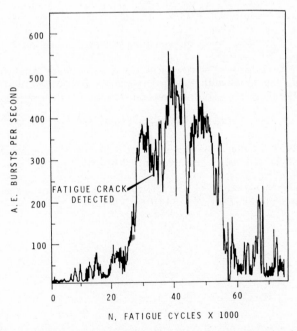

FIG. 23. A plot of acoustic emission from a specimen undergoing fatigue.

containers, (2) continuous surveillance of known or suspected flaw areas in an operating pressure system, (3) evaluation of weld quality from the standpoint of crack formation during weld cooling, and (4) study of fatigue crack growth characteristics under various loading conditions.

VI. Acknowledgements

The authors wish to gratefully acknowledge the support and assistance provided by other members of the NDT Methods Development Section of Battelle-Northwest. In particular, J. F. Dawson and D. M. Romrell, technicians assigned to the program, are to be recommended for their effort and contributions beyond that to be normally expected, and C. E. Fitch for his study of basic wave propagation and signal reconstruction.

REFERENCES

Day, C. K. (1968). "An Investigation of Acoustic Emission for Defect Formation in Stainless Steel Weld Coupons", BNWL-902.

Fitch, C. E. (1969). "Acoustic Emission Signal Analysis in Flat Plates", BNWL-1008.

Green, A. T., Lockman, C. S. and Steele, R. K. (1964). "Acoustic Verification of Structural Integrity of Polaris Chambers" (Society of Plastic Engineers, Atlantic City, N.J.) 1964.

Hutton, P. H. (1968a). Acoustic emission in metals as an NDT tool. *Mater. Evaluation* **26** (7), 125.

Hutton, P. H. (1968b). "Nuclear Reactor System Noise Analysis, Dresden I Reactor, Commonwealth Edison Company", BNWL-867.

Hutton, P. H. (1968c). "Acoustic Emission Detection in the Presence of Hydraulic Noise", BNWL-933.

Jolly, W. D. (1968). "An *In Situ* Weld Defect Detector, Acoustic Emission", BNWL-817.

Kaiser, J. (1953). Untersuchungen uber das Auftreten Gerauschen Beim Zugversuch (Ph.D. Thesis, Techn. Hoschh., Munchen, 1950); also (1953) *Arch. Eisenhüttenwesen* **24**, 43–45.

Mason, W. P., McSkimin, J. H. and Shockley, W. (1948). Ultrasonic observation of twinning in tin, *Phys. Rev.* **73** (10), 1213–1214.

CHAPTER 2

Ultrasonic Spectroscopy

O. R. GERICKE

Department of the Army, Army Materials and
Mechanics Research Center, Watertown, Massachusetts,
U.S.A.

I. INTRODUCTION

The heading of this chapter requires a clarification because the term "Ultrasonic Spectroscopy" has more than one connotation. It describes phenomena involving the diffraction of light by ultrasonically produced density gratings in transparent media, and it also denotes methods of analyzing the frequency components of ultrasonic signals used in nondestructive testing of materials. It is the latter interpretation that applies to the contents of this chapter.

A. Optical Analogy

Although the type of ultrasonic spectroscopy to be discussed here will not involve light waves, a reference to optics is considered helpful for explaining certain basic aspects of ultrasonic spectroscopy that are analogous to light spectroscopy. Such a comparison can also be justified by the fact that the well known successes of optical spectroscopy have actually stimulated the development of similar techniques in the ultrasonic field.

The origin of optical spectroscopy dates back almost 300 years to Newton's discovery that by using a glass prism white light can be split up into a series of colored light bundles. Today it is known that white light actually represents short pulses of electromagnetic radiation which according to Fourier's theorem can be regarded as arrays of various radiation frequencies called spectra.

Short pulses of vibrational energy used in ultrasonic testing have characteristics that are similar to white light. They too carry a spectrum of frequencies. Hence, a spectrum analysis of these pulses, conducted after they have been used to interrogate a test specimen, can be expected to yield results that are equally as important as data obtained by optical spectroscopy. Since ultrasound represents elastic instead of electromagnetic waves, however, ultrasonic spectroscopy of a specimen will yield different information.

This becomes immediately obvious if one compares the transmissivity of some materials to visible light with their transmissivity to ultrasound in the 1 to 25 MHz range commonly used for nondestructive testing.

Metals, for example, are opaque to light but frequently transmit ultrasound extremely well. Certain plastics, on the other hand, may be translucent but may nevertheless exhibit very strong ultrasonic attenuation.

B. Earlier Techniques

An interest in the frequency dependence of the ultrasonic response of materials has existed prior to the development of special spectroscopic equipment. Earlier studies of this type conducted with more conventional ultrasonic test equipment will now be discussed.

1. Manual Frequency Variation

Perhaps the oldest technique for obtaining the ultrasonic transmission characteristic of a material as a function of the frequency is manual tuning of a conventional pulse-echo test instrument. A prerequisite is, of course, that the instrument used is equipped with a tunable transmitter or receiver. Furthermore, it is desirable that the ultrasonic transducer employed for the test has a broad frequency response. Transducers with narrow response can also be used but have to be exchanged when the test frequency is varied.

The manual tuning technique is rather cumbersome and therefore seldom applied outside the laboratory. The technique is primarily used for determining the frequency dependence of the ultrasonic attenuation of materials (Roderick and Truell, 1952).

2. Frequency Modulation

Automatic tuning can be substituted for the manual frequency variation procedure and leads to frequency modulated test signals. Instruments employing such signals have been in use for many years for the purpose of thickness measurements. The technique is generally referred to as ultrasonic resonance testing because it determines those mechanical resonances of a test specimen that depend on its wall thickness. Detailed procedures have been described by Evans (1959).

The electronic equipment utilized for resonance testing usually does not provide a linear amplitude versus frequency read-out and is therefore only suited for the detection of resonance peaks in a spectrum.

3. Multiple Frequency

Multiple-frequency testing represents another approach to ultrasonic spectroscopy. The technique utilizes two or more tuned pulse-echo test instruments connected to a single transducer which must have a broad frequency response. The instruments have to be tuned to different frequencies and their outputs displayed on a single cathode-ray tube by means of a multitrace procedure. The obtained read-out is still an amplitude versus time function but contains the frequency as a parameter since each of the displayed traces represents a different frequency.

With a sufficient number of pulse-echo instruments and a corresponding read-out capability at hand, the technique approaches "real-time" spectroscopy. This means that the time dependence of the analyzed frequency components can be determined.

The main disadvantage of the procedure is the inherent bulk and cost of the equipment. For this reason it has thus far been implemented only for two test frequencies (Gericke, 1964).

4. Pulse Shape Examination

Another forerunner of ultrasonic spectroscopy is the technique of examining the shape of an ultrasonic pulse before and after it has interrogated a specimen. The technique has evolved from the observation that ultrasonic pulses undergo changes in contour when transmitted through an attenuating specimen or when reflected from certain discontinuities. These alterations of the pulse shape can be interpreted as changes in the spectral energy distribution of the ultrasonic signal occurring in the specimen (Posakony, 1959).

A theoretical and experimental study of ultrasonic pulse distortion conducted by the author (Gericke, 1963) has led to the conclusion that certain changes in the pulse spectrum can be recognized by a visual examination of the ultrasonic pulse shape. If the attenuation in a test specimen increases with the ultrasonic frequency, for example, an originally rectangular pulse, after having been transmitted through the specimen, will exhibit rounded edges that are quite evident.

More complicated changes in the spectral energy distribution of an ultrasonic pulse, however, may not be as easily deductable from pulse contour changes. Thus, it cannot be expected that a person can be trained to judge a pulse shape in a manner that is equivalent to a pulse spectrum analysis in all situations.

A more realistic approach would be to record and subsequently analyze the ultrasonic pulse by means of numerical Fourier analysis. Such a process could be computerized, but would nevertheless be tedious in comparison to the electronic techniques of spectrum analysis described below.

II. THEORY

It has already been mentioned in the Introduction that the main purpose of introducing the ultrasonic frequency as a new test variable is to render the ultrasonic inspection method more informative. While the techniques discussed earlier have pursued this goal to some extent, a full exploitation of this idea requires the development of new ultrasonic test concepts.

A. Basic Concepts of Current Techniques

Efforts to produce special procedures for ultrasonic spectroscopy could have been directed towards the development of acousto-optical means of analysis. But, since compatibility with conventional ultrasonic test equipment was deemed desirable, only electronic procedures of spectrum analysis were taken into consideration.

The result of this work is the development of three different basic techniques for electronic signal generation and analysis which employ piezoelectric transducers as the necessary link between the electronic and the acoustic domain.

The simplest technique is to inject frequency modulated ultrasound into the test specimen with a swept-frequency generator and to plot the received signals on an oscilloscope whose horizontal beam deflection is furnished by the frequency sweep signal of the generator.

The second procedure utilizes pulsed ultrasound which, similar to white light, contains a multitude of frequencies. An electronic spectrum analyzer is required at the receiver side to separate and display the various frequency components of the pulse spectrum.

The third procedure is pulsed frequency-modulation which is a combination of the two previous techniques.

B. Spectra of Test Signals

Extremely important for the success of ultrasonic spectroscopy is an accurate knowledge of the spectral energy distribution of the ultrasonic signals originally introduced into a test specimen. The two main factors determining this distribution are the output spectrum of the generator that excites the transducer and the frequency response characteristic of the transducer itself.

1. Generator Output Spectra

The transducer excitation furnished by an electronic generator is the primary factor determining the spectrum of the produced ultrasonic test signal.

In the frequency-modulation technique, the amplitude versus frequency function of the signal supplied to the transducer can be chosen quite arbitrarily. Normally, however, one prefers a signal having a constant amplitude over all frequencies. Under these conditions the generator output spectrum is represented by a single vertical line of constant height whose horizontal position changes in accordance with the momentary generator frequency.

In the case of pulse excitation, the spectrum of the signal fed to the transducer is more complicated. It contains numerous simultaneously produced frequency components (Brown and Gordon, 1967).

Under idealized conditions assuming an instantaneous rise and decay of the pulse, the pulse spectrum can easily be derived by Fourier analysis. For a single direct current pulse with rectangular shape, the spectral function is

$$A(f) = HT\frac{\sin \pi Tf}{\pi Tf} \tag{1}$$

where H denotes the pulse amplitude, T the pulse duration, and f the frequency.

In the case of an oscillatory pulse with rectangular envelope and a carrier frequency f_0, one obtains a modified equation for the spectral function

$$B(f) = \frac{HT}{2}\frac{\sin \pi T(f-f_0)}{\pi T(f-f_0)}. \tag{2}$$

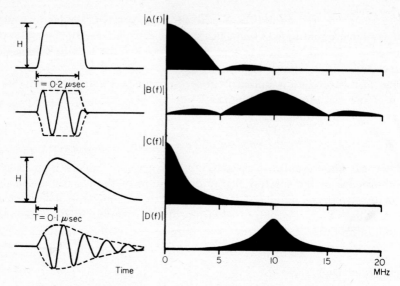

FIG. 1.   Various types of pulses and associated spectral functions.

Aside from a factor of 1/2, eqn. (2) differs from eqn. (1) only insofar as it indicates a shift of the spectral maximum towards higher frequencies by a constant quantity f_0 which represents the carrier frequency of the pulse.

Figure 1 shows diagrams of the absolute values of $A(f)$ and $B(f)$ plotted on a linear amplitude scale and over a frequency range of 0 to 20 MHz. To reflect a practical situation, a pulse duration of 0·2 μsec and finite rise and decay times were assumed. In the case of the oscillatory pulse, a carrier frequency f_0 of 10 MHz was chosen for the plot.

More common in ultrasonic testing are signals having exponential rise and decay as illustrated by the subsequent pulse traces in Fig. 1. In this case, the spectrum of the direct current pulse is given by

$$|C(f)| = eHT\frac{1}{1+4\pi^2T^2f^2} \quad \text{with } e = 2\cdot7183 \tag{3}$$

and the spectrum of the oscillatory pulse by

$$|D(f)| = \frac{eHT}{2}\frac{1}{1+4\pi^2T^2(f-f_0)^2}. \tag{4}$$

Figure 1 shows graphs of these functions, depicted to the right of the corresponding pulse trace. Again f_0 is made 10 MHz.

In Fig. 1 and in all figures that follow, plots of spectral functions are specially identified by coloring the space between the curve and the abscissa black. This type of diagram corresponds to a dense line spectrum as it is often observed on the cathode-ray tube screens of electronic spectrum analyzers. Unless otherwise stated, spectral amplitudes will always be plotted on a linear, arbitrary scale.

A comparison of the first two spectra of Fig. 1 indicates that the main spectral lobe, which contains most of the pulse energy, is twice as wide for the oscillatory pulse as for the direct current pulse. The width of the main lobe is, in addition, inversely proportional to the pulse duration T, as one can see from eqn. (1) by noting that the first null of $A(f)$, marking the end of the main lobe, is obtained when

$$f = 1/T. \tag{5}$$

Figure 2 illustrates the interdependence of pulse duration and spectral width by showing spectra of two pulses differing in length by a factor of two. A similar relationship of pulse duration to spectral width exists for the other discussed pulse forms governed by eqns. (2), (3), and (4).

Since the spectral functions of Fig. 2 are plotted on the same vertical scale, one can note that the shorter pulse has the lower spectral amplitude. This

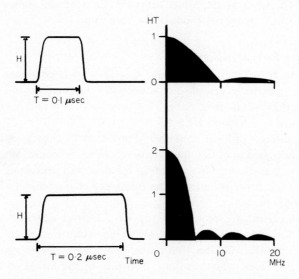

Fig. 2.　Influence of pulse duration on pulse spectrum.

is attributable to the fact that according to eqn. (1) the spectral amplitude is proportional to the pulse duration T.

From the foregoing it can be concluded that in order to obtain a very broad ultrasonic spectrum with large amplitudes, the pulse amplitude H must be made very high while the pulse duration T should be made as short as possible.

In practice a compromise value must be sought for T since H is limited either by the electrical breakdown of the transducer element or the maximum available pulse generator output voltage.

The pulse spectra shown in Figs. 1 and 2 apply to single pulses and are continuous except for the nulls. In a practical situation, repetitive pulses are needed in order to obtain a stationary pattern in the time domain and to supply sufficient signal energy to the electronic spectrum analyzer that is used. As a result of the pulse repetition, the continuous spectrum splits up into a line spectrum with frequency differences between two consecutive lines being equal to the pulse repetition rate. The spectral envelope, however, which is the curve connecting the peaks of all spectral lines remains unchanged.

Finally, a form of transducer excitation will be discussed which is obtained by pulsing a frequency-modulated signal. If the frequency modulation is slow compared to the pulse repetition rate and the pulse duration relatively long, the resultant spectrum is almost the same as with normal frequency modulation. Due to the pulsing, however, the spectrum becomes discontinuous and resembles in this respect the above-discussed repetitive pulse spectrum. The spectral amplitude function can again be chosen arbitrarily as with the other frequency-modulation technique.

2. Transducer Response Characteristics

The second factor determining the spectrum of a generated ultrasonic signal is the transducer response characteristic. Depending on the spectroscopic technique that is used, either two very similar or only a single transducer is needed (see Section III.B.).

If two transducers are utilized their individual frequency characteristics have to be multiplied to obtain the overall frequency response. When working with a single transducer, used for transmission and reception, the "loop response", which is the product of the frequency characteristic with itself, is the determining factor.

At present, discs of piezoelectric (or ferroelectric) material are used as transducers for ultrasonic spectroscopy. They vibrate in a compressional mode in accordance with the alternating electrical potential applied across their metallized faces. The response of a disc to the forces of the applied electrical field depends on its mechanical behavior. Thickness and elastic modulus in the direction normal to the disc surface are therefore important parameters. In addition, damping is a significant factor which can be caused by a backing, a wear plate attached to the front surface, and the transducer/specimen coupling conditions.

In order to obtain information on the frequency characteristic of a piezo-electric disc one can refer to the case of forced vibrations of a damped oscillator illustrated in Fig. 3 for amplitude resonance. The response functions shown in this figure involve various values of the quality factor Q which is inversely proportional to the amount of damping. The other important parameter is the resonance frequency of the undamped oscillator which is f_0.

Figure 3 indicates that a low Q will result in almost aperiodic behavior, a condition that would be highly desirable for ultrasonic spectroscopy. But it is also evident that the oscillation amplitude falls off sharply with decreasing Q. In other words, by lowering Q the response sensitivity is reduced.

Figure 3 shows further that the oscillator response approaches unity at frequencies below f_0. This means that a wider frequency coverage can be attained if a transducer element with a high resonance frequency is selected for ultrasonic spectroscopy. There is a practical limit to this approach, however, since increasing the resonance frequency requires decreasing the thickness of the transducer element and will therefore result in a more fragile component.

From the above, it is obvious that in designing transducers for ultrasonic spectroscopy one has to compromise between band width and sensitivity.

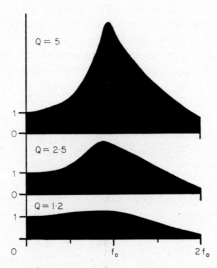

FIG. 3. Frequency response of a damped oscillator for various values of Q.

III. INSTRUMENTATION

The piezoelectric transducer used to convert electrical signals to mechanical vibrations and vice versa is a key component of the instrumentation developed for ultrasonic spectroscopy. A detailed treatment of transducer characteristics, prior to the discussion of the electronic equipment, is therefore warranted.

A. Transducers

General aspects of the characteristics of ultrasonic transducers are treated in detail by Goldman (1962) and by Filipczynski et al. (1966). In accordance with the previous section, the features to look for in a transducer to be used for the specific purpose of ultrasonic spectroscopy are a high resonance frequency of the undamped element and a relatively low Q. In addition, the piezoelectric conversion efficiency of the transducer must be high to obtain a good signal-to-noise ratio in spite of the strong damping that is required.

It is fortunate that these requirements are matched to a great extent by the demands of conventional ultrasonic pulse-echo testing. The high resolution sought in these tests calls for short ultrasonic pulses which in turn necessitate that the resonance frequency of the unmounted element be high and the Q of the transducer assembly be low. It is therefore possible to acquire transducers from commercial sources that are quite suitable for ultrasonic spectroscopy.

As examples, frequency characteristics of three commercially made transducers fabricated from different piezoelectric substances are shown in Fig. 4. The depicted characteristics were obtained by experimental techniques that are discussed below in Section IV.A.

The multiple peaks observed in the transducer characteristics are attributable to additional resonances caused by the wear plates with which the three transducers are fitted.

The frequencies noted next to the piezoelectric material in Fig. 4 represent the resonance frequencies of the unmounted transducer elements. Until recently, these frequencies were the only data that manufacturers would supply aside from specifying the piezoelectric material and the dimensions of the transducer element.

The transducer response characteristics shown in Fig. 4 are drawn on a linear amplitude scale and are normalized for better comparison. The actual transducer sensitivities given in arbitrary units are as follows:

Barium Titanate: 1 (Reference level);

Lead Metaniobate: 0·75;

Lithium Sulfate: 0·5.

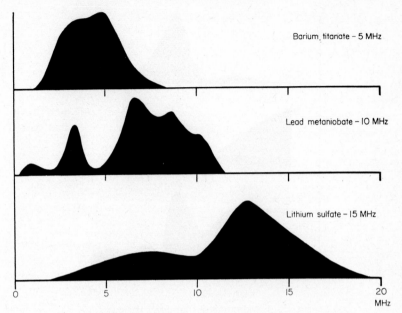

Barium titanate – 5 MHz

Lead metaniobate – 10 MHz

Lithium sulfate – 15 MHz

FIG. 4. Frequency responses of various piezoelectric transducers.

As predicted on page 39, these values indicate that a broadening of the spectral response is accompanied by a decrease in transducer sensitivity.

1. Response Equalization

Ultrasonic signals used for spectroscopy should have a spectrum that is more uniform than the response characteristics of Fig. 4 would have to offer. Hence, an equalization of the transducer response is desirable. It can be achieved by supplying an excitation voltage to the transducer which counteracts response fluctuations.

If one of the frequency-modulation techniques is used, an automatic variation of the generator output amplitude, in accordance with a function that is the inverse of the transducer response characteristic, will provide a uniform ultrasonic spectrum.

In the case of transducer excitation by direct-current pulsing, however, only a partial response compensation is possible. It is accomplished by adjusting the excitation pulse duration so that a null in the excitation pulse spectrum (see Fig. 1) coincides with the largest transducer response maximum.

Figure 5 illustrates this latter equalization procedure for the case of direct-current pulsing. The first graph indicates the loop response of the transducer. The second represents the relevant portion of an appropriately chosen

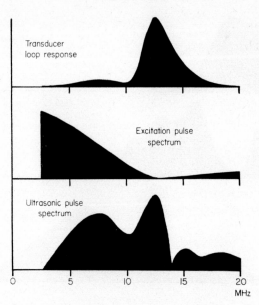

FIG. 5. Equalization of transducer response.

excitation pulse spectrum obtained by adjusting the pulse duration to $0 \cdot 075 \, \mu$sec. The last trace finally depicts the spectrum of the generated ultrasonic pulse which exhibits a fairly continuous coverage of the frequency range involved.

It must be stressed that frequency regions for which a transducer exhibits no useful response at all cannot be enhanced by electronic compensation techniques.

2. Coupling Conditions

A very important "component" of the ultrasonic spectroscope is the layer of the coupling medium introduced between the transducer and the test specimen to avoid the excessive acoustic impedance mismatch that would result from an airgap. Unless this layer is made extremely thin, it will exhibit resonances in the frequency range used for spectroscopy and will therefore interfere with the transmission of certain frequency components. Hence, particularly when very smooth planar surfaces are involved, the transducer must be effectively wrung down on the specimen to reduce the layer thickness of the couplant to a minimum.

Considering glycerin as a couplant, a layer thickness of 1 mm (0·040 in.) would lead to resonances occuring at about 1, 2, 3, 4,... etc. MHz which certainly are harmful. By reducing the layer thickness to 0·01 mm (0·0004 in.), however, the fundamental resonance of the layer would be increased to 100 MHz which is well beyond the range of frequencies currently used for ultrasonic spectroscopy.

B. Electronic Equipment

While, at present, complete instruments for ultrasonic spectroscopy are not available from commercial sources, all electronic subunits required to assemble them can be purchased.

The selection of the electronic apparatus depends primarily on the spectroscopic technique involved. In the following, schematic block diagrams will be given that show the basic features of the electronic instrumentation used with each technique. Each block on these diagrams will represent a commercially available device.

1. Frequency-Modulation Procedure

The frequency-modulation technique which involves a practically continuous transmission of test signals, in general requires the use of individual transducers for transmission and reception. If a reflection test procedure is employed these two transducers can be mounted together in the form of a single probe sometimes referred to as a transceiver search unit. An exception is the spectroscopy conducted solely for the purpose of determining the thickness resonances of a test specimen which requires only a single transducer.

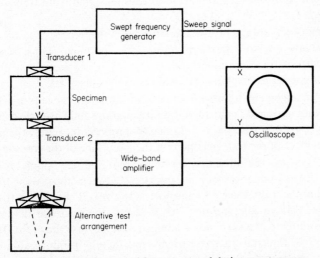

FIG. 6. Block diagram of frequency-modulation spectroscope.

Figure 6 illustrates equipment operating with two electrically isolated transducers. The heart of the instrument is a swept-frequency generator. It drives transducer 1, which functions as ultrasound transmitter, and simultaneously furnishes a sweep signal to the horizontal deflection electrodes of the oscilloscope beam.

In through-transmission testing, the ultrasonic signal traverses the specimen and is picked up by transducer 2 which reconverts it to an oscillatory voltage. In a different arrangement used for reflection testing, the transmitted ultrasound is beamed into the specimen at a slight angle and is received by transducer 2 after reflection from the back surface of the specimen or an internal discontinuity.

The electrical signal developed by transducer 2 is only a small fraction of the generator output and must be amplified by a wide-band amplifier to obtain a voltage sufficient for the vertical deflection of the beam of the cathode-ray tube oscilloscope. The trace displayed by the oscilloscope represents an amplitude versus frequency plot, i.e. a spectrum, because the horizontal deflection of the beam is furnished by the generator frequency sweep.

The circuit shown in Fig. 6 can be modified by inserting a detector and suitable filter between the wide-band amplifier output and the oscilloscope. Under these conditions, the oscilloscope trace will correspond to the envelope curve of the spectrum.

Another possibility is to substitute an electronically tuned radio-frequency amplifier for the wide-band amplifier. The tuning of this amplifier has to be effected by the sweep signal of the swept-frequency generator, delayed in accordance with the travel time of the ultrasonic signal from transducer 1 to transducer 2.

The primary advantage of amplifier tuning is the suppression of spurious signals that are due to reflections within the specimen. Thus, a spectrum obtained with a tuned amplifier will, for example, be free from amplitude fluctuations caused by thickness resonances in the specimen.

A third arrangement for the frequency-modulation procedure which has already been mentioned in the Introduction employs only a single transducer. As a result, a spectral read-out is obtained that represents merely the mechanical loading effect of the test specimen on the front face of the transducer.

Figure 7 shows the modified circuitry required for single transducer operation. An ohmic resistor is inserted between the swept-frequency oscillator and the transducer. Its purpose is to convert changes in transducer current into voltage fluctuations. As a result, the decrease in transducer current encountered at a specimen resonance will increase the radio-frequency potential across the transducer terminals and thereby produce an amplitude peak in the spectrum displayed by the oscilloscope. If more than one specimen

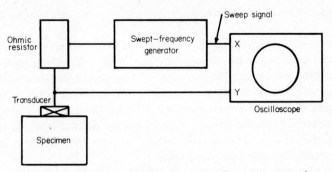

FIG. 7. Block diagram of spectroscope used for resonance testing.

resonance occurs within the range of the swept-frequency generator and the transducer response, the spectral trace will exhibit several such peaks.

Figure 8 gives examples of oscilloscope displays obtained with the circuit of Fig. 7. The first trace illustrates the response from an uncoupled transducer; the second is obtained for the same transducer when it is coupled to a 250 mm (10 in.) thick aluminium specimen exhibiting thickness resonances that are too closely spaced to be resolved. One can note that by coupling

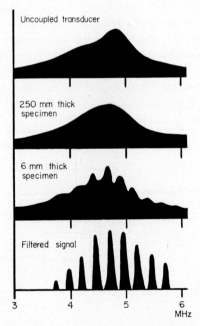

FIG. 8. Spectra obtained in resonance testing.

the transducer to the specimen its frequency response is changed to some degree.

The third trace shown in Fig. 8 is obtained with the transducer coupled to a 6 mm (0·25 in.) thick aluminium specimen. Since this specimen is relatively thin, resonance peaks are now well separated and can be observed riding on top of the response spectrum. By adding circuitry for rectification and filtering in front of the oscilloscope Y-input, the amplitude modulation can be extracted from the trace to yield indications of the type portrayed by the last trace in Fig. 8. The equidistant spectral lines seen in this plot represent solely the specimen resonances.

2. *Pulse Procedure*

The electronic instrumentation used for pulse spectroscopy differs considerably from the equipment discussed previously. Figure 9 illustrates the schematic block diagram of an ultrasonic pulse spectroscope which permits a simultaneous observation of a signal in the time domain and in the frequency domain. The indications obtained in the time domain are quite similar to those produced by conventional pulse-echo test instruments.

The circuit shown in Fig. 9 consists of a pulse generator for transducer excitation, a wide-band amplifier for the amplification of received echoes, an electronic spectrum analyzer for separating the various frequency components that are contained in a received signal, and an oscilloscope which displays the time domain in a way which is similar to conventional pulse-echo test instruments. The circuit further includes a time gate, inserted between the amplifier output and the electronic spectrum analyzer, for selecting the ultrasonic echo returns that one wants to analyze. In order to enable the identi-

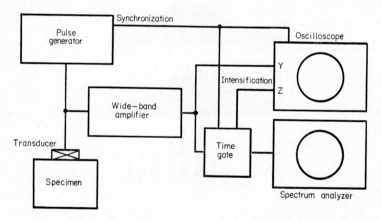

FIG. 9. Block diagram of a pulse-echo spectroscope.

fication of a selected echo on the time domain display, the time gate furnishes an appropriate intensifying signal to the oscilloscope.

The transducer excitation is periodic to produce a stationary time domain pattern on the oscilloscope screen and render echo signals suitable for processing by a spectrum analyzer that functions according to the frequency-scanning principle.   This type of analyzer is essentially a super-heterodyne radio receiver equipped with a swept-frequency local oscillator. An amplitude versus frequency display is obtained by using the oscillator sweep signal as the horizontal beam deflection voltage of the cathode-ray tube incorporated in the analyzer. A detailed discussion of these electronic spectrum analyzers can be found in a publication by Hewlett-Packard Company (1967) and in an article by Engelson (1969).

The circuit shown in Fig. 9 is very versatile. It has been described in greater detail in an earlier publication (Gericke, 1965).

3. *Pulsed Frequency-Modulation Procedure*

By combining the frequency-modulation with the pulse technique one arrives at the circuitry illustrated in Fig. 10 which, similar to the schematic of Fig. 6, does not require a special electronic spectrum analyzer. The time gate contained in the circuit opens and closes periodically at a rate that is considerably higher than the frequency sweep rate and thereby breaks the frequency-modulated signal up into a series of pulses each of which is followed by a transmission pause. During transmission pauses the transducer is able to function as a receiver for echoes returning from the specimen.

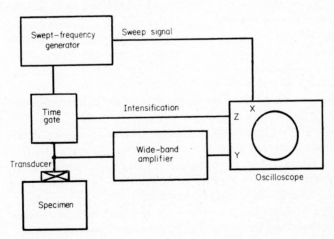

FIG. 10. Block diagram of pulsed frequency-modulation spectroscope.

The spectrum attributable to a particular echo signal is presented on the oscilloscope screen by decreasing its normal trace intensity and using the suitably delayed time gating pulse to brighten the trace momentarily during the arrival of the selected echo at the transducer. The frequency sweep rate, the gating pulse repetition rate, and its duty cycle have to be matched to the delay between the pulse transmission and the echo return, which in turn depends upon the distance traveled by the pulse in the specimen and on the ultrasonic velocity.

Compared with the previously discussed pulse procedure, pulsed frequency-modulation requires longer pulse durations. The technique is therefore not suited for applications requiring optimum resolution in the time domain. Thus far it has primarily been used for the purpose of determining transducer response characteristics (Gericke, 1966a).

IV. Applications

As already pointed out at the beginning, the principle purpose of ultrasonic spectroscopy is to enhance the effectiveness of the ultrasonic method of nondestructive testing. In the following, a number of examples will be discussed that illustrate the practical potential of the spectroscopic method in ultrasonic inspection work.

A. Testing Transducer Frequency Response

A fundamental requirement of ultrasonic spectroscopy is that one has an accurate knowledge of the frequency response of the transducer or the transducers that are involved. Techniques for determining transducer response characteristics are therefore of prime importance for the spectroscopic method. But they are also valuable for conventional time domain inspection techniques in cases where the test instrument utilizes transducer shock-excitation and wide-band amplification because, under these conditions, the transducer response is the only factor determining the ultrasonic frequencies involved in a test.

It is known that the results of conventional ultrasonic inspection can vary considerably from transducer to transducer even if element size and resonance frequency (measured on the unmounted element) remain the same. To avoid this problem, one must control both the transducer beam profile and its response characteristic during fabrication.

In order to determine the frequency response of a transducer, both the pulse and the pulsed frequency-modulation technique can be used. The test is carried out by analyzing the first back echo obtained from a specimen whose ultrasonic attenuation exhibits a negligible degree of frequency dependence. In addition, the specimen must be relatively thin to avoid errors due to the divergence of the ultrasonic beam (often referred to as geometric attenuation) which strongly depends on the frequency.

The frequency-modulation technique yields the transducer "loop" response directly, provided that the transducer excitation is constant over the frequency range concerned. The direct-current pulse technique, on the other hand, requires the use of extremely short generator pulses to broaden the first lobe of the transducer excitation spectrum as much as possible (see Section II.B.). Even then, results have to be corrected for the drop-off in amplitude towards higher frequencies which is present in the excitation pulse spectrum.

From the "loop" response the regular response characteristic can be derived by plotting the computed square root of the measured amplitude values. The response characteristics shown earlier in Fig. 4 were obtained in this manner.

More details on transducer response measurements using the pulsed frequency-modulation technique have been given in an earlier publication (Gericke, 1966a).

B. Measuring Thickness

As mentioned previously, frequency-modulation equipment that is designed for single transducer operation (Fig. 7) is used predominantly for ultrasonic thickness measurements. While this technique, which is usually referred to as resonance testing, may not be the most accurate ultrasonic procedure now available for thickness measurements, it has nevertheless gained wide acceptance and is therefore well documented elsewhere (Evans, 1959).

A problem encountered with resonance testing is the uncertainty in deciding whether an observed spectral line represents the fundamental or a harmonic resonance of the specimen. Unless one has a rough idea of the specimen thickness, it is therefore advisable to test over a relatively wide frequency range which yields at least two consecutive spectral lines. For two consecutive specimen resonances, f_n and f_{n+1}, the thickness d can unambiguously be computed from

$$d = \frac{v}{2(f_{n+1} - f_n)} \tag{6}$$

in which v represents the ultrasonic velocity of the tested material.

C. Examining Microstructure

The attenuation of ultrasound in polycrystalline and amorphous substances is dependent upon the microstructure of the material and the frequency of the ultrasonic signal (Truell and Elbaum, 1962). Relative differences in microstructure can therefore be determined by obtaining ultrasonic attenuation spectra for the materials involved.

Prior to the availability of spectroscopic instruments, special but basically conventional pulse-echo test equipment was used for attenuation studies. This equipment, which is still being manufactured, requires manual tuning and an exchange of transducers if measurements are to be conducted at various frequencies. It further involves the observation of multiple back echoes since the attenuation is determined as the rate of amplitude decay in a train of multiple back echoes. In contact testing this is a serious disadvantage because echo amplitudes of higher order than the first are dependent upon the difficult to measure loading effect of the transducer on the specimen entrance surface.

In addition to producing quicker results, the spectroscopic method offers the benefit not to involve multiple echoes since an analysis of the first back echo reflected from the free surface of the test specimen suffices. Should a reference level be required, the spectral amplitude obtained for the low frequency end of the back-echo spectrum can be used for that purpose.

1. Polycrystalline Metals

In the frequency range of 1 to 25 MHz, the attenuation of ultrasound in polycrystalline metals with random orientation of crystallites is caused to a large extent by scattering at grain boundaries. The elastic anisotropy of the single crystal in conjunction with the average grain size determine magnitude and frequency dependence of the ultrasonic attenuation.

The degree of anisotropy is small in magnesium and aluminum and polycrystalline specimens made from these metals therefore exhibit little attenuation in the 1 to 25 MHz range. Hence, they are suitable as test blocks for determining the response characteristics of transducers (see also Section IV.A.).

Copper, on the other hand, represents a material with a relatively strong anisotropy and can serve to demonstrate the frequency dependence of losses due to ultrasound scattering.

This is illustrated in Fig. 11 which shows back-echo spectra obtained for a 9·5 mm (0·375 in.) thick aluminum plate and a 50 mm (2 in.) thick copper block. Since the attenuation in a thin plate of aluminum is very low, the first spectrum is mainly determined by the excitation signal and the "loop" response of the transducer used in the experiment. The spectrum of the much thicker copper specimen, on the other hand, demonstrates the losses

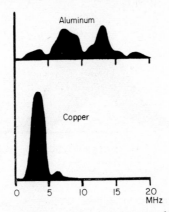

FIG. 11. Influence of specimen microstructure on ultrasonic spectrum.

encountered by the ultrasonic signal in polycrystalline copper having an average grain diameter of 0·004 mm (0·00016 in.). To permit an easier comparison with the aluminum spectrum, the relatively small echo return from the copper block was amplified 15 dB more than the aluminum back-echo.

The results of this test indicate that ultrasonic frequencies above 8 MHz are very strongly attenuated in the copper specimen and are therefore completely eliminated from the back-echo spectrum.

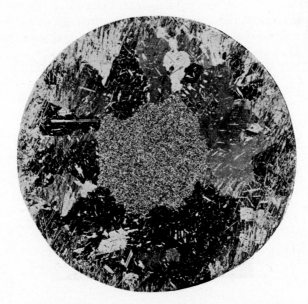

FIG. 12. Etched cross section of copper specimen with dual grain structure

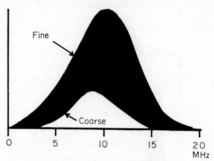

FIG. 13. Influence of grain size on spectral maximum.

To study the characteristics of copper further, a specimen with a dual grain structure was made. Fig. 12 shows the etched cross section of the cylindrical specimen whose diameter and length are both 50 mm (2 in.). While the core of the specimen exhibits a grain structure with an average diameter of about 0·004 mm (0·00016 in.), its peripheral region consists of extremely coarse grains.

Figure 13 shows the back-echo spectra observed for the two regions of the specimen over a frequency range that was narrowed down to 3–7 MHz. The upper edge of the blackened area corresponds to the spectrum of the fine grain and the lower edge to the spectrum of the coarse grain. In addition to an overall reduction in spectral amplitude observed for the coarse region, a shift of the spectral maximum towards lower frequencies can be noted. A similar observation has been made with graphite specimens by Serabian (1967).

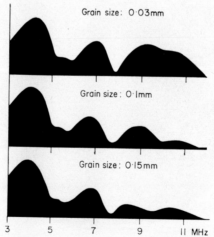

FIG. 14. Spectra of steel specimens with different grain sizes.

As a further example of the effect of the microstructure on the ultrasonic spectrum, Fig. 14 shows back-echo spectra obtained for steel specimens that have different grain sizes (Gericke, 1965).

In contrast to earlier figures, the spectral amplitudes have been plotted on a logarithmic scale. As a result, differences in the ordinate observed for these spectra are actually representative of amplitude ratios and, when plotted as functions of frequency, represent directly the relative increases in attenuation encountered with rising ultrasonic frequency. Figure 15 illustrates these functions, derived from the spectra of Fig. 14 and plotted with the grain size as parameter.

2. Amorphous Materials

Ultrasonic attenuation in glass, plastics, and other amorphous materials is caused by loss mechanisms other than scattering at grain boundaries but the ultrasonic spectra obtained for such materials are nevertheless quite characteristic.

This is illustrated by Fig. 16 which shows back-echo spectra for a glass, a Lucite (a thermoplastic material similar to Plexiglass or Perspex), and a clear, transparent specimen cast from a polyester resin. As a reference, Fig. 16 shows, in addition, the spectrum of the ultrasonic signal used to examine

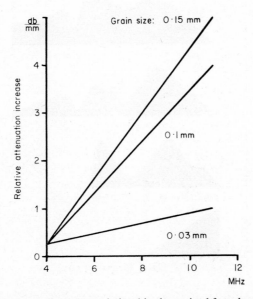

FIG. 15. Attenuation *vs.* frequency relationship determined from back-echo spectra.

these specimens. Spectral amplitudes are plotted on a linear scale and are normalized for easier comparison. Due to spectrum analyzer limitations the lowest frequency observable in this experiment was 2·5 MHz.

One notes that glass passes all frequency components of the interrogating ultrasonic signal almost equally well with only a slight relative amplitude reduction at frequencies above 10 MHz. Lucite, on the other hand, conducts ultrasound only up to about 14 MHz where the material becomes practically opaque. The cast resin exhibits the most inferior ultrasonic transmission characteristics with a complete cutoff occurring already at about 9 MHz.

The frequency limitations exhibited by cast polyester resins must be remembered when wedges for ultrasonic angle beam probes are fabricated from such substances.

D. Analyzing Defect Geometry

While conventional ultrasonic pulse-echo testing is a superior nondestructive method for the detection of small internal discontinuities, it often does not provide sufficient information for deciding whether the inspected item should be accepted or rejected.

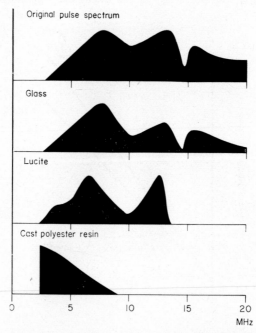

Fig. 16. Spectra obtained for amorphous materials.

In assessing the magnitude of an ultrasonically exposed discontinuity, the height of the ultrasonic echo is sometimes considered as a reliable guidance. Unfortunately, such an interpretation of the test data is frequently erroneous since the echo amplitude is influenced by several factors of which the size of the discontinuity is only one. Orientation, geometry, and acoustic impedance, for example, are equally as important for the magnitude of the echo return as the mere size of a discontinuity.

One way to obtain more information from an ultrasonic test is to beam at the discontinuity from various angles, a procedure which is only feasible if the specimen has a suitable shape. Another approach is to vary the wavelength of the interrogating ultrasonic beam. This is accomplished by changing the ultrasonic frequency and leads into ultrasonic spectroscopy.

In order to demonstrate the practical value of applying the spectroscopic method for defect assessment, results of conventional pulse-echo testing will be contrasted with spectroscopic data. The pulse-echo spectroscope outlined in Fig. 9 is ideally suited for this purpose because it yields read-outs in both the frequency and the conventional time-domain.

1. Small Isolated Defects

Considering first single defects, experimental results shall be presented which have been obtained for artificial flaws of very closely controlled geometry. To minimize the influence of frequency-dependent attenuation, the specimens containing these defects were fabricated from aluminum.

Figure 17 shows a cross section and side views of the fabricated cylindrical specimens which are 50 mm (2 in.) high and 50 mm (2 in.) in diameter. Illustrated also are three different categories of artificial defects. One is a cylindrical hole drilled parallel to the surface to which the transducer is coupled. The second is a crack-like defect represented by the bottom of a machined slot which is oriented parallel to the test surface. The third type of defect is also

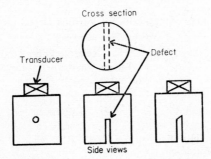

FIG. 17. Test specimens with artificial defects.

represented by the bottom of a slot but its short dimension is now angulated with respect to the test surface. Three defects of the last category are provided with angulations of 10°, 20°, and 40°. A total of five different defect configurations is thus available.

In all five cases, the rectangular projection of the defect onto the test surface is the same and measures 50 mm (2 in.) by 4 mm (0·15 in.).

Table I lists the data obtained for these five specimens when defect echo amplitudes are determined by a conventional procedure using a high-resolution (short pulse) transducer. Although all defects have essentially the same magnitude, one notes large differences in measured echo height which amount to as much as a ratio of 50 to 1 or 34 dB. Hence, a defect assessment based solely on the echo height would lead to significant error.

It will now be shown that by subjecting the defect echoes of the above specimens to a spectrum analysis, certain clues can be gained that shed light on the existing differences in defect geometry.

Figure 18 shows the spectra that are obtained for the five specimens when echo amplitudes are normalized. The first spectrum, representing 0° defect orientation, roughly resembles the spectrum of the ultrasonic pulse originally injected into the test specimen (compare the first trace of Fig. 11). This result is plausible since a sizeable discontinuity oriented parallel to the test surface should reflect all frequency components of the impinging ultrasonic pulse equally as well.

The second spectrum shows that a small defect angulation of 10° affects the return of low frequencies in the spectrum less than the return of high frequencies since the latter are better collimated and therefore more effectively reflected away from the transducer.

The next spectrum, obtained for 20° angulation, exhibits a pronounced spectral substructure and lacks, in addition, the higher frequency components. A possible explanation for the substructure is that signal interference effects

TABLE I. Echo Amplitudes for Various Defect Geometries

Defect Geometry	Relative Echo Amplitude Linear Scale (arbitrary units)	Logarithmic Scale (in dB)
Crack with 0° orientation	100 (Reference level)	0 (Reference level)
Crack with 10° orientation	15·8	−16
Crack with 20° orientation	5·6	−25
Crack with 40° orientation	2·0	−34
Cylindrical hole	25	−12

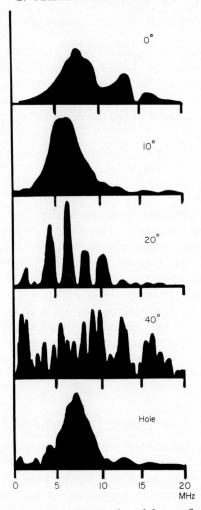

FIG. 18. Spectra obtained for various defect configurations.

are encountered because, due to the greater defect angulation, ultrasonic energy is returned from different depth levels.

For the largest angle of 40°, the maxima and minima in the spectrum are more closely spaced. This can again be explained as an interference phenomenon if one takes the now greater depth range of the defect into account.

The last spectrum depicted in Fig. 18 is obtained for the specimen containing a cylindrical hole. The spectrum is different from all the others and exhibits its largest amplitudes in an intermediate frequency range centering at 7·5 MHz.

The quintessence of the presented results is that a comparison of defect echo heights for the purpose of size assessment is not advisable if the spectral signatures of the defect echoes show significant differences.

This statement can be rephrased by saying that a good correlation of spectral signatures is essential for the quantitative interpretation of pulse-echo test data.

Other experimental results leading to similar conclusions have been reported elsewhere (Gericke, 1966b, 1967a).

2. *Multiple Defects*

Occasionally an ultrasonic beam penetrating into a test specimen will encounter more than one defect on its path. If such defects are well separated in depth, an individual defect-echo analysis can still be carried out by an appropriate adjustment of the time gate of the ultrasonic spectroscope (see Fig. 9). But if indications in the time domain are of the type shown in Fig. 19 they cannot be further resolved.

The spectrum of such an indication is not a smooth curve. It exhibits a substructure similar to that encountered in some of the spectra of Fig. 18. Figure 20 gives examples showing spectra obtained for two different multiple defects. The traces are contrasted with the spectrum of the ultrasonic pulse that is initially injected into the specimens.

V. FUTURE DEVELOPMENTS

Efforts to improve the method of ultrasonic spectroscopy could follow along two different lines. One approach would be to gain more practical experience by comparing the spectral signatures of a larger number of test specimens with the results of destructive testing. Such a project should eventually lead to the compilation of numerous spectra in the form of an atlas to which one could refer to interpret the spectral traces obtained from unknown specimens.

FIG. 19. Echo return from multiple defect.

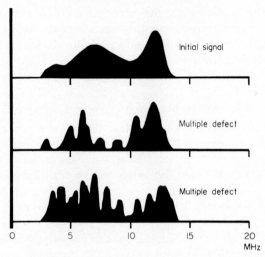

FIG. 20. Spectra obtained for multiple defects.

The other approach would be to aim for a further improvement of the equipment used for ultrasonic spectroscopy. Some thoughts in this latter area will now be presented.

A. *Improvement of Transducers*

A definite improvement of the ultrasonic spectroscope would be an extension of its frequency range. A coverage of 1–100 MHz, for example, would be desirable for the examination of steel microstructure.

While electronic circuitry providing such frequency coverage can be constructed without too much difficulty, the design of a suitable piezoelectric transducer might pose a problem. Hence, other techniques of ultrasonically energizing and probing a test specimen should be considered.

In order to inject an ultrasonic signal into a test specimen, mechanical impacting techniques could be used. According to the statements made in Section II.B., generating an ultrasonic spectrum of 100 MHz width would require imparting a mechanical impulse of not more than 0·01 μsec duration to the test specimen. This, unfortunately, is by no means a simple task. It calls for a "hammer" of very small mass to be brought down on the specimen with an extremely high velocity.

One feasible method is the acceleration of liquid particles by a powerful electric spark, another the inductive propulsion of a small metal ring by an adjacent pulse excited coil. The elastic wave resulting from the thermal shock produced by a laser beam offers a further possibility.

The second step in eliminating the piezoelectric transducer is to introduce a new device for sensing ultrasound. If the specimen is metallic, a probe based on a capacitive effect or a semiconductor contact could be used (Gericke, 1967b). Both these devices would exhibit less frequency dependence than piezoelectric transducers.

B. *Improvement of Electronic Circuitry*

Retaining the piezoelectric transducer as the link between the specimen and the electronic equipment has the advantage of avoiding the potentially greater complexity of novel devices. Thus, the question arises as to how the ultrasonic spectroscope could be improved by modifying its electronic circuitry.

According to Section II.B.1. the voltage pulse exciting a piezoelectric transducer should be as high and as short as possible. While the maximum pulse amplitude is ultimately limited by the electrical breakdown of the transducer element, pulsers currently available for generating very short signals do not reach this critical limit, which is of the order of 1000 volts.

The design of a generator capable of supplying a pulse of up to 1000 volts amplitude into a 10 ohm load would constitute a definite improvement. The pulse duration should be less than 0·05 μsec with rise and decay times of not more than 0·005 μsec. Using such equipment the desired spectral range of 100 MHz might be attainable with piezoelectric transducers.

1. *Spectrum Analyzer*

From the point of view of the signal-to-noise ratio, electronic spectrum analyzers employing frequency scanning techniques are not ideal. They sample the amplitude of each spectral component contained in the input signal only for a short moment and are therefore affected by statistical signal fluctuations.

A better S/N ratio would be available from a so-called "real-time" spectrum analyzer which utilizes an individual filter circuit for each frequency. But even a quasi-continuous coverage of the wide frequency range desired for ultrasonic spectroscopy would require an extremely large number of filters and would therefore probably result in a prohibitive cost and bulk of the equipment.

A less expensive improvement of the spectrum analyzer S/N ratio can be achieved by signal averaging procedures which take advantage of the fact that the input signal fed to the spectrum analyzer is repetitive.

Signal averaging of the spectrum analyzer output can be accomplished in several ways. Photographic registration or a storage-type cathode-ray tube used as a read-out will integrate the spectral trace over a longer period of time and thus even out statistical fluctuations. Another possiblity is to connect an electronic signal averager of the type described by Nittrover (1968) to the output of the spectrum analyzer.

An improvement of the S/N ratio provided by signal averaging would permit the use of transducers having lower Q and thus help to increase the frequency range of the ultrasonic spectroscope.

REFERENCES

Brown, B. and Gordon, D. (1967). "Ultrasonic Techniques in Biology and Medicine". Iliffe, London.
Engelson, M. (1969). *Microwaves* **8** (3), 52–55.
Evans, D. T. (1959). *In* "Nondestructive Testing Handbook " (R. C. McMaster, ed.). Vol. 2, p. 4420. The Ronald Press, New York.
Filipczynski, L., Pawlowski, Z. and Wehr, J. (1966). "Ultrasonic Methods of Testing Materials". Butterworths, London.
Gericke, O. R. (1963). *J. acoust. Soc. Am.* **35**, 364–368.
Gericke, O. R. (1964). *J. acoust. Soc. Am.* **36**, 313–322.
Gericke, O. R. (1965). *Mater. Res. Stand.* **5**, 25–30.
Gericke, O. R. (1966a). *Mater. Evaluation* **24**, 409–411.
Gericke, O. R. (1966b). *J. Metals, N.Y.* **18**, 932–937.
Gericke, O. R. (1967a). *In* "Proc. 6th Symp. on Nondestructive Evaluation of Aerospace and Weapons Systems Components and Materials" pp. 1–16. Amer. Soc. of Nondestructive Testing, San Antonio, Texas.
Gericke, O. R. (1967b). *In* "Physics and Nondestructive Testing" (W. D. McGonnagle, ed.) pp. 257–282. Gordon and Breach, London, New York.
Goldman, R. (1962). "Ultrasonic Technology". Chapman and Hall, London.
Hewlett-Packard Company (1967). "Spectrum Analysis", Application Note No. 63. Hewlett-Packard, Palo Alto, California.
Nittrover, C. A. (1968). *Electron. Instrum. Dig.* **4** (9), 12–19 and (5), 28–32.
Posakony, G. T. (1959). *In* "Nondestructive Testing Handbook" (R. C. McMaster, ed.) Vol. 2, p. 4420. The Ronald Press, New York.
Roderick, R. L. and Truell, R. (1952). *J. appl. Phys.* **23**, 267–279.
Serabian, S. (1967). *J. acoust. Soc. Am.* **42**, 1052–1059.
Truell, R. and Elbaum, Ch. (1962). *In* "Encyclopedia of Physics" (S. Fluegge, ed.) Vol. 11/2, pp. 163–172. Springer-Verlag, Heidelberg.

CHAPTER 3

Ultrasound Imaging Systems

J. E. JACOBS

BioMedical Engineering Center, Technological Institute,
Northwestern University,
Evanston, Illinois, U.S.A.

I. INTRODUCTION

This chapter is concerned with detecting ultrasound energy over an extended area. Ultrasound imaging systems of this type closely parallel those utilizing light and X-ray.

This type of imaging system is distinct from the other types of ultrasound display systems in which a geometrically related image is obtained by a combination of mechanical scanning and relative time measurement.

There are a number of methods of obtaining an area display of an ultrasound field. These may be classified by the basic processes used in their production as follows:

(a) Photographic and Chemical Effects;

(b) Thermal Effects;

(c) Optical and Mechanical Techniques;

(d) Electronic Techniques.

Each of these methods produces an image of the distribution of ultrasound energy in a plane. Each method differs in characteristics, which in turn limits the application in a practical nondestructive testing system. In the material that follows, the theory underlying such systems, as well as the practical applications, will be detailed.

A. Photographic and Chemical Systems

The first report of the detection of ultrasonic energy using a photographic emulsion was that of Marinesco and Truillet (1933). Since that time numerous studies to ascertain the physical mechanism responsible for the conversion of ultrasound energy into a photographic image have been reported; however, as of this date, a well defined, acceptable explanation for the conversion has not been identified (Arkhangelskii, 1967). The method most widely used involves pre-exposing the photographic plate in a uniform light field followed by the introduction of this pre-exposed plate into a tank of developing solution in which the ultrasound to be detected is being propagated. This indirect method of detecting ultrasound with a photographic emulsion is based on the observation that a photographic plate undergoing development will have its development process speeded up by the application of the ultrasound. It is thought that the role of the ultrasound is to increase the developing speed through mechanical agitation at the developer-film emulsion interface. The sensitivity of this system has been reported to be in the order of one watt per square centimeter.

Another method for detecting ultrasound which may be classified as photographic, although, in the strictest sense, it should be a chemical reaction, is based on the fact that ultrasound will release iodine from potassium iodide solutions. This effect has been used in connection with a photographic emulsion. In another method, advantage is taken of the fact that iodine is readily detectable visually in a starch solution, hence this technique has been reported as a method of forming images of the ultrasound field (Bennett, 1952). The sensitivity of the iodine release, as well as a photographic emulsion system, is in the order of one watt per square centimeter.

These photographic and chemical methods are at best relatively insensitive and rather cumbersome in comparison to those other methods available for the visualization of extended area ultrasound fields. As near as can be told at this date, they are of historic interest only and do not constitute a practical or widely accepted method of ultrasound imaging.

B. Thermal Methods

Ultrasound energy when absorbed is converted to thermal energy. The availability of ultrasound related thermal energy forms the basis for the second method of ultrasound imaging.

One system, formerly in relatively widespread use, is based on the fact that a fluorescent phosphor once excited and subsequently subjected to ultrasound energy, will have the fluorescence in the region subjected to ultrasound energy quenched or darkened. Through the use of persistent phosphors, it is possible to design a system that will form images of the ultrasound field.

In a working system, a phosphor screen is continuously excited by a suitable light source, usually ultraviolet. The ultrasound field is caused to impinge on the phosphor and, depending on the phosphor used, produces a quenching or enhancing effect on the fluorescence. The sensitivity of such systems has been reported to be in the order of one tenth of one watt per square centimeter (Ermst and Hoffman, 1952).

Another effect which has been reported, that lends itself to the detection of ultrasound fields, is the effect of thermal energy on a photo-emissive surface. The sensitivity of such a system is in the order of a tenth of a watt per square centimeter which severely limits its applicability to a practical system (von Ardenne, 1955).

The direct detection of the thermal effects produced by the ultrasound field by means of thermal sensitive detectors such as thermo-couples or thermistors has been postulated as the basis for an ultrasound imaging system. Again, as in the case with other thermal systems, the sensitivity is in the order of a tenth of a watt per square centimeter, and falls far short of many of the other systems that have been demonstrated for ultrasonic imaging (Dunn and Fry, 1959).

II. OPTICAL AND MECHANICAL TECHNIQUES

Ultrasound fields may be detected by a number of mechanical methods of which probably the most elegant is that described by Pohlman (1948).

Of all the mechanical and optical methods used, probably the Pohlman cell has found the widest application. Pohlman's original method made use of a detector cell in which small aluminum flakes are suspended in xylene.

This cell has recently been improved in efficiency by using water as the dispersing medium. Figure 1 is a schematic representation of the image converting system based on the Pohlman principle using a water suspended aluminum flake system (Ogura et al., 1968).

When an acoustic image of the object under inspection is focused by the acoustic lens on the surface of the light source side of the aluminum suspension cell, a force is exerted on the aluminum flakes dispersed in the water

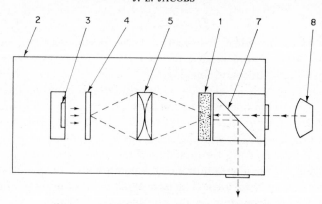

FIG. 1. Pohlman cell ultrasound imaging system. Al-suspension cell is 1, 2 is water filled tank, 3 is ultrasound source, 4 is object inspected, 5 is lens system, 7 is half-silvered plate, 8 is light source to illuminate Al-suspension cell.

and they become parallel with the wave surface of the ultrasound wave like a small Rayleigh plate. Under the force of the ultrasound energy the direction of each flake becomes uniform. On the aluminum suspension cell surface, that part subjected to the ultrasonic wave contains aluminum flakes whose direction is aligned, while the part free of the ultrasonic energy is not affected and has the aluminum flakes facing in all directions due to Brownian movement. When this aluminum-flake cell is illuminated by the light source, in those areas where the ultrasound wave is present the aluminum flake direction is uniform and becomes bright, while in those portions where ultrasound energy is not present, the randomness of the aluminum flakes is such that reflection does not occur. An image obtained from such a cell is presented in Fig. 2. The principle disadvantage of the system is the time of response which may approach the order of 10 to 20 seconds.

Another method that has been used to visualize ultrasound fields involves the optical shadowing of a liquid surface deformed by the ultrasound radiation image with subsequent detection of the light variation caused by the density changes or acoustic birefringence within the ultrasound transmitting medium (Schuster, 1951). The deformation of a liquid surface is widely used as a simple method of examining the field patterns of ultrasound transducers (Piguleuskii, 1958).

A method of examining the field pattern of transducers is that termed the Schlieren method. This method, shown schematically in Fig. 3, is based on the principle that a light ray is refracted and therefore deviated laterally when it passes through a block of material whose refractive index differs from that of the surrounding medium, or when it passes through a medium with

FIG. 2.   Image of gear obtained using system of Fig. 1.

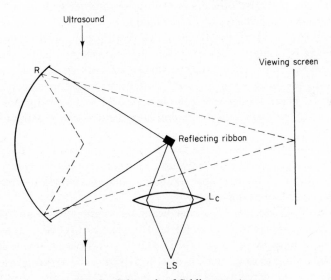

FIG. 3.   Schematic of Schlieren system.

a variable refractive index. An ultrasound beam consists of a rapid progression of compressions and rarefactions, and therefore the refractive index of a medium traversed by the ultrasound beam is constantly changing in the path of the beam. A practical Schlieren system normally uses mirrors instead of the usual lens system in order to avoid chromatic aberrations.

III. ELECTRONIC SYSTEMS

One of the most sensitive methods of detecting an extended ultrasound field is to make use of the piezoelectric effect in solids. The piezoelectric effect describes the ability of certain materials, for example quartz, to convert the mechanical energy present in an ultrasound field into an electric potential.

The voltage produced on such materials by reflected or transmitted ultrasound may be observed in a variety of ways. For example, if the transducer is moved and information related to its movement and electrical output is presented on an oscilloscope, an image of the ultrasound field may be obtained. The oscilloscope presentation produces an image of the ultrasound transmitted or reflected by the test object by correlating the mechanical scan of the transducers with the electronically processed output of the piezoelectric detector (Suckling and MacLean, 1955). The necessity for the mechanical movement of the transducers or the object in order to form an image is one of the major drawbacks of this technique.

This problem is effectively eliminated by detecting the ultrasound induced charge image by using a single extended area piezoelectric plate. The point-to-point voltage generated on the transducer will correspond in intensity and geometry to the ultrasound intensity pattern striking the extended piezoelectric plate. These voltage variations may be observed by using point probes on the back of the transducer or by scanning the transducer with an electron beam.

The latter approach, that is the use of an electronic beam for the scanning of the extended area piezoelectric transducer, has been extensively investigated and forms the basis for the most sensitive system developed to date for the detection of ultrasound images.

The use of an electron scanning beam, in conjunction with the piezoelectric plate, to display ultrasound field intensities over an extended area was first postulated by Sokoloff (1939). The Sokoloff type of image converter has reached a stage of development such that reliable and versatile instruments are available commercially. Numerous investigations made over the past decade have resulted in a complete understanding of the performance characteristics of the systems using the Sokoloff image converter tube as well as an elucidation of the difficulties associated with the use of such a system.

The original Sokoloff disclosure envisaged a television camera tube consisting of a quartz plate which formed the face plate of an iconoscope type of pick-up tube, one side of the quartz plate being subjected to the impinging ultrasound field, which in turn produced a piezoelectric potential distribution corresponding both in intensity and spatial distribution to the impinging field.

These ultrasound induced potentials modulate the secondary emission electrons produced by a high velocity electron scanning beam. This modulation of the secondary emission induced by the scanning beam produces the image-forming signal which is subsequently amplified and displayed using conventional television techniques.

Many alternatives to the Sokoloff tube have been proposed for conversion of the ultrasound image into an electrical signal for subsequent display; however, as near as can be told at this date, the Sokoloff tube is the configuration which forms the basis of all practical and working systems (Oschepkov et al., 1955; Freitag et al., 1959; Smyth et al., 1963; Haslett, 1966; Grasyuk et al., 1965; Fotland, 1960; Smyth and Sayers, 1960; Sheldon, 1958; Prokhorov, 1964).

The sensitivity of the modern Sokoloff type ultrasound image converter tube is limited by the ability of the piezoelectric voltage on the input surface to modulate the secondary emission electrons. The first tubes developed were limited by the noise of the subsequent amplifiers. However, modern versions, such as in Fig. 4, utilize a secondary emission electron multiplier which has such low noise characteristics that the sensitivity limitation becomes that given above. The minimum recognizable potential on the piezoelectric surface limits the sensitivity of the modern ultrasound image tube to power levels in the order of 10^{-7} watts per square centimeter (Jacobs et al., 1963).

The detail resolvable by the ultrasound visualization system is limited to a large extent by the wavelengths of the sound frequencies themselves. These wavelengths in water, a commonly used coupling medium, vary from six hundredths of an inch at a frequency of 1 MHz down to six thousandths of an inch at a frequency of 10 MHz. These considerations, however, do not completely determine the limiting resolution of a practical system. Since the velocity of sound in the commonly used piezoelectric material is considerably greater than that of the commonly used liquid coupling media, the longer wavelength limitation occurs in the solid conversion layer material and becomes the determining factor in resolvable detail of the practical system. Although the image converter tubes will respond across a wide frequency range, their sensitivity and resolution capabilities are increased at odd multiples of a resonant frequency determined by the thickness of the input surface. The tubes have a maximum sensitivity at the frequency where the

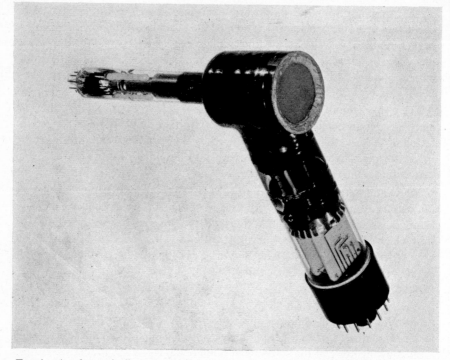

FIG. 4. An electronically scanned ultrasound image converter tube. Operation of the tube is as follows: A high velocity electron scanning beam stabilizes the scanned surface of the quartz plate at a potential approximately 3 volts positive relative to the stabilizing mesh. The secondary emission electrons produced by the scanning beam are modulated by the piezoelectrically-induced potential resulting from the impinging ultrasound field. The secondary electron cloud within the stabilizing mesh region forms a virtual cathode whose electrons are directed into the electron multiplier by the accelerating mesh. The use of the electron multiplier eliminates for all practical purposes stray signal inputs to the system. At 2 MHz, input level of 10^{-7} W/cm² produces a discernible signal output from the tube. Seal is of an indium based solder.

input conversion plate thickness is equal to a half wavelength of the incident sound field in the solid plate material.

The difference in velocity of sound in the solid material of the plate in contrast to the liquid coupling medium, and the increased sensitivity of the plate at odd multiples of half wavelength, dictate a situation such that the detail resolvable is approximately three times that one would predict based on the wavelength of the ultrasound in a liquid medium. Stated another way, the minimum resolvable detail is a cylinder whose diameter is equal to the thickness of the plate. There also exists a frequency dependence of the maximum detail resolution obtainable on the plate. This results in maximum resolution occurring at a frequency approximately 94 per cent of that of the half-wave resonance point (Hartwig, 1959). At this point, the resolution of the system is given by eqn. (1):

$$\text{Minimum diameter resolvable} = \frac{2 \cdot 86}{\text{Freq. in MHz}} \text{ mm.} \tag{1}$$

Investigations as to the effect of an added optical system have shown that, due to the large acoustic impedance discontinuity existing between the normally used liquid coupling media and the solid piezoelectric image conversion plate, a limited angle of incidence must be used. This seriously limits the aperture of the optics with the result that the use of an optical coupling system with an image conversion system of the type described here results in a degradation of the resolvable detail. Equation (2) describes the effect of adding an optical system (Jacobs, 1967):

$$d_{\min} = \frac{4 \cdot 34}{\text{Freq. in MHz}} \text{ mm.} \tag{2}$$

Experimental verification of these equations is shown in Figures 5–7.

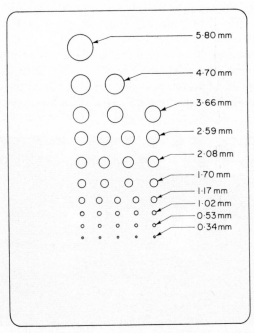

FIG. 5. Test object used for resolution studies. Material is tungsten loaded epoxy which attenuates incident sound 60 db.

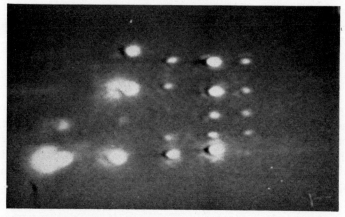

FIG. 6. Image of resolution test. Ultrasound frequency 7 MHz, object in contact with face plate. Smallest hole resolved is 0·52 mm. Resonant frequency of plate 7 MHz.

A. *Image Display Techniques*

The majority of the systems reported to date utilize continuous wave ultrasound sources and conventional closed circuit display techniques. These systems, shown in Fig. 8, at frequency ranges of 1 to 4 MHz are plagued by standing wave patterns.

The standing wave problem may be solved by a pulsed ultrasonic image system. Such systems are widely used in the C-scan method involving the mechanical movement of an ultrasonic transducer. Pulsed television systems have also been proposed, and one has been demonstrated (Sayers, 1966;

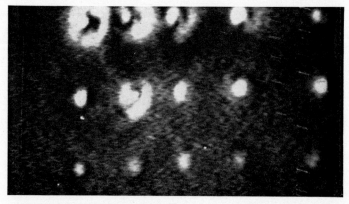

FIG. 7. Image obtained with resolution test object in direct contact with face plate. Ultrasound frequency 10 MHz, smallest resolved hole 0·34 mm.

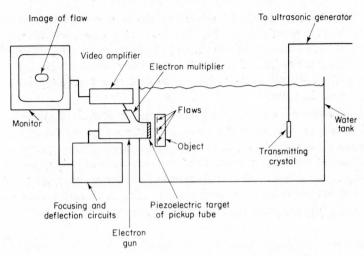

FIG. 8. Schematic representation of complete visualization system with monochrome television display utilizing a tube of the type shown in Fig. 4.

Goldman, 1962; Lawrie, 1964). The demonstrated system has proved that such systems can be made. However, the slow frame time of this particular system, about 4 seconds for typical operations, is a disadvantage because inspection object movement capability is limited. This slows the inspection rate and also decreases flaw detection capability as compared to the fast, continuous-wave system, with which advantage is taken for detecting moving inhomogeneity indications.

Pulsed systems can offer a number of advantages in addition to standing wave problem elimination. One can use high intensity ultrasonic pulses in order to obtain greater signal strengths. For short pulse lengths, greater ultrasonic intensity can be tolerated without cavitation. In practice, however, threshold sensitivity is not an important problem with the ultrasound television system. A more important advantage is that one can obtain uniform ultrasonic illumination without going to large transmitter-to-detector distances in order to obtain the far-field pattern, because transducer near-field patterns do not have time to form. Another significant advantage is that reflection methods would be more meaningful. A gated system might aid in obtaining useful images from different depths in the object, thereby contributing a useful technique for examining thick objects.

Recently the application of holography to the problem of acoustic imaging has received widespread attention (Halsted, 1968). Most of these methods rely on the holographic techniques developed for optical fields, in which the phase and amplitude are recorded on a photographic film as modulations in density and spacing of a spatial reference carrier. This is accomplished by photographing, in some plane, the interference pattern between the reference beam and the light scattered from the object under study. The sound waves require levels in the order of 1 W/cm^2 to register directly on the photographic plate; therefore, several methods have been devised for making visible the distribution of the sound intensity in a particular plane.

An entirely different method which has no analogy in optics is the sampling method of holography. Here the sound field is scanned with a transducer which gives an electrical output corresponding in amplitude and phase to the local sound pressure. A phase reference may be added electronically or by using a conventional acoustic reference beam. In either case, the resulting electrical signal is further processed and recorded as density variations on photographie film in synchronism with the scanning.

The holograms obtained by these various methods are viewed in visible light. Because of the difference between light and sound wavelengths, the reconstructed visible image is, in general, not identical with the acoustically recorded one. The typical distortion encountered is one in depth unless the hologram is scaled down by the ratio of sound to light wavelength. (cf. Chapter 5, p. 133).

It is reported that the expected real and virtual images are present as well as the expected three-dimensional depth properties. The difficulty encountered is that the great difference in sound and light wavelengths results in a reconstruction geometry different from that familiar in much of holography. In the report cited, the objects used were foam plastic figures several centimeters in size located 10 to 25 centimeters from the aperture. The ultrasound frequency used was 10 MHz with an aperture 4 centimeters by 5 centimeters. It is reported that the resolution achieved equalled the diffraction limit imposed by the sound wavelength and aperture size (Smith, 1967).

All of the holographic methods for image display reported above involve the use of photographic film. A new technique reported holds promise of removing this time-consuming step to permit image display in real time (Boutin *et al.*, 1967).

B. Display of Amplitude and Phase Characteristics of the Ultrasound Image Using the NTSC Color System

The readily available equipment used in the NTSC commercial television system offers the opportunity to achieve a simultaneous display of both the phase and amplitude characteristics of the resultant sound system.

The adaptation of the equipment and techniques utilized in the NTSC color system to the display of ultrasound images gives an improvement in the sensitivity of the system to acoustic impedance changes in the order of some 20 to 40 times over that obtainable utilizing monochrome displays.

The NTSC color system is based on a signal processing method that results in the color vs. signal phase relationship shown in Fig. 9 (Barstow, 1955). If the ultrasound signals received have uniform amplitude, but are delayed in time with regard to the reference signal, this variable delay will be displayed using a monitor operating on the NTSC standards as a particular color. Some idea of the differential color sensitivity of the eye may be ascertained from an examination of the Munsell color circle shown in Fig. 10. The Munsell system for categorizing color has come into widespread use as an industrial standard.

In a practical system utilizing the NTSC technology for the display of the ultrasound image, the reference signal is generated by a quartz crystal oscillator as shown in Fig. 11. The entire system is synchronized at 3·58 MHz both in terms of line and frame scanning frequency as well as phase demodulation characteristics (Jacobs, 1968a; Jacobs et al., 1968).

The color display system reported here is capable of detecting changes in acoustic impedance in the order of one part in 10^8 (Jacobs et al., 1968). Such

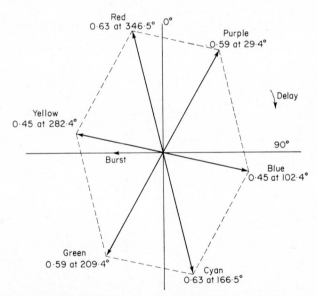

FIG. 9. Vector diagram showing relationship between 3·58 MHz carrier amplitude and phase relative to the reference carrier of the NTSC color display system. The position of the reference carrier (i.e. yellow-green) relative to the various hues was chosen by the NTSC to permit the design of simplified receivers. It should be recognized that the signal output of the color demodulation system is combined with the luminance or brightness information to produce the complete display. This is illustrated schematically in Fig. 11.

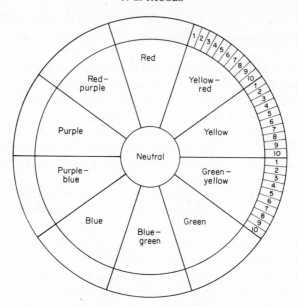

Fig. 10.   One of the ten value sectors of the Munsell color solid.   There are ten major hue sectors each having ten subdivisions specified as well as sixteen degrees of saturation from the center to the edge of each sector.

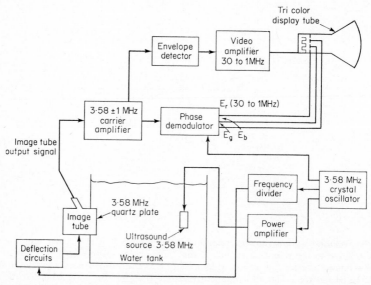

Fig. 11. Schematic representation of ultrasound image system utilizing the NTSC color system for display of the image.

a sensitivity has understandably widespread application, an example of which is the ability to determine differential electron population in semiconductors as described below. Using the ultrasound frequency of 3·59 MHz, the limiting detail of the system is shown by eqn. (1) to be in the order of 0·8 mm.

IV. Considerations in Ultrasonic Inspection Systems

Ultrasound energy is of particular interest in the nondestructive testing of materials due to the fact that it is propagated through the material by means of mechanical vibrations and hence interacts with material in a manner that differs fundamentally from that of ionizing radiation. With ultrasound energy the principal parameters affecting its propagation are the acoustic impedance, velocity and absorption coefficients of the medium. The ratio of the acoustic pressure to particle velocity is defined as the specific acoustic impedance. The product of the acoustic impedance times the velocity of the sound in a medium is termed the characteristic impedance for the particular medium. This one quantity is probably the most significant factor influencing the reflection and transmission of sound waves.

Attenuation of the sound waves arises from (1) deviation of energy from the parallel beam by regular reflection, refraction and scattering; and (2) absorption by which the mechanical energy is converted into heat. The absorption losses are characteristic of the material through which the waves travel and can yield valuable information about the physical properties of that medium.

When the sound source vibrates in the direction of the wave motion, longitudinal waves are propagated. These waves are termed compression waves. Where the motion of the source is at right angles to the direction of wave motion, transverse or shear waves are propagated. Except in special cases, shear waves can only be propagated in solids.

The percentage of incident energy reflected when a sound beam encounters an abrupt change in medium depends upon (1) the impedance mismatch, and (2) the angle of incidence. The impedance ratio or mismatch factor between the two media is a convenient base for determining the intensity relationships. This ratio is in the order of magnitude of 20 for liquids to metals which results in approximately 80 per cent reflection, and about 10^5 for gases to metals which gives virtually 100 per cent reflection.

If an ultrasound beam is incident at angles other than normal at an interface between two materials having unequal sound velocities, the transmitted beam will assume a new direction of propagation given by the angle of refraction.

In the most general case, additional beams may be generated by mode conversion. The angular relationship for plane wave propagation is given by Snell's Law.

In addition to the simpler interface conditions of normal incidence between two media and the angular reflection at a free surface, several more complicated and important cases of angular incidence arise in practice. In each case the reflected and transmitted beams may include both longitudinal and shear waves, unless (1) critical angles are exceeded, or (2) the medium is a nonviscous liquid which can support only longitudinal waves.

In an extended solid medium acoustic energy propagates in three principal modes as longitudinal, shear and/or surface waves. Each propagates at a characteristic velocity for a specific material. When the beam strikes an interface between materials of different acoustic velocity or impedance properties at other than normal incidence, some energy may be converted into other modes of vibration during reflection and refraction. The net result of this interaction of the sound energy with the material is to give an extremely complex propagation characteristic.

Ionizing radiation inspection systems depend for the most part on the absorption of the radiation within the material, hence slight irregularities, generally speaking, will not be detected. In contrast, the ultrasound energy propagates through the material in such a manner that these discontinuities represent an abrupt change in transmission with the result that small defects are readily discerned.

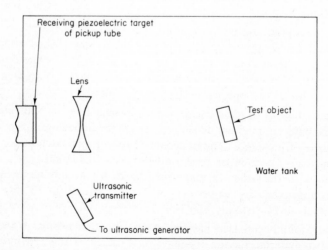

FIG. 12. Top view of the television ultrasonic image system arranged for reflection testing.

The problem of uniform illumination of the input surface of the tube is most easily solved by use of the far field of the exciting ultrasound transducer. Experience has shown that if the distance from the exciting transducer to the input surface of the image converter is 16 times the diameter of the driving transducer, then for all practical purposes the sound field consists of plane waves and is uniform. In many applications, since the system response is effectively instantaneous, some degree of non-uniformity of the sound field may be tolerated. In these situations the object being inspected is moved in the field, thus permitting relative movement of the object to be used as an aid in discerning non-uniformities of the object proper. The system is normally used in the transmission mode which, depending upon the degree of detail desired, may or may not use optics. In the reflection mode the object is illuminated by ultrasound in a manner which permits the reflected energy to be used to form the image. In this configuration optics are necessary to focus the reflected energy on to the input surface of the tube. Figure 12 illustrates such a system.

V. APPLICATIONS OF ULTRASOUND IMAGING SYSTEMS

To date the ultrasound image converter has been applied most successfully to problems associated with the nondestructive testing of certain metal configurations, and to a limited extent in studies relating to biological organisms (Jacobs, 1968b). In the nondestructive testing of metal structures the image converter system has proved to be of particular value in the detection of "non-bonds". These non-bonds occur in many metallurgical processes associated with the cladding of base metal with a protective coating of another metal sheet. Due to the nature of this cladding process the lack of mechanical bond between the two metals represents such a small change in total density of material as to essentially rule out the use of X-ray absorption measurements which are based upon the density and absorption characteristics of the intervening material. However, since the transmitted or reflected ultrasound energy is dependent upon the acoustical impedance of the path between the sending transducer and the receiver, the abrupt change in acoustic impedance associated with the non-bond results in a reflection at the non-bond area of essentially 100 per cent of the transmitted energy. Berger has reported successful inspection of nuclear fuel plates using an ultrasound camera and properly oriented specimens (Berger, 1967). A second area in which the ultrasound image converter has demonstrated applicability is associated with the detection of microporosity within otherwise uniform metal sections. In this instance the microporosity evidences itself by an abrupt change in acoustical impedance, and in fact becomes an energy scattering source which is most readily detected by the reflection technique. The major portion of the sound incident

energy is transmitted through the object leaving only that portion scattered by
the defect to be detected. It has been found in practice that at a frequency of
2 MHz it is possible to reliably detect microporosity in the order of 100
microns within aluminum plate having a thickness of one inch (Jacobs *et al.*,
1964). Figures 13–16 illustrate the performance of the system in these applications.

The principle advantage of the use of ultrasound as a means of inspecting
weldments is that a major property of a material most characteristic of its
strength is its acoustic impedance (Sharpe, 1965). The acoustic impedance of
the material is, in addition, an extremely sensitive function of the degree of
stress or strain that exists within the material proper. The welding process
inherently is susceptible to the creation of regions of extreme stress in
materials. These highly stressed conditions are not detectable using the conventional X-ray absorption techniques. It is, however, possible under certain

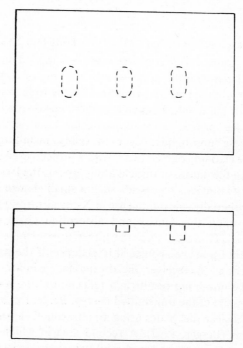

Fig. 13. An aluminum non-bond test object. The large block is 1 by 1½ by 4 inches. The
cover plate is ⅛ by 1½ by 4 inches. The three oval areas milled into the surface of the larger
block are ¼ by ½ inch and 0·002, 0·005 and 0·010 inch deep. The cover plate was
bonded to the larger block with Eastman 910 adhesive. The three oval areas were not
bonded, thereby providing three well defined aluminum–air interfaces within the test
object.

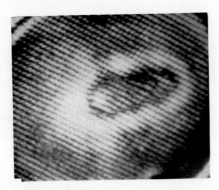

FIG. 14. Frame of a motion picture recording of a kinescope presentation of the ultrasonic image of one of the oval non-bond areas of the test piece shown in Fig. 13. The detection system was that pictured in Fig. 8. The oval non-bond area can be seen in the central portion of each frame. The active diameter of the pick-up tube piezoelectric target used for this test was 1·5 in. The outline of this target is shown by the circular shape immediately surrounding the oval non-bond image. The images outside that circle are other objects (wire, glass, etc.) inside the pick-up tube.

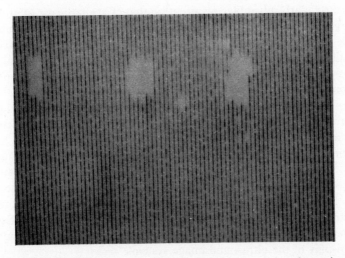

FIG. 15. Reproduction of a recorder trace from a through-transmission ultrasonic, mechanically scanned inspection system using two transducers showing the three oval areas in the aluminum test block described in connection with Fig. 13. Good bonding is indicated in the remainder of the block.

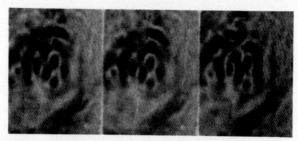

FIG. 16. Ultrasonic image of one of the non-bond areas of the test piece shown in Fig. 13. This print showing three frames of a kinescope recording, was taken using the system pictured in Fig. 8, except that a reflection technique was used. Reflection from the oval non-bond area appears as a white, " 8 "-shaped area. Some additional reflections from the aluminum surface are also detected. Fig. 12 is the test arrangement.

conditions, through the use of X-ray crystallography, to determine the degree of stress in the material. The use of ultrasound imaging, with its resultant measurement of the acoustic impedance of the material, permits these stress patterns to be visualized.

An ultrasound image system as in Fig. 11 is readily able to detect acoustic impedance variations. At the frequencies routinely used with such systems, variations in the acoustic impedance cause reflection or transmission discontinuities that are easily visible.

One particularly advantageous situation that exists in the use of ultrasound for image formation in weldment inspection is that the acoustic impedance between vacuum or air inclusions and the surrounding metal is such that essentially 100 per cent reflection occurs at these interfaces. The result is that extremely fine porosity, which in terms of the absorption of penetrating radiation would not appreciably change the absorption of the total object, will, with ultrasound energy, constitute a total reflection point within the material (Beecham, 1966). A change in the crystalline structure of the material, particularly in the case of microstructure, will so influence the acoustic properties of the material that a major discontinuity in the transmission is created. In many of the metals commonly encountered, the crystalline structure associated with the welding process is of such a size that effective scattering sources for the wave front of the ultrasound are created. The end result is that changes in the crystalline structure of a metal, that are totally undetectable through the use of penetrating radiation, become in a sense "black and white" in the display of the ultrasound image systems.

In a consideration of the various types of weldment to be used as a problem for the application of ultrasound image converters, it was decided that the most difficult and yet illustrative type of weld to inspect would be the spot weld (Matting and Wilkens, 1965). The reason for choosing this type of weld is that in the majority of cases a second metal is not present. The presence of a second metal will in many instances simplify the inspection problem, particularly when penetrating radiation is used.

The types of defects encountered in spot welding are (1) porosity in the spot, (2) so-called stick welds, which occur quite commonly in aluminum where the oxide of the aluminum is actually welded to the exclusion of the metal itself, (3) expulsion welds where the major nugget of the weld has been blown out with a resultant loss of integrity, and (4) cracks in the spot.

In terms of variation in acoustic impedance, the stick weld differs very little from a good weld; however, the use of the color display technique permits the identification of this type of weld.

The performance of the ultrasound image system with various welds, both good and defective in the various types of materials, is shown by Figs. 17–20. Conditions of the welds were verified by both sectioning of the weld and pull test (Jacobs, 1968c).

As may be seen from these figures, the system is capable of reliably detecting the commonly encountered spot welding defects. This is not totally unexpected since a good weld is characterized by a minimum of acoustic discontinuity

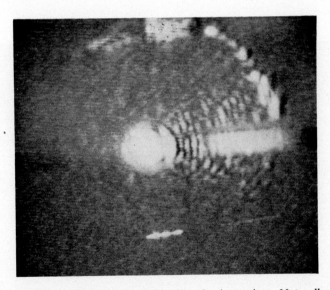

FIG. 17. Image of good weld between 0·0051 in. 24st aluminum plates. Note: all weld images are prints of 16 mm cinéscope recordings of image produced by system detailed in Fig. 8. Line through center of weld is the supporting structure for quartz input plate, frequency 10 MHz.

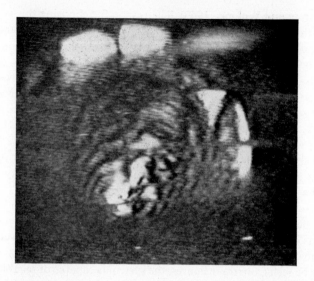

FIG. 18. Image of expulsion welds between 0·051 in. 24st aluminum plates. Freq. 10 MHz.

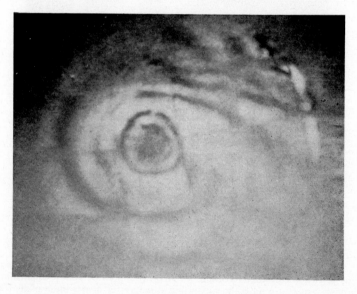

FIG. 19. Image of porosity in weld between 0·125 in. 24st aluminum plates. Freq. 10 MHz.

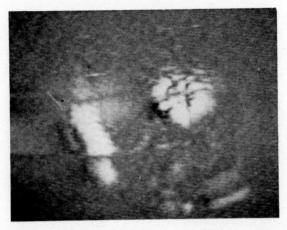

FIG. 20. Image of cracks in weld between 0·125 in. 24st aluminum plates. Freq. 10 MHz.

between the two plates, which results in the maximum transmission of the incident sound. Any deviation in this conducting path, of course, is readily discerned.

In biological applications the ultrasound image converter system encounters difficulties due to a multiplicity of scattering sources within the material itself. To date the only applications which have been demonstrated on a routine basis are those associated with the visualization of rather gross anatomical structures. It has been demonstrated, using experimental animals, that it is possible to ascertain the bony structure from the surrounding soft tissue, and in the case of certain small animals to demonstrate the cardio-vascular flow patterns within the intact animal (Jacobs, 1965). Figures 21 and 22 demonstrate the performance of the systems available today.

Quite aside from the problems associated with the absorption and scattering within the biological organism itself, a major problem exists due to the acoustic impedance discontinuities of the various gas filled regions of the body. This very characteristic which makes ultrasound so attractive for the inspection of certain metals, i.e. the ready detection of gas interfaces, is an overpowering obstacle to ultrasound visualization of many regions of the intact biological organisms.

A major deterrent to the widespread use of the ultrasound camera in biology today is the limited attainable detail associated with the acoustic propagation characteristics of the intact animal. However, the camera does have many applications in those problems associated with specialized studies related to fluid flow, investigations of blood vessel size, and studies relating to the gross structure of the various organs of the body. It is anticipated that

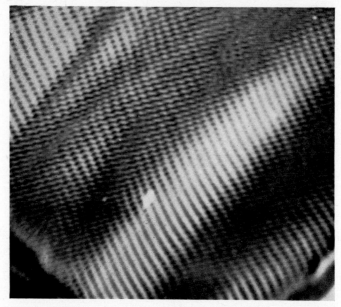

Fig. 21. Print from one frame of 16 mm cinéscope recording of image obtained through an adult forearm. Input frequency 1 mc. Diameter of sensitive area 4 inches. Note discrimination of bone from surrounding muscle.

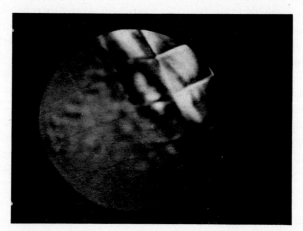

Fig. 22. Image obtained from soft tissue tumor in the forearm. Tumor is oval area in upper left quadrant surrounded by normal tissue exhibiting uniform acoustic impedance. In the original color display from which this black and white illustration was made the lipoma center was displayed as bright blue and surrounded by a red ring. The remainder of the area was a low intensity blue.

holographic techniques may be of use in improving the performance of the systems in the visualization of biological organisms (Thurstone, 1967).

Utilizing the color display system described above, it is possible to detect the changes in electron population in the order of $10^{13}/cm^3$ in semiconductors due to the change in velocity of the sound propagated through the medium. This ability to measure conductivity changes over a large area holds promise of extending the use of the ultrasound image converters into many problems where previously the sensitivity of the system has constituted a serious limitation (Gulyaev, 1967).

A demonstrated and rather unique application of the system is the determination, under dynamic conditions, of the diffusion constants of liquid solutions. The acoustic impedance of a solution is a function of concentration and the degree of ionization of the salt (Kannuna, 1948; Hueter and Bolt, 1955). Since the system provides an extended area display, rather fine gradations in the process may be readily discerned (Jacobs and Buss, 1968). This application holds great potential in many studies of this general type.

VI. SUMMARY

Experience to date has demonstrated that the ultrasound imaging systems have reached a state of development such that they are relatively easy to use. The systems have demonstrated usefulness in the inspection of metallic materials for bond, voids and homogeneity. The exact configuration of the inspection system used depends upon the particular problem at hand.

The development of holographic techniques holds promise of extending the usefulness of the systems in applications relating to the biological organism. Experiments in acoustic holography made to date are disappointing, due to the necessity for geometric scaling. When the real time systems proposed for holographic imaging become a reality, it is felt that this type of ultrasound image converter display will find widespread use.

Regarding future developments, it is anticipated that the use of thin film piezoelectric transducers will increase the sensitivity of the detector systems to levels that will permit practical operation in the 50 to 100 MHz range. It should be recognized, however, that the systems presently available exhibit detail resolution equal to many of the currently available X-ray image intensification systems, when these systems are operated at maximum sensitivity. It is strongly felt that many of the early difficulties associated with the widespread use of the ultrasound imaging systems were of a technical nature. The problems have been solved with the result that ultrasound imaging should find widespread use in many nondestructive testing applications where the unique characteristics of ultrasound propagation are best utilized.

REFERENCES

Arkhangelskii, M. E. (1967). Action of ultrasound on the processes of photographic development and fixing. *Soviet Phys. Acoust.* **12** (3), 241–249.

Barstow, J. M. (1955). The ABC's of colour television. *Proc. IRE* **43** (11), 1574–1579.

Beecham, D. (1966). Ultrasonic scatter in metals; its properties and its application to grain size determination. *Ultrasonics* **4**, 67–76.

Bennett, C. S. (1952). A new method for a visualization and measurement of ultrasonic fields. *J. acoust. Soc. Am.* **24**, 470.

Berger, H. (1967). An ultrasonic imaging Lamb-wave system for reactor fuel-plate inspection. *Ultrasonics* **5**, 39–42.

Boutin, H., Maron, E. and Mueller, R. K. (1967). "Real Time Display of Sound Holograms by *KD*P* Modulation of a Coherent Light Source," Abstract K4. Program of the 74th Meeting of the Acoustic Society of America, Miami Beach, Florida, Nov. 1967.

Dunn, F. and Fry, W. J. (1959). Ultrasound absorption microscope. *J. acoust. Soc. Am.* **31**, 632–633.

Ermst, P. J. and Hoffman, C. W. (1952). New methods of ultrasonoscopy and ultrasonography. *J. acoust. Soc. Am.* **24**, 207.

Fotland, R. A. (1960). "Ultrasonic Inspection Devices." U.S. Patent No. 2,919,574, filed June 12, 1957, serial No. 665,228, issued January 5, 1960.

Freitag, W., Martin, H. F. and Schellbach, G. (1959). "Descriptions and Results of Investigations of an Electronic Image Converter." Proc. 2nd Int. Conf. on Medical Electronics, Paris (C. N. Smyth, ed.) pp. 373–379. Iliffe, London.

Goldman, R. (1962). "Ultrasonic Technology," pp. 211–222. Reinhold Publishing Corp., New York.

Grasyuk, D. S., Oschepkov, P. K., Rozenberg, L. D. and Semennikov, Yv. B. (1965). An ultrasound introscope with the new U-55 electron-acoustic image converter. *Soviet Phys. Acoust.* **11** (4), 438–441.

Gulyaev, Yu. V. (1967). Acousto density effect in semiconductors. *Soviet Phys. solid St.* **9** (2), 328–330.

Halsted, J. (1968). Ultrasound holography. *Ultrasonics* **6** (2), 79–87.

Hartwig, K. (1959). Uber das Raeumliche Aufloesungsuer mogen von Barium-Titanat-Und Quarzplatten bei der Ultraschallabbildung. *Acustica* **9**, 109–117.

Haslett, R. W. G. (1966). An ultrasonic to electronic image converter tube for operation at 1·20 MhZ. *The Radio and Electronic Engineer* **21**, 161.

Hueter, T. F. and Bolt, R. H. (1955). "Sonics," pp. 436–439. John Wiley, New York.

Jacobs, J. E. (1965). *In* "Biomechanics and Related Bio-Engineering Topics " (R. M. Kenedi, ed.). Pergamon Press, New York and Oxford.

Jacobs, J. E. (1967). Performance of the ultrasound microscope. *Mater. Evaluation* **25** (3), 41–46.

Jacobs, J. E. (1968a). An ultrasound imaging system, utilizing colour television display techniques. *Bio-Med. Eng.* **3** (11), 504–509.

Jacobs, J. E. (1968b). Present status of ultrasound image converter system. *Trans. N.Y. Acad. Sci.* Series II, **30** (3), 444–456.

Jacobs, J. E. (1968c). "Imaging Systems for Weld Inspection." Proc. 1967 Symposium on Nondestructive Testing of Welds (T. F. Drovillard, ed.) pp. 134–152. American Society for Nondestructive Testing, Evanston, Illinois.

Jacobs, J. E. and Buss, L. E. (1968). "The Determination of Diffusion Constants Over an Extended Area by Means of the Ultrasound Camera". Proceedings of the 23rd Annual Instrument Society of America Conference, Vol. I, October 30th, 1968, New York. (Instrument Society of America, Pittsburgh).

Jacobs, J. E., Berger, H. and Collis, W. J. (1963). An investigation of the limitations to the maximum attainable sensitivity in acoustic image converters. *Trans. IEEE* UE-10, 83–88.

Jacobs, J. E., Collis, W. J. and Berger H. (1964). An evaluation of an ultrasonic inspection system employing television techniques. *Mater. Evaluation* 22, 209–212.

Jacobs, J. E., Keimann, K. and Buss, L. (1968). Use of colour display techniques to enhance sensitivity of the ultrasound camera. *Mater. Evaluation* 26 (8), 155–159.

Kannuna, M. (1948). Methode zur Diflusionsmesung. zweier Flussigkeiten vermittels Ultraschallwellen. *Helv. phys. Acta* Bd. 21, S93/116.

Lawrie, W. E. (1964). "Ultrasonic Nondestructive Coding Evaluation". Proc. 1964 Symposium on Physics and Nondestructive Testing. Illinois Institute of Technology Press, Chicago, Illinois.

Marinesco, N. and Truillet, J. J. (1933). Action of supersonic waves upon a photographic plate. *C.r. hebd. Séanc. Acad. Sci., Paris*, 196, 858.

Matting, A. and Wilkens, G. (1965). Testing of spot welds. *Ultrasonics* 3, 161–165.

Ogura, T., Kojima, T. and Vesugi, N. (1968). Ultrasonic image converter. *Japan Electron. Engng.* 2, 66–68.

Oschepkov, P. K., Rosenberg, L. D. and Semennikov, Yv. B. (1955). Electronacoustic converter for the visualization of acoustic images. *Soviet Phys. Acoust.* 1, 362–391.

Piguleuskii, E. D. (1958). The sensitivity and resolution of acoustic-optical image conversion at a liquid surface. *Soviet Phys. Acoust.* 4, 359.

Pohlman, R. (1948). Internal examination of material by ultrasonic optical pictures. *Z. angew. Phys.* 1, 181.

Prokhorov, V. G. (1964). Use of an electrokinetic target for the reception of an ultrasonic image. *Soviet Phys. Acoust.* 14 (1) 112–114.

Sayers, J. F. (1966). The future of ultrasonic camera in industrial inspection. *Ultrasonics* 4, 92–93.

Schuster, K. (1951). Ultrasonic-optical imaging by means of surface relief. *Jenaer Jb.* 217–220.

Sharpe, R. S. (1965). Low power ultrasonic techniques for materials inspection. *Ultrasonics* 3, 25–29.

Sheldon, E. E. (1958). "Apparatus for Supersonic Examination of Bodies". U.S. Patent No. 2,848,890, filed May 7, 1952, serial No. 286,521, issued August 26, 1958.

Smith, R. B. (1967). "Ultrasonic Imaging Using a Scanned Hologram Method", Paper 17, IEEE Symposium on Sonics and Ultrasonics, Vancouver, Canada, Oct. 1967.

Smyth, C. N. and Sayers, J. F. (1960). " Improvements in or Relating to Ultrasonic Viewing Devices ". British Patent No. 841,025, filed June 24, 1957, application No. 19,830,157. Complete specification published July 13, 1960.

Smyth, C. N., Poynton, F. Y. and Sayers, J. F. (1963). The ultrasound image camera. *Proc. Instn elect. Engrs* **110** (1), 16–28.

Sokoloff, S. (1939). " Means for Indicating Flaws in Materials ". U.S. Patent No. 2,164,125, filed August 21, 1937, issued June 27, 1939.

Suckling, E. E. and MacLean, W. R. (1955). Method of transducing ultrasonic shadowgraph or image for display on oscilloscope. *J. acoust. Soc. Am.* **27**, 297–300.

Thurstone, F. L. (1967). "Three-dimensional Imaging by Ultrasound Hologrphy". 1967 Digest 7th Int. Conf. on Medical and Biological Engineering (B. Jacobson, ed.), Stockholm, Sweden: Royal Acad. Engrg. Sciences.

von Ardenne, M. (1955). On making sound visible using an electron-optical image converter. *Nachrichten technik* **5**, 49.

CHAPTER 4

Ultrasonic Critical Angle Reflectivity†

F. L. BECKER‡ AND R. L. RICHARDSON§

*Battelle Memorial Institute, Pacific Northwest Laboratory,
Richland, Washington, U.S.A.*

I. LIST OF SYMBOLS

σ	stress (eqns. 3 and 4)
κ_1	constants relating stress to strain
κ_2	constants relating stress to time rate of strain
ε	strain
$\partial\varepsilon/\partial t$	time rate of strain
ω	angular frequency

† This paper is based on work performed under United States Atomic Energy Commission Contract AT (45–1)–1830.

‡ Scientist, Nondestructive Testing Department, Pacific Northwest Laboratory, operated by Battelle Memorial Institute for the United States Atomic Energy Commission, Richland, Washington.

§ Senior Research Scientist, Applied Mathematics Department, Pacific Northwest Laboratory, operated by Battelle Memorial Institute for the United States Atomic Energy Commission, Richland, Washington.

t	time
σ_{ij}	components of stress tensor (eqn. 5)
$C_{ij}^{\kappa l}$	constants relating stress to strain
$D_{ij}^{\kappa l}$	constants relating stress to time rate of strain
$\varepsilon_{\kappa l}$	components of strain tensor
$\varepsilon_{\kappa l,t}$	components of rate of strain tensor with respect to time
λ, μ	Lamé constants relating stress to strain (eqn. 6)
λ', μ'	constants relating stress to time rate of strain
δ_{ij}	Kronecker delta: 1 if $i = j$, 0 if $i \neq j$
ρ	density (eqn. 7)
$\sigma_{ij,j}$	components of spatial rate of stress summed on j
U_i	components of displacement
$U_{i,tt}$	components of acceleration
$U_{i,j}$	components of spatial rate of displacement (eqn. 8)
$\mathbf{U}$	the displacement vector (eqn. 9)
∇	the vector gradient operator
$\mathbf{U}_{,t}$	the velocity vector
$\mathbf{U}_{,tt}$	the acceleration vector
ϕ	longitudinal wave displacement potential (eqn. 10)
$\phi_{,t}$	longitudinal wave velocity potential
$\phi_{,tt}$	longitudinal wave acceleration potential
$\mathbf{0}$	the null vector
ψ	shear wave displacement potential (eqn. 11)
$\psi_{,t}$	shear wave velocity potential
$\psi_{,tt}$	shear wave acceleration potential
v	either μ or $\lambda + 2\mu$ (eqn. 12)
E	either μ' or $\lambda' + 2\mu'$
$\mathbf{R}$	the position vector
$f(\mathbf{R}, t \| \omega)$	a function of $\mathbf{R}$ and t given ω
Z	complex amplitude function (eqn. 13)
i	$\sqrt{(-1)}$
$\mathbf{N}$	complex unit normal vector
c	complex propagation constant
e	base of natural system of logarithms (2·718)
v	medium velocity of plane waves (eqn. 16)

d	a factor proportional to damping
α_n	attenuation factor (nepers/radian) in the direction of wave motion
x, y, z	distances along coordinate axes (eqn. 18)
$\mathbf{i}, \mathbf{j}, \mathbf{k}$	unit vectors along coordinate axes
γ	complex valued angle (eqn. 19)
A	real part of γ (eqn. 20)
B	imaginary part of γ
$\mathbf{N}r$	real part of $\mathbf{N}$ (eqn. 21)
$\mathbf{N}i$	imaginary part of $\mathbf{N}$
ζ, η	wave coordinates to which the (x, y) coordinates were rotated
λ, μ	temporary parameters defined by eqns. (30) and (31) (eqn. 23)
$\mathbf{U}_n$	a unit vector in the direction of wave motion (eqn. 27)
$\mathbf{U}_c$	a unit vector in the plane of the wave (eqn. 28)
v^*	phase velocity of generalized plane waves (eqn. 34)
α_n	attenuation in the direction of wave motion (eqn. 35)
α_c	attenuation in the plane of the wave (eqn. 36)
α	angle between generalized plane wave motion and x axis (eqn. 39)
ϕ_{ij}	longitudinal wave displacement potentials (eqn. 40)
ψ_{ij}	shear wave displacement potentials
σ_{yy}	direct stress on the plane $y = 0$ (eqn. 42)
σ_{xy}	shear stress on the plane $y = 0$
λ_i, μ_i	Lamé constants relating stress to strain
λ_i', μ_i'	Lamé constants relating stress to rate of strain
$\mathbf{N}_{ijk}$	unit normal vectors (eqn. 44)
c_{ij}	media propagation constants
$\gamma_{ijk}, \gamma_{ij}$	complex valued angles
π	$\sim 3 \cdot 141592654$
ρ_i	media densities
Z_{ijk}	complex valued wave amplitudes
α_{nij}	wave attenuation in direction of motion (eqn. 61)
d_{ij}	media factors proportional to damping
v_{ij}	media wave velocities
f	Re $\cos \gamma_{ij}$ (eqn. 63)
g	Im $\cos \gamma_{ij}$
h	$f + ig$

α, β temporary parameters (eqn. 69)

γ, δ temporary parameters (eqn. 71)

ζ temporary parametric angle (eqn. 72)

σ Re $\{\cos \gamma/c\}$

τ Im $\{\cos \gamma/c\}$

α_1 angle between Re $(\mathbf{N}/c)$ and $\mathbf{i}$ (eqn. 76)

α_2 angle between Im $(\mathbf{N}/c)$ and $\mathbf{i}$

v' velocity of attenuation (eqn. 77)

Δ lateral beam displacement (eqn. 79)

II. Introduction

Ultrasonic energy incident on a liquid–solid interface at the Rayleigh critical angle has been shown to be an important nondestructive testing method by Fitch (1963), Dixon (1967), Hunter (1969) and Rollins (1966). The method is effective in detecting changes in near-surface material properties of the reflecting media. There is need for a better understanding of the basic mechanisms involved, and the establishment of quantitative relationships which can relate responses to material properties.

Classical non-attenuative wave models such as Brekhovskikh (1960) fail to predict the nature of the Rayleigh critical angle reflection factor. This is demonstrated in Fig. 1. The observed reflection factor (solid line) differs

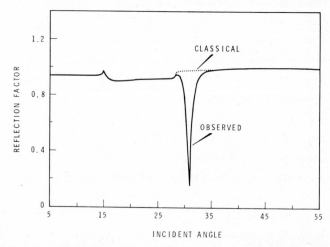

Fig. 1. Reflected amplitude versus angle of incidence for a water/stainless steel interface. The observed reflection factor (solid line) differs from that calculated by non-attenuation wave models (dashed line) in the area of the Rayleigh critical angle.

substantially from that predicted by the classical models (dashed line) near the Rayleigh critical angle (30·8 degrees for this water/stainless steel interface). The failure of classical models to predict critical angle phenomena emphasizes the need for further modelling of the mathematical relationships involved in the propagation of ultrasound. We shall generalize the classical non-attenuative model to obtain a model which includes the effects if attenuation. Next, the predictions for values of the reflection factor near the Rayleigh critical angle will be examined for agreement with measured data. Using this model we then investigate the advantages, limitations and sensitivity of the test method for various applications.

III. THEORY

Models which require dissipation of the energy in a stress wave as it propagates have been discussed in detail. Thompson (1933) discussed the relative merits of models accounting for the two inconsistencies of Hooke's law:

$$\text{Damping of vibrations in solids,} \tag{1}$$

$$\text{Hysteresis between stress and strain.} \tag{2}$$

Earlier, Kimball and Lovell (1927) had performed experiments on shafts forced to rotate in a strained position. Their experiments showed the hysteresis to be independent of frequency for a number of materials. Knopoff and MacDonald (1958) discussed classical modifications of Hooke's law in which they attempted an explanation of (1) and (2). They discussed the experiments and observations pertaining to (2) for constituents of the earth's crust and concluded, with Kimball and Lovell, that these materials have constant hysteresis (specific dissipation function) over a wide range of frequencies (10^{-2}–10^{7} Hz). They also conclude that a one-dimensional partial differential equation cannot be made to satisfy the requirements (1) and (2) with the latter being independent of frequency. This requirement of hysteresis being independent of frequency is demonstrated graphically by the fact that the ellipse relating stress to strain is unaltered by changes in frequency. Again, this requirement is demonstrated algebraically if we examine the linear relationship (3) between stress $\sigma(\omega, t)$ and strain $\varepsilon(\omega, t)$

$$\sigma = \kappa_1 \varepsilon + \frac{\kappa_2}{\omega} \frac{\partial \varepsilon}{\partial t}. \tag{3}$$

Here, ω is the regular frequency, κ_1 and κ_2 are constants and $\partial/\partial t$ is a linear differential operator with respect to time. Suppose $\varepsilon = \sin \omega t$. Then $\sigma = \kappa_1 \sin \omega t + \kappa_2 \cos \omega t$.

Eliminating time between the two equations yields

$$\sigma^2 + \left(\frac{\sigma - \kappa_1 \varepsilon}{\kappa_2}\right)^2 = 1, \tag{4}$$

an ellipse independent of ω. From (4) we may conclude that (3) has properties which satisfy the requirements (1) and (2) with (2) independent of frequency, as Knopoff and MacDonald desired.

The analysis proceeds with a generalization of (3) to three dimensions

$$\sigma_{ij} = C_{ij}^{\kappa l}\, \varepsilon_{\kappa l} + \frac{D_{ij}^{\kappa l}}{\omega}\, \varepsilon_{\kappa l, t}, \tag{5}$$

where $C_{ij}^{\kappa l}$ and $D_{ij}^{\kappa l}$ are arbitrary constants.

In (5), the Einstein convention for tensor notation is employed. Supposing the medium to be isotropic, (5) can be written with two pairs of constants where the unprimed constants of (6) are the Lamé constants and the primed constants follow in a manner parallel to that given in Love (1944) Chapter III

$$\sigma_{ij} = \lambda \delta_{ij} \varepsilon_{\kappa\kappa} + 2\mu \varepsilon_{ij} + \frac{\lambda'}{\omega} \delta_{ij} \varepsilon_{\kappa\kappa,t} + \frac{2\mu'}{\omega} \varepsilon_{ij,t}. \tag{6}$$

Neglecting body forces, the dynamic equilibrium conditions per unit volume in terms of density ρ and the components of displacement U_i are given by

$$\sigma_{ij,j} = \rho U_{i,tt}. \tag{7}$$

The strain-displacement equations are

$$\varepsilon_{ij} = \tfrac{1}{2}(U_{i,j} + U_{j,i}). \tag{8}$$

We obtain the equation of particle motion by eliminating the strains from (6) and (8) and using the derivatives of the result to eliminate the stresses from (7)

$$\mu \nabla^2 \mathbf{U} + (\lambda + \mu) \nabla (\nabla \cdot \mathbf{U}) + \frac{\mu'}{\omega} \nabla^2 \mathbf{U}_{,t} + \frac{(\lambda' + \mu')}{\omega} \nabla(\nabla \cdot \mathbf{U}_{,t}) = \rho \mathbf{U}_{,tt}. \tag{9}$$

In eqn. (9), the notation $\mathbf{U}$ is the vector notation for the displacement components and ∇ is the vector gradient operator. Equation (9) is similar to a special case of the equation of particle motion, given by Newlands (1954), with the exception that eqn. (9) allows ω to take on complex values. In addition, the author remarks that the various wave velocities, compression, shear, and Rayleigh have complex values. We intend to reveal a physical interpretation for these complex valued velocities. The investigation of eqn. (9) continues by assuming, as in the lossless case, that the displacement vector (**U**) is made up of the superposition of two independent displacements **L** and **S**. Since eqn. (9) is linear, each component of **U** yields an independent equation. If **L** = $\nabla\phi$ then

$$\nabla\left[(\lambda+2\mu)\,\nabla^2\phi+\frac{\lambda'+2\mu'}{\omega}\nabla^2\phi_{,t} - \rho\phi_{,tt}\right]=\mathbf{0}. \tag{10}$$

If **S** = $\nabla \times \boldsymbol{\psi}$ then

$$\nabla \times \left[\mu\nabla^2\boldsymbol{\psi}+\frac{\mu'}{\omega}\nabla^2\boldsymbol{\psi}_{,t} - \rho\boldsymbol{\psi}_{,tt}\right]=\mathbf{0}. \tag{11}$$

In (10) the quantity in brackets must be equivalent to a scaler constant which is equated to zero for convenience. Similarly, in eqn. (11) the constant vector field is chosen to be null. An equation of type (10) or (11) will yield specific dissipation functions independent of frequency, and may be represented by a single equation

$$\left(v+\frac{E}{\omega}\frac{\partial}{\partial t}\right)\nabla^2 f(\mathbf{R}, t|\omega) = \rho\,\frac{\partial f(\mathbf{R}, t|\omega)}{\partial t^2}\ , \tag{12}$$

where v and E take on appropriate values.

Assume a solution be of the form

$$f(\mathbf{R}, t|\omega) = Z(\omega)\exp\left[i\omega\left(\frac{\mathbf{N}.\mathbf{R}}{c} - t\right)\right], \tag{13}$$

where **N** is a unit vector and c is a real valued propagation constant. In an article by Newlands (1954), it has been shown that c must be complex for $E \neq 0$ in eqn. (12). In addition, we find examples of analyses where **N** takes on complex values in the expansion of a spherical wave potential function into a system of plane waves.

If we substitute eqn. (13) into eqn. (12), we obtain

$$\left(\frac{i\omega}{c}\right)^2\mathbf{N}.\mathbf{N}\left[v+\frac{E}{\omega}(-i\omega)\right]=\rho(-i\omega)^2,$$

which imposes two generalizations

$$\mathbf{N}.\mathbf{N} = 1 \text{ for all complex valued } \mathbf{N}, \tag{14}$$

$$c^2 = \frac{v - iE}{\rho}. \tag{15}$$

At least two questions arise. First, what is the direction of motion of wave crests? Second, how fast are wave crests moving? These questions have been answered by Fitch and Richardson (1967). However, a more heuristic approach is to inspect directions of $\mathbf{R}$ in $f(\mathbf{R}, t|\omega)$, given by a generalization of $\mathbf{N}/c$ in eqn. (13). Again, it is customary to define a relation between $\mathbf{N}.\mathbf{R}/c$ and t which makes the exponent vanish. This relation then yields the velocity of wave crests, commonly called the phase velocity. Here, we are required to choose real values of the position vector $\mathbf{R}$ and real values of time. Consequently, the relation must be real valued.

Consider first the generalization of medium velocity given by eqn. (15) and its effect on the wave potential eqn. (13). The parameters which determine c^2 in eqn. (15) are inconvenient, and so new parameters v and d are defined such that

$$\left. \begin{array}{l} c = v/(1+id) \text{ where } d, v \geqslant 0, \\[2mm] \text{consequently, } d = E/[v + (v^2 + E^2)^{\frac{1}{2}}], \\[2mm] \text{and } v = \{2(v^2 + E^2)/\rho[v + (v^2 + E^2)^{\frac{1}{2}}]\}^{\frac{1}{2}}. \end{array} \right\} \tag{16}$$

The substitution of eqn. (16) into eqn. (13) yields

$$f(\mathbf{R}, t|\omega) = Z(\omega) \exp\left(-\frac{\omega d}{v}\mathbf{N}.\mathbf{R}\right) \exp\left[i\omega\left(\frac{\mathbf{N}.\mathbf{R}}{v} - t\right)\right]. \tag{17}$$

From eqn. (17) we see immediately that generalizing c yields a plane wave which has the usual attenuation factor $\alpha_n = \omega d/v$ (in nepers per unit distance) in the direction of $\mathbf{N}$ (i.e. the direction of wave propagation). These choices of v and d impose physical limitations on the propagating wave which require that it attenuate with d nepers per radian as it propagates with velocity v. By taking limits as E goes to zero, d goes to zero and v becomes the usual undamped plane wave velocity in the medium.

In addition to the previous generalization of c, the generalization of $\mathbf{N}$ to complex values is still to be accomplished. In a regular coordinate system with unit vectors $\mathbf{i}$ in the positive x direction and $\mathbf{j}$ in the positive y direction, we let

$$\mathbf{R} = \mathbf{i}x + \mathbf{j}y. \tag{18}$$

The unit normal ($\mathbf{N}$) takes the form

$$\mathbf{N} = \mathbf{i} \cos \gamma + \mathbf{j} \sin \gamma, \tag{19}$$

where γ is positive when measured counter-clockwise from the positive x axis. We generalize $\mathbf{N}$ by requiring

$$\gamma = A + iB, \tag{20}$$

where $i = \sqrt{(-1)}$ and is distinct from the unit vector $\mathbf{i}$ by the indicated notation.

Substituting eqn. (20) into eqn. (19), and expanding

$$\mathbf{N} = \mathbf{N}_r \cosh B + i\mathbf{N}_i \sinh B, \tag{21}$$

where

$$\mathbf{N}_r = \mathbf{i} \cos A + \mathbf{j} \sin A,$$

$$\mathbf{N}_i = -\mathbf{i} \sin A + \mathbf{j} \cos A.$$

Note that $\mathbf{N}$ obeys the usual requirements $\mathbf{N}.\mathbf{N} = 1$ when the usual rules of vector dot product and the product of complex numbers are applied. Also, the unit vectors $\mathbf{N}_r$ and $\mathbf{N}_i$ form a right-handed coordinate system where the vector product $\mathbf{N}_r \times \mathbf{N}_i = \mathbf{k}$, $\mathbf{k}$ being a unit vector in the direction of the positive z axis. Proceeding with the generalization of the unit normal we substitute eqn. (21) into eqn. (13) which yields

$$f(\mathbf{R}, t | \omega) = Z(\omega) \exp \left(-\frac{\omega \sinh B}{c} \mathbf{N}_i.\mathbf{R} \right) \exp \left[i\omega \left(\frac{\cosh B}{c} \mathbf{N}_r.\mathbf{R} - t \right) \right]. \tag{22}$$

Plane wave motion in eqn. (22) is in the direction of $\mathbf{N}_r$, whereas attenuation is perpendicular to $\mathbf{N}_r$. Consequently, the amplitude in the plane of the wave is not uniform, but varies exponentially, increasing or decreasing in the direction of $\mathbf{N}_i$ depending on the sign of B. Schoch (1950) and Brekhovskikh (1960) called these generalized plane waves "inhomogeneous" plane waves. Much confusion has resulted from over-use of the term homogeneous in analysis. It seems better to use the term "generalized" plane wave since it is more descriptive. Not only is the generalized plane wave non-uniform in amplitude, but it also moves in the direction of $\mathbf{N}_r$ at a velocity less than that of sound in the medium. The physical example of this phenomenon is given by the Rayleigh wave, and the Stonely waves. More study of the relationship between this generalized plane wave representation and the structure of boundary waves is anticipated. Now that each factor c and $\mathbf{N}$ of plane wave generalizations has been studied separately, the factors are now considered simultaneously. As has occurred in each previous case, we expect the wave to move in the direction given by $\mathrm{Re}(\mathbf{N}/c)$ and to be attenuated in the direction $\mathrm{Im}(\mathbf{N}/c)$. However, $\mathrm{Im}(\mathbf{N}/c)$ has a meaningless direction unless it is interpreted as before, by examining components in the direction of motion and perpendicular to the direction of motion. We need to resolve $\mathrm{Im}(\mathbf{N}/c)$ into components in the direction $\mathrm{Re}(\mathbf{N}/c)$ and in the direction perpendicular to $\mathrm{Re}(\mathbf{N}/c)$. To carry out the generalization, we need to rotate from the rectangular (x, y) coordinates to the rectangular (ζ, η) coordinates, and from the unit vectors $\mathbf{i}$ and $\mathbf{j}$ to the unit vectors $\mathbf{U}_N$ and $\mathbf{U}_c$. Here, $\mathbf{U}_N$ is a unit vector in the direction of $\mathrm{Re}(\mathbf{N}/c)$ and $\mathbf{U}_c$ is such that $\mathbf{U}_N \times \mathbf{U}_c = \mathbf{k}$, where $\mathbf{k}$ is the unit vector in the positive z direction. The geometric parameters (γ, A, B) given in eqn. (20) and the medium parameters (c, v, d) given in eqn. (16) are used to provide values for $\mathbf{U}_N$ and $\mathbf{U}_c$, from which the physical character of the propagating generalized plane waves may be determined

$$\frac{\mathbf{N}}{c} = \mathrm{Re}\left(\frac{\mathbf{N}}{c}\right) + i\,\mathrm{Im}\left(\frac{\mathbf{N}}{c}\right) = \left|\mathrm{Re}\left(\frac{\mathbf{N}}{c}\right)\right|\mathbf{U}_N + i(\lambda\mathbf{U}_N + \mu\mathbf{U}_c). \tag{23}$$

Here λ and μ are parameters to be evaluated.

$$\mathrm{Re}\left(\frac{\mathbf{N}}{c}\right) = \frac{\mathbf{N}_r \cosh B - \mathbf{N}_i d \sinh B}{v}. \tag{24}$$

$$\mathrm{Im}\left(\frac{\mathbf{N}}{c}\right) = \frac{\mathbf{N}_r d \cosh B + \mathbf{N}_i \sinh B}{v}. \tag{25}$$

$$\left|\mathrm{Re}\left(\frac{\mathbf{N}}{c}\right)\right| = \frac{(\cosh^2 B + d^2 \sinh^2 B)^{\frac{1}{2}}}{v}. \tag{26}$$

$$\mathbf{U}_N = \frac{\mathbf{N}_r \cosh B - \mathbf{N}_i d \sinh B}{(\cosh^2 B + d^2 \sinh^2 B)^{\frac{1}{2}}}. \tag{27}$$

$$\mathbf{U}_c = \frac{\mathbf{N}_i \cosh B + \mathbf{N}_r d \sinh B}{(\cosh^2 B + d^2 \sinh^2 B)^{\frac{1}{2}}}. \tag{28}$$

The $\mathbf{U}_N$ and $\mathbf{U}_c$ given in eqn. (27) and eqn. (28) fulfill the requirements for a right-handed coordinate system. The vector $\mathbf{R}$ is transformed from the (x, y) coordinate system to the (ζ, η) coordinate system where:

x is measured positive in the direction $\mathbf{i}$,

y is measured positive in the direction $\mathbf{j}$,

ζ is measured positive on the direction $\mathbf{U}_N$, and

η is measured positive in the direction $\mathbf{U}_c$.

Thus,

$$\mathbf{R} = \zeta\mathbf{U}_N + \eta\mathbf{U}_c. \tag{29}$$

The value of λ is found from the dot product of eqn. (25) with eqn. (27)

$$\lambda = \frac{d}{v(\cosh^2 B + d^2 \sinh^2 B)^{\frac{1}{2}}} . \tag{30}$$

The value of μ is found from the dot product of eqn. (25) with eqn. (28)

$$\mu = \frac{(1+d^2)\sinh 2B/2v}{(\cosh^2 B + d^2 \sinh^2 B)^{\frac{1}{2}}} . \tag{31}$$

With eqn. (29) and eqn. (23) we are able to evaluate

$$\mathbf{R}.\left(\frac{\mathbf{N}}{c}\right) = \left|\mathrm{Re}\left(\frac{\mathbf{N}}{c}\right)\right|\zeta + i(\lambda\zeta + \mu\eta). \tag{32}$$

Substituting eqn. (32) into eqn. (13) we are able to interpret the character of the generalized plane wave

$$f(\mathbf{R}, t|\omega) = Z(\omega)\exp(-\omega\lambda\zeta)\exp(-\omega\mu\eta)\exp\left[i\omega\left(\left|\mathrm{Re}\left(\frac{\mathbf{N}}{c}\right)\right|\zeta - t\right)\right]. \tag{33}$$

The phase velocity of the wave $f(\mathbf{R}, t|\omega)$ is found by setting

$$\zeta = t\left/\left|\mathrm{Re}\left(\frac{\mathbf{N}}{c}\right)\right|\right..$$

Thus the phase velocity of the generalized plane wave is called v^*, and is equal to the coefficient of t

$$v^* = 1\left/\left|\mathrm{Re}\left(\frac{\mathbf{N}}{c}\right)\right|\right. = v/(\cosh^2 B + d^2 \sinh^2 B)^{\frac{1}{2}}. \tag{34}$$

Curiously, generalized plane waves do *not* travel at medium velocity v even when no damping ($d = 0$) occurs, but plane waves ($B = 0$) always travel at medium velocity v as given in eqn. (16). Note that the medium velocity v, for a viscoelastic medium, is not the same as the one derived from the lossless wave equation for an elastic medium. This result is consistent with the findings of eqn. (17) and eqn. (22).

The attenuation of the generalized plane wave in the direction of motion is $\omega\lambda$, as shown in eqn. (33), and is called the normal attenuation α_n. The use of eqn. (34) in eqn. (30) simplifies α_n

$$\alpha_n = \omega d v^*/v^2. \tag{35}$$

Crest attenuation of the generalized wave is $\omega\mu$, as shown in eqn. (33), and is called α_c. Similarly, the use of eqn. (34) in eqn. (31) simplifies α_c

$$\alpha_c = \omega(1+d^2)\,(v^*/2v^2)\sinh 2B. \tag{36}$$

From (35) it is obvious that a non-attenuating medium has no normal attenuation α_n. But eqn. (36) shows that the non-attenuating medium will support plane waves with exponentially varying crest amplitude. These results were anticipated in eqn. (17) and eqn. (22). Consequently, we have evaluated the phase velocity of waves in the medium.

To answer the second question, we must find the real direction of motion of the generalized plane waves with respect to the (x, y) coordinates, i.e. the direction of $\mathbf{U}_N$ in terms of the unit vectors $\mathbf{i}$ and $\mathbf{j}$. From eqn. (28) and eqn. (34)

$$\mathbf{U}_N = \frac{(\mathbf{N}_r \cosh B - \mathbf{N}_i d \sinh B)\, v^*}{v}. \tag{37}$$

Substituting the values of $\mathbf{N}_r$ and $\mathbf{N}_i$ given in eqn. (21) into eqn. (37) yields

$$\mathbf{U}_N = \mathbf{i}[\cos A \cosh B + d \sin A \sinh B] \quad (v^*/v)$$

$$+ \mathbf{j}[\sin A \cosh B - d \cos A \sinh B] \quad (v^*/v). \tag{38}$$

Hence, the plane waves move in the direction which makes a positive angle α with respect to the positive x axis

$$\alpha = \tan^{-1}\left\{\frac{\tan A - d \tanh B}{1 + d\tan A \tan B\,\mathrm{h}}\right\}. \tag{39}$$

The values of d and tanh B are expected to be small. The value of α is approximately A, with the possible exception of waves colliding with the plane $x = 0$, for which the value of tan A becomes quite small.

The foregoing analysis of generalized plane waves is an attempt to convince one of the validity of eqn. (5) as a linear approximation to the solution of the original questions concerning attenuation of stress waves. A more convincing example is given by seeking an explanation of the anomaly given by Brekhovskikh (1960), page 33, concerning the dip in the magnitude of the reflected ultrasonic wave at critical incidence. The irregularities between observed and computed data are eliminated by using eqn. (5) in the solution of the problem of reflection of a plane wave at oblique incidence from a plane boundary between two dissimilar attenuating media. The analysis leading to the reflection factor is given by Fitch and Richardson (1967) with anticipated use of the digital computer. Next we will use the common procedures to compute a solution, in an attempt to indicate the similarity of this solution to those in the existing literature. (See, for example, Cooper (1967) for the frequency dependent case.) Figure 2 reveals the geometric pattern of response waves anticipated from an incident longitudinal wave in medium 1.

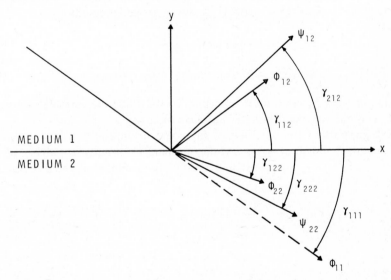

Fig. 2. Geometry of response wave potentials from an incident longitudinal wave potential.

The wave potentials ϕ and ψ have been described in eqn. (10) and eqn. (11). The superposition of an incident longitudinal displacement potential ϕ_{11}, a reflected longitudinal displacement potential ϕ_{22}, and a reflected shear displacement potential ψ_{22} make up the total displacement potential in the incident medium, where the subscripts will be discussed below. The superposition of a refracted longitudinal displacement potential ϕ_{22} and a refracted shear displacment potential ψ_{22} make up the total displacement potential in the refracting medium. At the plane boundary ($y=0$) separating the two media, the potentials in medium 1 have the following relationships to the potentials in the second medium:

(i) The two components of displacement are equal;

(ii) The direct and shear components of stress normal to the plane are equal.

We have used superposition to resolve the displacement vector according to $\mathbf{U} = \nabla\phi + \nabla \times \boldsymbol{\psi}$, which is always possible for any vector field. In the sequel, it is natural to use Cartesian coordinates (x, y, z) with respective unit vectors $\mathbf{i}$, $\mathbf{j}$, $\mathbf{k}$. The following analysis is not violated by assuming z independence for the potentials $\boldsymbol{\psi} = \mathbf{k}\Psi$ (x, y, $t|\omega$) and $\phi = \phi(x, y, t|\omega)$, where ϕ and ψ have the format of $f(\mathbf{R},t|\omega)$ in eqn. (13). Continuity, (i) above, of the x and y components of the displacement vector $\mathbf{U}$ yields the respective equations

$$\phi_{11,x} + \phi_{12,x} + \psi_{12,y} = \phi_{22,x} + \psi_{22,y} \tag{40}$$

$$\phi_{11,y} + \phi_{12,y} - \psi_{12,x} = \phi_{22,y} - \psi_{22,x} \tag{41}$$

at the interface between the two media ($y=0$), where the variable subscripts following the comma denote partial differentiation.

The generalized relation between stress and strain, eqn. (6), may be used to equate direct and shear stresses at the interface. Continuity, (ii) above, of σ_{yy} at the boundary ($y=0$) is written in terms of the potentials ϕ and ψ in eqn. (6)

$$\left[\lambda_1 + 2\mu_1 + \frac{\lambda_1' + 2\mu_1'}{\omega}\frac{\partial}{\partial t}\right]\nabla^2(\phi_{11} + \phi_{12}) - 2\left[\mu_1 + \frac{\mu_1'}{\omega}\frac{\partial}{\partial t}\right]$$

$$\times (\phi_{11,xx} + \phi_{12,xx} + \psi_{12,xy})$$

$$= \left[\lambda_2 + 2\mu_2 + \frac{\lambda_2' + 2\mu_2'}{\omega}\frac{\partial}{\partial t}\right]\nabla^2\phi_{22} - 2\left[\mu_2 + \frac{\mu_2'}{\omega}\frac{\partial}{\partial t}\right](\phi_{22,xx} + \psi_{22,xy}). \tag{42}$$

Continuity, (ii) above, of σ_{xy} at the boundary ($y=0$) is written in terms of the potentials ϕ and ψ in eqn. (6)

$$\left[\mu_1 + \frac{\mu_1'}{\omega}\frac{\partial}{\partial t}\right]\{\nabla^2\psi_{12} + 2[\phi_{11,xy} + \phi_{12,xy} - \psi_{12,xx}]\}$$

$$= \left[\mu_2 + \frac{\mu_2'}{\omega}\frac{\partial}{\partial t}\right]\{\nabla^2\psi_{22} + 2[\phi_{22,xy} - \psi_{22,xx}]\}. \tag{43}$$

Using the following subscript notation in eqn. (13),

$$Z_{ijk}(\omega)\exp\left[i\omega\left(\frac{N_{ijk}\cdot R}{c_{ij}} - t\right)\right] = \left\{\begin{array}{l}\phi_{jk}(R, t|\omega) \text{ if } i = 1 \\ \psi_{jk}(R, t|\omega) \text{ if } i = 2\end{array}\right\}_{; j,k \,=\, 1,2.} \tag{44}$$

In the boundary conditions, eqns. (40)–(43), we carry out the indicated linear operations on the functions given by eqn. (44). Consequently setting $y = 0$, we find a function

$$\exp\left[i\omega\left(\frac{x\cos\gamma_{ijk}}{c_{ij}} - t\right)\right]$$

common to every term. Obedience of these potentials to the boundary condition demands that the coefficients of x be equivalent

$$\frac{\cos\gamma_{ijk}}{c_{ij}} = \text{a complex constant}; \ i, j, k = 1, 2. \tag{45}$$

Equation (45) is the generalized form of Snell's law resulting from the generalization of the equation relating stress to strain. Here, we have used the definitions

$$N_{ijk} = i\cos\gamma_{ijk} + j\sin\gamma_{ijk} \tag{46}$$

and

$$R = ix + jy. \tag{47}$$

The index notation is given in detail for clarification:

Longitudinal wave parameters have subscript $i = 1$;

Shear wave parameters have subscript $i = 2$;

Incident medium parameters have subscript $j = 1$;

Refracting medium parameters have subscript $j = 2$;

Incident wave parameters have index $k = 1$;

Response wave parameters have index $k = 2$.

It is important to adopt the policy that the angle of reflection is the negative of the angle of incidence when angles are measured positively in the counterclockwise direction from the positive x axis (even if they are complex). The respective boundary eqns. (40)–(43) can be simplified by omitting the common function of time and the function of x (made common by the generalized Snell's law). In addition we can delete the index k by writing all angles in terms of response angles ($k=2$), then completing the operations indicated by the symbols ∇ and $\partial/\partial t$, we obtain

$$(Z_{111}+Z_{112})\frac{\cos\gamma_{11}}{c_{11}}+Z_{212}\frac{\sin\gamma_{21}}{c_{21}}=Z_{122}\frac{\cos\gamma_{12}}{c_{12}}+Z_{222}\frac{\sin\gamma_{22}}{c_{22}}, \quad (48)$$

$$(-Z_{111}+Z_{112})\frac{\sin\gamma_{11}}{c_{11}}-Z_{212}\frac{\cos\gamma_{21}}{c_{21}}=Z_{122}\frac{\sin\gamma_{12}}{c_{12}}-Z_{222}\frac{\cos\gamma_{22}}{c_{22}}, \quad (49)$$

$$[\lambda_1+2\mu_1-i(\lambda_1'+2\mu_1')]\left(\frac{Z_{111}+Z_{112}}{c_{11}^2}\right)-2(\mu_1-i\mu_1')\Bigg[(Z_{111}+Z_{112})$$

$$\times\left(\frac{\cos^2\gamma_{11}}{c_{11}^2}\right)+Z_{212}\frac{\sin\gamma_{21}\cos\gamma_{21}}{c_{21}^2}\Bigg]=[\lambda_2+2\mu_2-i(\lambda'+2\mu_2')]\frac{Z_{122}}{c_{12}^2}$$

$$-2(\mu_2-i\mu_2')\left(Z_{122}\frac{\cos^2\gamma_{12}}{c_{12}^2}+Z_{222}\frac{\sin\gamma_{22}\cos\gamma_{22}}{c_{22}^2}\right), \quad (50)$$

$$(\mu_1-i\mu_1')\left\{\frac{Z_{212}}{c_{21}^2}+2\left[(-Z_{111}+Z_{112})\frac{\sin\gamma_{11}\cos\gamma_{11}}{c_{11}^2}-Z_{212}\frac{\cos^2\gamma_{21}}{c_{21}^2}\right]\right\}$$

$$=(\mu_2-i\mu_2')\left\{\frac{Z_{222}}{c_{22}^2}+2\left[Z_{122}\frac{\sin\gamma_{12}\cos\gamma_{12}}{c_{12}^2}-Z_{222}\frac{\cos^2\gamma_{22}}{c_{22}^2}\right]\right\}. \quad (51)$$

The addition of the subscripts to eqn. (15) yields

$$\rho_j c_{ij}^2 = \begin{cases} \lambda_j+2\mu_j-i(\lambda_j'+2\mu_j') & \text{if } i = 1 \\ \mu_j-i\mu_j' & \text{if } i = 2 \end{cases}_{,j\,=\,1,2}. \quad (52)$$

Incorporating eqn. (52) in eqns. (48)–(51) and re-arranging, we obtain the following simultaneous equations

$$\frac{\cos \gamma_{11}}{c_{11}}Z_{112} - \frac{\sin \gamma_{22}}{c_{22}}Z_{222} - \frac{\cos \gamma_{12}}{c_{12}}Z_{122} + \frac{\sin \gamma_{21}}{c_{21}}Z_{212} = -\frac{\cos \gamma_{11}}{c_{11}}Z_{111},$$

(53)

$$\frac{\sin \gamma_{11}}{c_{11}}Z_{112} + \frac{\cos \gamma_{22}}{c_{22}}Z_{222} - \frac{\sin \gamma_{12}}{c_{12}}Z_{122} - \frac{\cos \gamma_{21}}{c_{21}}Z_{212} = \frac{\sin \gamma_{11}}{c_{11}}Z_{111},$$

(54)

$$\rho_1\left(1 - 2\frac{c_{21}^2}{c_{11}^2}\cos^2\gamma_{11}\right)Z_{112} + \rho_2 \sin 2\gamma_{22}\, Z_{222} - \rho_2\left(1 - 2\frac{c_{22}^2}{c_{12}^2}\cos^2\gamma_{12}\right)Z_{122}$$

$$- \rho_1 \sin 2\gamma_{21}Z_{212} = -\rho_1\left(1 - 2\frac{c_{21}^2}{c_{11}^2}\cos^2\gamma_{11}\right)Z_{111},$$

(55)

$$2\rho_1\frac{c_{21}^2}{c_{11}^2}\sin \gamma_{11} \cos \gamma_{11}\, Z_{112} - \rho_2(1 - 2\cos^2\gamma_{22})\, Z_{222} +$$

$$-2\rho_2\frac{c_{22}^2}{c_{12}^2}\sin \gamma_{12} \cos \gamma_{12}\, Z_{122} + \rho_1(1 - 2\cos^2\gamma_{21})\, Z_{212}$$

$$= 2\rho_1\frac{c_{21}^2}{c_{11}^2}\sin \gamma_{11} \cos \gamma_{11}\, Z_{111}.$$

(56)

Equations (53)–(56) yield the response amplitudes of waves rebounding from an interface between two dissimilar media that do not slip. At the water–metal interfaces where non-slip conditions prevail, local test results indicate the amount of energy going into shear waves in the liquid is small. Neglecting the small effect of liquid shearloss, we obtain a simplification of the analysis. The simplification is parallel to that given by Ewing *et al.* (1957), page 78, where c_{21} goes to zero. Since the generalized Snell's law must always be true, we have

$$\lim_{c_{21} \to 0} \frac{\cos \gamma_{21}}{c_{21}} = \text{complex constant.}$$

Hence, $\gamma_{21} \to \pm \frac{1}{2}\pi$, $Z_{212} \to 0$, and eqn. (54) is no longer valid since the horizontal component of displacement is not continuous at $y = 0$ where slip is allowed. The resulting simplified equations are, respectively,

$$\frac{\sin \gamma_{11}}{c_{11}} Z_{112} + \frac{\cos \gamma_{22}}{c_{22}} Z_{222} - \frac{\sin \gamma_{12}}{c_{12}} Z_{122} = \frac{\sin \gamma_{11}}{c_{11}} Z_{111}, \quad (57)$$

$$\rho_1 Z_{112} + \rho_2 \sin 2\gamma_{22} Z_{222} - \rho_2 \left(1 - 2\frac{c_{22}^2}{c_{12}^2} \cos^2 \gamma_{12}\right) Z_{122} = -\rho_1 Z_{111}, \quad (58)$$

$$-\rho_2 (1 - 2 \cos^2 \gamma_{22}) Z_{222} - 2\rho_2 \frac{c_{22}^2}{c_{12}^2} \sin \gamma_{12} \cos \gamma_{12} Z_{122} = 0. \quad (59)$$

Using Cramer's rule to solve the simultaneous eqns. (57), (58) and (59) for Z_{112},

$$\frac{Z_{112}}{Z_{111}} = \frac{\cos^2 2\gamma_{22} + \dfrac{c_{22}^2}{c_{12}^2} \sin 2\gamma_{12} \sin 2\gamma_{22} + \dfrac{\rho_1 \, c_{11} \sin \gamma_{12}}{\rho_2 \, c_{12} \sin \gamma_{11}}}{\cos^2 2\gamma_{22} + \dfrac{c_{22}^2}{c_{12}^2} \sin 2\gamma_{12} \sin 2\gamma_{22} - \dfrac{\rho_1 \, c_{11} \sin \gamma_{12}}{\rho_2 \, c_{12} \sin \gamma_{11}}}. \quad (60)$$

The result is seen to agree with that of Schoch (1950) when losses are neglected. The evaluation of eqn. (60) requires the use of the generalized Snell's law, eqn. (45).

In experimental work, one can measure the plane wave to a degree of approximation determined by the equipment. We can measure only the real angles of incidence. Consequently, we can measure the values of medium velocity v, medium attenuation in the direction of wave propagation α_n, and the angle γ_{11} that the incident wave makes with the positive x axis (where positive angles are measured counter-clockwise from the positive x axis to the arrow end of the vector pointing in the direction of propagation). In addition, the single angular frequency ω can be measured.

To evaluate the response angles, first we can find the parameter d from eqn. (17), where

$$\alpha_{nij} = \omega d_{ij}/v_{ij}. \quad (61)$$

Next, c_{ij} is evaluated from eqn. (16),

$$c_{ij} = v_{ij}/(1 + i \, d_{ij}) \quad (62)$$

and then the ratio $c_{ij} \cos \gamma_{11}/c_{11}$ is evaluated as a complex constant, which is used in the generalized Snell's law, eqn. (45)

$$\cos \gamma_{ij} = \frac{(1+id_{11})}{(1+id_{ij})} \frac{v_{ij}}{v_{11}} \cos \gamma_{11} = f_{ij} + ig_{ij}. \tag{63}$$

We use eqn. (20) to evaluate $A + iB = \gamma$, where the subscripts have been dropped for convenience

$$\cos (A+iB) = f + ig = h. \tag{64}$$

Expand the cosine function

$$\cos A \cosh B - i \sin A \sinh B = f + ig$$

and equate real and imaginary parts

$$\cos A \cosh B = f$$
$$\sin A \sinh B = -g. \tag{65}$$

Using the trigonometric identity $\cos^2 A + \sin^2 A = 1$, and the hyperbolic identity $\cosh^2 B - \sinh^2 B = 1$, we get

$$\frac{f^2}{\cosh^2 B} + \frac{g^2}{\sinh^2 B} = 1$$
$$\frac{f^2}{\cos^2 A} - \frac{g^2}{\sin^2 A} = 1. \tag{66}$$

Solve eqn. (66) by eliminating $\sinh^2 B$, and by eliminating $\sin^2 A$

$$\cosh B = \left\{ \frac{1+|h|^2}{2} + \left[\left(\frac{1+|h|^2}{2} \right)^2 - f^2 \right]^{\frac{1}{2}} \right\}^{\frac{1}{2}}$$
$$\cos A = \left\{ \frac{1+|h|^2}{2} - \left[\left(\frac{1+|h|^2}{2} \right)^2 - f^2 \right]^{\frac{1}{2}} \right\}^{\frac{1}{2}}. \tag{67}$$

Use of the identities yields

$$\sinh B = \pm \left\{ \frac{|h|^2 - 1}{2} + \left[\left(\frac{1+|h|^2}{1} \right)^2 - f^2 \right]^{\frac{1}{2}} \right\}^{\frac{1}{2}}$$
$$\sin A = \mp \left\{ \frac{1-|h|^2}{2} + \left[\left(\frac{1+|h|^2}{2} \right)^2 - f^2 \right]^{\frac{1}{2}} \right\}^{\frac{1}{2}}. \tag{68}$$

The signs of A and B are determined from the geometry and the sign of g. First, cosh $B \geqslant 1$ for all B. Second, every wave geometry can be chosen to make A an acute angle. Consequently, cos $A > 0$. If A is measured counterclockwise from the positive x axis, $A > 0$. It follows from eqn. (65) that B has a sign opposite the sign of g. Conversely, if A is measured clockwise from the positive x axis, $A < 0$ and eqn. (65) requires that B have the same sign as the sign of g.

Computation of eqn. (67) and eqn. (68) is made more precise if we introduce parameters α and β such that

$$\cosh B = \sqrt{\alpha} + \sqrt{\beta}$$
$$\cos A = \sqrt{\alpha} - \sqrt{\beta}. \tag{69}$$

Equation (65) shows that $\alpha - \beta = f$. Equation (67) is used to show

$$\alpha + \beta + 2(\alpha\beta)^{\frac{1}{2}} = \frac{1+|h|^2}{2} + \left[\left(\frac{1+|h|^2}{2} \right)^2 - f^2 \right]^{\frac{1}{2}}$$

and

$$\alpha + \beta - 2(\alpha\beta)^{\frac{1}{2}} = \frac{1+|h|^2}{2} - \left[\left(\frac{1+|h|^2}{2} \right)^2 - f^2 \right]^{\frac{1}{2}}.$$

Addition of these equations yields a result which together with the result from eqn. (65) leads to a solution for α and β

$$\alpha = \tfrac{1}{4}[(1+f)^2 + g^2]$$
$$\beta = \tfrac{1}{4}[(1-f)^2 + g^2]. \tag{70}$$

Again, introducing parameters γ and δ, we let

$$\sin A = \pm (\sqrt{\gamma} + \sqrt{\delta})$$
$$\sinh B = \pm i(\sqrt{\gamma} - \sqrt{\delta}). \tag{71}$$

It follows from eqn. (65) that $\gamma - \delta = ig$. Equation (68) is used to show

$$\gamma + \delta + 2(\gamma\delta)^{\frac{1}{2}} = \frac{1-|h|^2}{2} + \left[\left(\frac{1+|h|^2}{2} \right)^2 - f^2 \right]^{\frac{1}{2}}$$

and

$$-\gamma - \delta + 2(\gamma\delta)^{\frac{1}{2}} = -\frac{1-|h|^2}{2} + \left[\left(\frac{1+|h|^2}{2} \right)^2 - f^2 \right]^{\frac{1}{2}}.$$

Subtraction yields a result which together with $\gamma = \delta - ig$ leads to a solution for γ and δ

$$\gamma = \tfrac{1}{2}\left[\left(\frac{1-|h|^2}{2}\right) + ig\right]$$

$$\delta = \tfrac{1}{2}\left[\left(\frac{1-|h|^2}{2}\right) - ig\right].$$

However, γ and δ are not easy to compute. To aid in the computation, we introduce a parameter

$$\xi = \tan^{-1}[2g/(1-|h|^2)]. \tag{72}$$

Then γ and δ become

$$\gamma = \tfrac{1}{2}\left[\left(\frac{1-|h|^2}{2}\right)^2 + g^2\right]^{\frac{1}{2}} e^{i\xi}$$

$$\delta = \tfrac{1}{2}\left[\left(\frac{1-|h|^2}{2}\right)^2 + g^2\right]^{\frac{1}{2}} e^{-i\xi}.$$

Noting that

$$\left(\frac{1-|h|^2}{2}\right)^2 + g^2 = \left(\frac{1+|h|^2}{2}\right)^2 - f^2 = 4\alpha\beta,$$

we are able to write eqn. (68) with α, β, and ξ

$$\sin A = \pm 2(\alpha\beta)^{\frac{1}{4}} \cos (\xi/2)$$
$$\sinh B = \mp 2(\alpha\beta)^{\frac{1}{4}} \sin (\xi/2). \tag{73}$$

The signs of A and B were discussed previously following equations (68).

Conventionally, Snell's law relates the wave direction-cosines to the velocities of sound in the media. An examination of the generalized Snell's law, eqn. (45), reveals a modification of the convention. The generalized Snell's law can be rewritten by introducing the parameters σ and τ equal to the real and imaginary parts of the complex constant

$$\frac{\cos \gamma}{c} = \sigma + i\tau. \tag{74}$$

Expanding with $\gamma = A + iB$ and $c = v/(1 + id)$,

$$\frac{\cos A \cosh B + d \sin A \sinh B}{v} = \sigma$$

$$\frac{d \cos A \cosh B - \sin A \sinh B}{v} = \tau. \tag{75}$$

Suppose we compare the result of eqn. (75) with some important direction cosines obtained from $\mathbf{N}/c$ of eqn. (23). First, we denote the angle between the vectors $\mathbf{i}$ and $\mathrm{Re}/(\mathbf{N}/c)$ as α_1. The cosine of that angle can be written

$$\cos \alpha_1 = \mathbf{i} \cdot \mathbf{U}_N = \frac{\cos A \cosh B + d \sin A \sinh B}{(\cosh^2 B + d^2 \sinh^2 B)^{\frac{1}{2}}}.$$

Noting that v^* is the generalized plane wave velocity, the first equation in eqn. (75) may be written

$$\frac{\cos \alpha_1}{v^*} = \sigma \tag{76}$$

for all waves.

Next we denote the angle between $\mathbf{i}$ and $\mathrm{Im}(\mathbf{N}/c)$ by α_2, and obtain its cosine from

$$\cos \alpha_2 = \mathbf{i} \cdot \mathrm{Im}(\mathbf{N}/c)/|\mathrm{Im}(\mathbf{N}/c)|$$

$$= \frac{d \cos A \cosh B - \sin A \sinh B}{(d^2 \cosh^2 B + \sinh^2 B)^{\frac{1}{2}}}.$$

If we define a velocity v' as

$$v' = \frac{v}{(d^2 \cosh^2 B + \sinh^2 B)^{\frac{1}{2}}} \tag{77}$$

then

$$\frac{\cos \alpha_2}{v'} = \tau \quad \text{for all waves.} \tag{78}$$

Thus, there are two physical requirements imposed by the generalization of Snell's law. The first relates generalized plane-wave direction-cosines to generalized wave-velocities. The second relates the direction cosines of generalized plane-wave attenuation to a type of velocity v' which usually exceeds medium velocity v. Perhaps this mechanism accounts for the diffusion associated with acoustical losses.

IV. EXPERIMENT

A. Model Confirmation

The first step in confirming the predictions of the theoretical wave model is to measure the physical parameters. Velocities and densities are easily measured to acceptable accuracies by standard methods. Measurement of attenuation is somewhat more difficult.

Longitudinal attenuation may be measured by several methods and the errors due to diffraction can be calculated. No such calculations are available for shear waves, although an experimental method has been devised by Fitch (1966). In this experimental method, which we shall use, two transducers are separated by a fixed water path. The amplitude of a signal transmitted through fused quartz is compared with one through the sample, when each is inserted into the fixed path. The thickness of the quartz is chosen to give the same diffraction as occurred in the sample. Knowing the velocities and densities of both materials and the attenuation in the quartz, the diffraction-corrected attenuation in the sample can be calculated. At oblique incidence, the mode-converted shear waves can be used to measure the shear wave attenuation.

For the purpose of confirming the predictions of the wave model, we have chosen 304 stainless steel as the test material. Attenuation versus frequency curves obtained by the above method are shown in Fig. 3. The measured velocities and density are shown in Table I.

TABLE I

Material	(g/cm^3)	Longitudinal Velocity (cm/sec)	Shear Velocity (cm/sec)
Water	1·0	1·49 × 10^5	—
304 Stainless Steel	7·932	5·84 × 10^5	3·13 × 10^5

The apparatus for these experiments is shown in Fig. 4. Movable transducer holders are mounted on the two semicircular tracks in a position for reflection factor measurements. When measuring attenuation the receiving transducer is positioned below the table. The track on the left, on which the receiver is mounted, may be translated in all three directions. The relative amplitude of the received signal is measured using an oscilloscope and a digital voltage source as a reference.

To approximate the plane wave conditions of the model, transducers with an active area one inch in diameter are used. This results in a maximum diffraction angle of 0·85 degrees at 5 MHz when the angle is calculated by

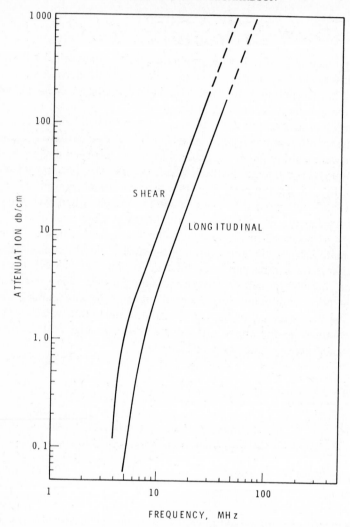

Fɪɢ. 3. Attenuation 304 stainless steel.

the method of Hueter and Bolt (1955). The angle is inversely proportional to frequency and is 0·4 degrees at 10 MHz and only 0·2 degrees at 20 MHz. The resultant error in the reflection factor is very small for frequencies greater than 10 MHz.

A second and more important consequence of using bounded beams to approximate a plane wave is the lateral beam displacement which occurs at the Rayleigh critical angle. To account for this displacement when measuring

FIG. 4. Apparatus used in experimental work.

reflection factors, the receiving transducer was translated by an amount given by the formula of Schoch (1950)

$$\Delta = \lambda \frac{2}{\pi} \frac{\rho_f}{\rho} \frac{r(r-s)}{s(s-1)} \frac{1+6s^2(1-q)-2s(d-2q)}{s-q} \tag{79}$$

in which

λ is the wavelength in the liquid,

ρ is the liquid density,

ρ_f is the solid medium's density,

$q = (c_s/c_d)^2$, with c_s the shear wave velocity, and c_d the longitudinal wave velocity in the solid medium,

$r = (c_s/c)^2$, with c the velocity of sound in the liquid medium,

$s = (c_s/v_R)^2$, with v_R the Rayleigh surface wave's velocity in the solid medium.

The reflection factors of eqn. (60) differ only slightly from the classica non-attenuative case except in the region near the Rayleigh critical angle. We shall therefore limit our experiments to this region. Reflection factors were measured for frequencies of 5–36 MHz at the water–304 stainless steel interface. The measured reflection factor versus incident angle is shown in Fig. 5 for a frequency of 10 MHz. The slight error shown here, in the angular

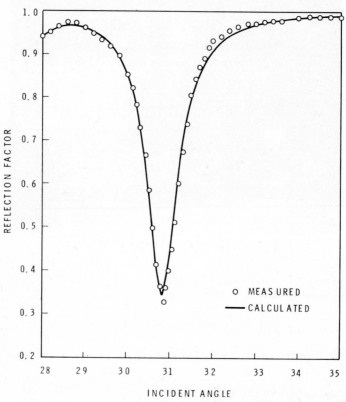

FIG. 5. Comparison of measured and calculated reflection factors at a water/304 stainless steel interface at 10 MHz.

position of the minimum and its depth, could be attributed to the fact that we have used the properties of the bulk material to approximate the reflecting surface. Aside from this slight error, the model is quite successful in predicting the reflection minimum at the critical angle.

The minimum critical-angle reflection factor given by eqn. (60) is shown in Fig. 6 for frequencies up to 40 MHz. Measured values for frequencies of 5–36 MHz are also shown. The indicated error bars correspond to the limits of resolution. The measured and calculated values are in good agreement with the possible exception of the 5 MHz measurement. At this frequency, diffraction and the large lateral beam displacement contribute significantly to this error.

The frequency of least reflection occurs at 15 MHz for this material. This frequency varies greatly with material properties of the reflecting medium and will be discussed more thoroughly in following sections.

With the large differences in amplitude which result when attenuation is added to the model, one might also expect changes in the phase of the reflected wave. This is indeed the case. The phase shift at the interface is the arc tangent of the ratio of the imaginary part to the real part of eqn. (60). Phase shifts expected at the water–304 stainless steel interface are shown in Fig. 7 for

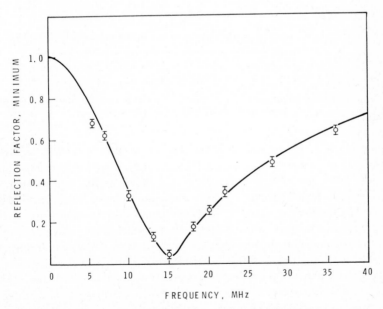

FIG. 6. Minimum critical-angle reflection versus frequency for a water/stainless steel interface. Measured values are shown from 5 to 36 MHz.

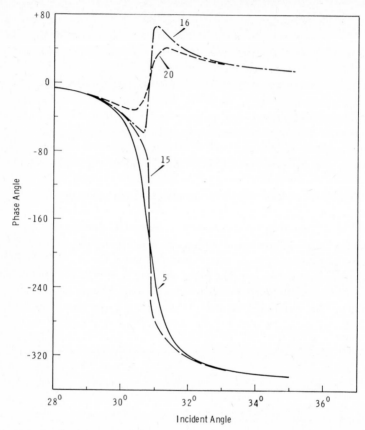

Fig. 7. Calculated relative phase of the reflected wave at a water/stainless steel boundary for frequencies of 5, 15, 16 and 20 MHz.

frequencies of 5, 15, 16 and 20 MHz. The phase shift is plotted versus the incident angle in the vicinity of the critical angle. Three important charac-teristics of the critical angle phase shift are shown in this figure. First, the slope increases with frequency up to the frequency of least reflection (15 MHz). Second, for frequencies above 15 MHz the phase shift no longer approaches −360 degrees but goes positive and then approaches zero degrees with its slope decreasing with further increase in frequency. Third, for frequencies below 15 MHz the phase shift is −180 degrees at the critical angle.

The relative phase of the reflected wave was measured for the 5 and 16 MHz cases and is shown in Fig. 8. These measurements were made by first adjusting the sample table so that a change in angle from 15 to 25 degrees produced no detectable phase change. Using a variable delay line and an oscillator tuned

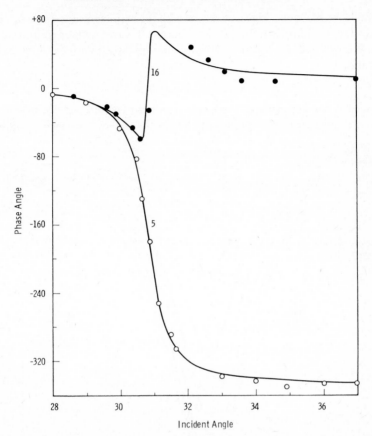

Fɪɢ. 8. Measured and calculated relative phase for frequencies of 5 and 16 MHz at a water/stainless steel boundary.

to the test frequency, a signal with the same phase as the reflected signal was compared to a fixed reference signal. The phase was then measured using Lissajous figures.

Measured and predicted phase angles are in good agreement for the 5 MHz case. The accuracy of the 16 MHz case is not as good as might be desired, but it is sufficient to confirm that the predicted phase reversal above the frequency of least reflection does occur. This has also been confirmed using short pulses to assure that an error of 360 degrees was not made in the measurment. The reduced accuracy at the higher frequency can be attributed to the smaller wavelength and the limitations of angular resolution and mechanical stability of the apparatus.

Agreement shown by Figs. 5, 6, and 8 with the predicted values substantiates the accuracy and applicability of the theoretical wave model and its assumptions. We will now use this model to further investigate critical angle reflectivity effects.

B. Critical Angle Effects

Now that the accuracy of the model has been established, we would like to use it to investigate the effect that changes in material properties will have

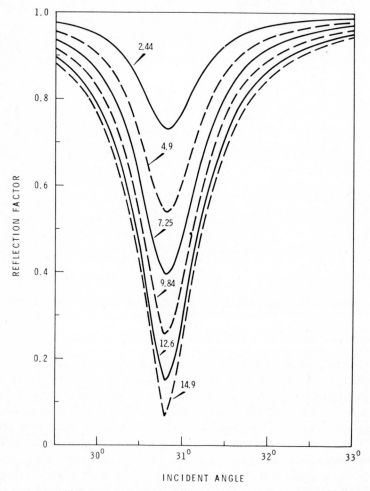

FIG. 9. Calculated critical-angle reflection factors for shear wave attenuations of 2·44, 4·9, 7·25, 9·84, 12·6 and 14·9 dB/cm. Longitudinal attenuation was constant at 1·25 dB. The test frequency was 8 MHz.

on the critical angle reflectivity. The effects of changes in attenuation and velocities will be studied. The same sample of 304 stainless steel will be used to demonstrate these effects.

To establish the relationships between attenuation and critical angle reflectivity, we shall first vary shear wave attenuation and calculate its effects while holding all other parameters constant. The measured attenuations at 8 MHz for 304 stainless steel are 4·9 dB/cm for shear waves and 1·25 dB/cm for longitudinal waves. Reflection factors for shear attenuations of 2·44, 4·9, 7·25, 9·84, 12·6 and 14·9 dB/cm were calculated and are shown in Fig. 9. It is easy to see that shear wave attenuation has considerable effect on the depth of the minimum reflection factor. In addition, it appears from these calculations that attenuation has no effect on the angular position of the minimum.

The same type of calculations were accomplished for frequencies of 15 and 25 MHz. and the results are shown in Fig. 10, along with those for 8 MHz.

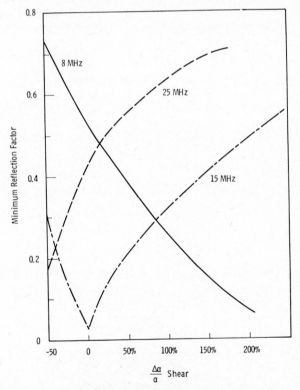

FIG. 10. Calculated minimum critical-angle reflection factors plotted versus the percentage change in shear wave attenuation for frequencies of 8, 15 and 20 MHz. Longitudinal attenuation was held constant.

In this figure the minimum value of the reflection factor is plotted versus the percentage change in shear wave attenuation which was used for the calculation. Fifteen MHz is the frequency of least reflection which explains the minimum in that curve. Larger changes in attenuation would have also resulted in minimums in the other two curves.

Similar calculations for changes in longitudinal attenuation, while shear attenuation was held fixed, were accomplished and are shown in Fig. 11. It is clear from these figures that shear wave attenuation is the predominant factor governing the depth of the minimum. The slope for changes in shear waves is as much as 17 times larger than that for longitudinal waves. For frequencies much greater or much less than the frequency of least reflection, the sensitivity to changes in attenuation is greatly reduced.

Sensitivities similar to those shown in Figs. 10 and 11 can also be expected for other materials. The principle difference will be in the frequency of least reflection, where maximum sensitivity will be obtained.

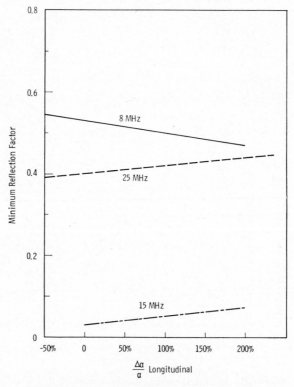

Fig. 11. Same condition as Fig. 10 except that longitudinal attenuation was varied while shear attenuation was held fixed.

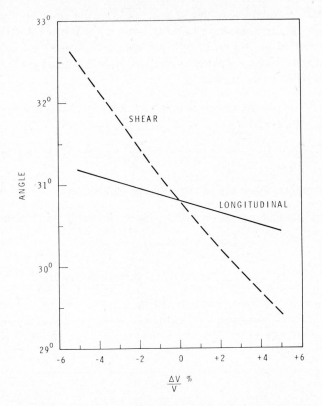

FIG. 12. Changes in the angular position of the critical angle minimum which result from percentage changes in longitudinal or shear wave velocities.

Next, the quantitative effects of variations in velocity on the angular position of the minimum will be investigated. The same procedure as above will be used here also. First, the shear wave velocity is varied from $+5$ to -5% of its measured value while all other parameters are held constant. Longitudinal velocity is then varied in the same fashion. The resultant change in the critical angle is shown for both cases in Fig. 12. These changes in velocity did not produce any change in amplitude. Likewise, changes in frequency from 5 to 30 MHz had no effect on the angular position of the minimum. From Fig. 12, it is easy to see that changes in shear wave velocity are most influential in determining the angular position of the minimum. A change in shear wave velocity of $0\cdot10\%$ results in the critical angle minimum of $0\cdot3$ degrees. A similar change in longitudinal velocity results in an angular change of only $0\cdot07$ degrees.

Using the information of this section, we can now analyze some of the critical angle effects which have been observed in nondestructive testing, such as those reported by Rollins (1966). We have also shown their quantitative relationships to changes in material parameters. In this analysis the following general statements can be made:

1. The depth of the minimum depends on attenuation, principally that of shear attenuation;

2. Angular position of the minimum depends on velocity, principally that of shear velocity;

3. The two effects listed above appear to be independent;

4. Changes in angular position with frequency probably result from velocity gradients with depth.

C. *Frequency of Least Reflection* (*FLR*)

The frequency of least reflection (FLR) is important to nondestructive testing since maximum sensitivity is obtained in the vicinity of this frequency. We have already seen that shear wave attenuation is an important parameter

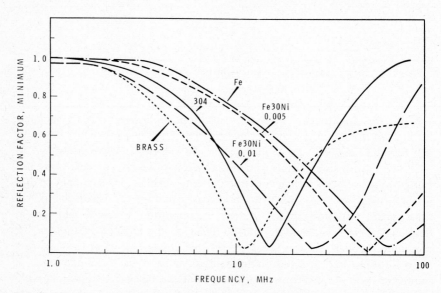

FIG. 13. Minimum critical-angle reflection factors versus frequency, for 304 ss, brass, Fe and Fe 30Ni with grain sizes of ·01 and ·005 inches.

in determining the depth of the reflection minimum. Next we shall consider its relationship to the FLR.

To explore the effects of losses on the reflection factor and frequency of least reflection, attenuation coefficients for both longitudinal and shear waves must be known over a range of frequencies. Experimental data for both types of waves are available for only a few materials. Of these limited data, none have apparently been corrected for shear wave diffraction effects.

It has been shown by Papadakis (1965) that it is possible to obtain a relatively accurate estimate of attenuation, due to scattering from grain size information. Scattering is the principal source of losses in granular materials at the frequencies of interest. For four hypothetical materials, this method will be used to predict the frequency versus attenuation curves, which will then be used to calculate the reflection factors. The materials chosen for this study are listed as follows, along with the average grain diameters which were used: Brass (30% Zinc), 0·002 in.; Fe, 0·01 in.; Fe–30Ni, 0·01 and 0·005 in.; Ni, 0·01, 0·005 and 0·001 in. Scattering factors necessary for the calculation of the attenuation were taken from Papadakis (1965). Other parameters necessary for the reflectivity calculations were obtained from Papadakis (1965) and Gray (1957).

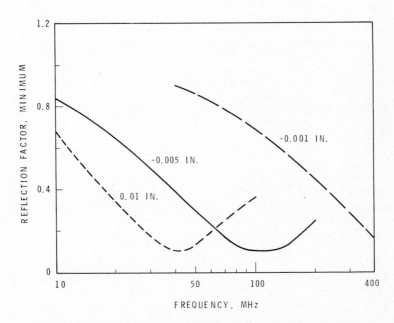

FIG. 14. Minimum critical-angle reflection factors versus frequency, for 0·01, 0·005 and 0·001 inch grain sizes of Ni.

TABLE II

Material	Grain Size (in.)	FLR (MHz)	Longitudinal Attenuation (dB/cm)	Shear Attenuation (dB/cm)	Shear Attenuation per wavelength (dB)	Shear Wave Attenuation at 20 MHz (dB/cm)
304 ss	·003	15	9·35	29·8	·624	109·2
Brass	·002	11	8·53	75·5	1·36	295·0
Fe	·01	65	9·1	114·9	·568	11·8
Fe 30Ni	·01	25	1·35	52·5	·58	31·1
Fe 30Ni	·005	50	5·4	105·0	·581	16·7
Ni	·01	42·5	49·2	110·0	·542	17·2
Ni	·005	120	29·0	220·0	·546	7·86
Ni	·001	>400	—	—	—	·985

Critical angle reflection factors have been calculated over a wide range of frequencies. The reflection minima versus frequency for these five materials and 304 stainless steel are shown in Figs. 13 and 14. Pertinent factors concerning the frequency of least reflection are listed in Table II. The varied responses shown in these figures help to illustrate the large range of responses which can be expected and the need for further analysis.

Analyzing Table II and the figures, several characteristics of the FLR can be deduced. It is readily apparent that the FLR does not occur at any particular frequency or attenuation for different materials or grain sizes. Positive conclusions are more difficult to deduce although some general statements can be made. Except for brass the shear wave attenuation per wavelength falls within a small range, ·542 to ·624 dB, even when the FLR varies from 15 to 120 MHz. The second important factor is, that for different grain sizes of the same material, the shear attenuation per wavelength is very nearly constant. Further investigation will be required to determine if this is a consequence of the manner in which the attenuation was approximated or is a general material property. One general rule, for which no exception has yet been found, is that the FLR is inversely proportional to the magnitude of the attenuation at some particular frequency. That is the larger the attenuation, for instance at 20 MHz, the lower the FLR will be (see last column of Table II).

D. Application Possibilities

The possibility of applying critical angle reflectivity appears very promising in the light of the sensitivities demonstrated in the previous section. Any changes in physical properties which affect attenuation or velocity can be detected using this method. Some specific examples will be considered next.

The difference in amplitude sensitivity between longitudinal and shear attenuations, shown in Figs. 10 and 11, suggests the possibility of measuring attenuation by reflection. We have shown previously, in Fig. 6, that the wave model accurately predicts the minimum value of the reflection factor at the critical angle. If the densities, velocities and an approximate value of longitudinal attenuation are known, the attenuation of shear waves can be measured. This is done by first measuring the minimum value of the critical angle reflection factor. This is then compared to a graph similar to Fig. 10 from which the attenuation can be found.

The uncertainty introduced by errors in longitudinal attenuation is small. If longitudinal attenuation is known within $\pm 25\%$ this would introduce a maximum error of only $\pm 2\cdot5\%$ in the shear wave measurement. Likewise an error of $\pm 100\%$ in the longitudinal attenuation results in an uncertainty of only 10%. These relationships hold for all three frequencies except the areas of shallow slope on the 25 MHz curve in Fig. 10. The accuracy of the measurements shown in Fig. 6 would represent an error of less than $\pm 5\%$ in shear wave attenuation for all but the 5 MHz case.

This technique is naturally limited to the surface layer only. The effective depth of penetration is approximately one Rayleigh wavelength, as shown by Viktorov (1967). This technique could be very effective over a selected frequency range in the following instances:

1. When it can be assumed that the surface layer adequately represents the bulk properties of the material;

2. When material geometry does not allow the use of pulse echo or through-transmission techniques;

3. When material attenuation or thickness is too large;

4. To evaluate the difference between bulk and surface layer properties.

One particular application sensitive to attenuation is the measurement of grain size. From Fig. 14 it is readily apparent that the difference in attenuation due to grain size could be detected. For this case, a test frequency of 40 MHz would yield signal ratios of 1, 4 and 9 for grain sizes of 0·01, 0·05, and 0·001 inch, respectively.

Like shear attenuation, shear wave velocity can also be easily measured by critical angle reflectivity. Given the longitudinal velocities, densities and approximate values for attenuations, the shear velocity may be measured. Assuming an angular resolution of 0·05 degrees, the shear wave velocity could be measured within 0·16%. The approximate values of attenuation are needed only to find an approximate frequency of least reflection, which will give the best angular resolution. The attenuations do not affect the accuracy of the measurement.

The differences in slopes for shear and longitudinal velocity changes, shown in Fig. 12, reduce the sensitivity to errors in longitudinal velocity. An error of 1·0% in the longitudinal velocity introduces an error of only 0·25% in the measured shear wave velocity. The velocity of the incident liquid medium must, of course, be known with good accuracy.

Measurement of shear velocity by this method would become practical under the same conditions previously listed for shear wave attenuation measurements. In addition, this method has the advantage of not requiring shear wave transducers which must be bonded to the sample.

The effect of stress on the velocity of ultrasonic waves has been studied in considerable detail by Smith (1963) and others. Through the use of Murnaghan constants (third-order elastic constants) the quantitative effects of stress on the velocity of longitudinal waves and shear waves of various polarizations can be found. The sensitivity of velocity to stress varies with each material. Three reported values of stress necessary to create a 0·1% change in shear velocity are listed as follows:

1. 200·0 bars or 2·0 ksi in 6061-T6 aluminum, reported by Smith (1963);

2. 3·1 ksi for polarization normal to stress and 6·25 ksi for polarization parallel to the stress, reported by Crecraft (1962);

3. 19·0 ksi for polarization parallel to stress in nickel–steel, reported by Shahbender (1961). Shear waves polarized normal to the stress experienced a change of only 0·025% at this same compressive stress.

Resolutions of 0·1% in shear wave velocity, which amounts to approximately 0·03° in Fig. 12, are possible with suitable equipment. Frequencies near the frequency of least reflection would give optimum results.

This method possesses all the advantages listed previously for velocity and attenuation measurements plus the ability to resolve surface stresses and stress gradients. The standard shear wave birefringence and sing-around methods of relating changes in velocity to stress require substantial path lengths to achieve their high resolution capabilities. Measured values then represent the average properties over the entire path length. The effective path for this critical angle reflectivity method is approximately one Rayleigh wavelength deep with a path length equivalent to the lateral beam displacement. It also has the additional advantage of not requiring the use of bonded shear wave transducers. This then allows the use of scanning techniques.

Attenuation also varies with strain as shown by Hikata *et al.* (1956). The sensitivity of critical angle reflectivity to this method of strain detection has not yet been investigated. If the attenuation changes prove to be detectable, (10% to 20%), the test can be of some value.

V. Conclusions

We have shown that this theoretical wave model accurately predicts the nature of the reflection factor in the vicinity of the critical angle. This is the only model which possesses this capability. The demonstrated accuracy of the model lends considerable support to the assumptions and conclusions of Section II.

The nature of the reflection minimum at the Rayleigh critical angle depends on the material properties of the reflecting medium. The depth of the minimum depends principally on shear wave attenuation while its angular position depends predominantly on shear wave velocity.

Several nondestructive testing applications were investigated and shown to be feasible. In addition, a unique method of shear wave attenuation measurement has been demonstrated. The ability to calculate the sensitivity of a proposed test using the wave model should prove to be extremely useful in exploiting the full potential of this method.

VI. Acknowledgement

Acknowledgement must be made for the efforts of C. E. Fitch, Jr., in guiding the initial development of this work. The theoretical content of this document owes its existence to the helpful suggestions of Dr. R. Y. Dean. The help of Dr. W. R. McSpadden in preparation of the manuscript is gratefully acknowledged.

REFERENCES

Brekhovskikh, L. M. (1960). "Waves in Layered Media," Academic Press, New York.

Cooper, Jr., H. F. (1967). *J. acoust. Soc. Am.* **42** (5), 1064–1069.

Crecraft, D. I. (1962). *Nature, Lond.* **195**, 1193–1194.

Dixon, N. E. (1967). "Proceedings of the Sixth Symposium on Nondestructive Evaluation of Aerospace and Weapons Systems Components and Materials," pp. 16–35.

Ewing, W. M., Jardelsky, W. S. and Press, F. (1957). "Elastic Waves in Layered Media," McGraw-Hill, New York.

Fitch, Jr., C. E. (1963). "Critical Angle Ultrasonic Tests," HW-79928, AEC Research and Development Report.

Fitch, Jr., C. E. (1966). *J. acoust. Soc. Am.* **40** (5), 989–997.

Fitch, Jr., C. E. and Richardson, R. L. (1967). *In* "Progress in Applied Materials Research" (E. G. Stanford, J. N. Fearon and W. J. McGonnagle, eds.) Vol. 8, pp. 79–120. Iliffe Books, London.

Gray, D. E. (ed.) (1957). "American Institute of Physics Handbook," pp. 3, 74–88. McGraw-Hill, New York.

Hikata, A., Truell, R., Granato, A., Chick, B. and Lücke, K. (1956). *J. appl. Phys.* **27** (4), 396–404.

Hueter, T. F. and Bolt, R. H. (1955). "Sonics," 1st Edn., p. 65. John Wiley, New York.

Hunter, D. O. (1969). "Ultrasonic Velocity and Critical Angle Changes in Irradiated A302-B and A542-B Steel," BNWL-988, AEC Research and Development Report.

Kimball, A. L. and Lovell, P. E. (1927). *Phys. Rev.* **30**, 948–959.

Knopoff, L. and MacDonald, J. F. (1958). *Rev. mod. Phys.* **30** (4), 1178–1192.

Love, A. E. H. (1944). "A Treatise on the Mathematical Theory of Elasticity," Dover, New York.

Newlands, M. (1954). *J. acoust. Soc. Am.* **26**, 434–448.

Papadakis, E. P. (1965). *J. acoust. Soc. Am.* **37** (4), 703–710.
Rollins, Jr., F. R. (1966). *Mater. Evaluation* **24**, 683–689.
Schoch, A. (1950). *Ergebn. exakt. Naturw.* **23**, 127–234.
Shahbender, R. A. (1961). IRE Trans. on Ultrasonic Engineering, March, pp. 19–22.
Smith, R. T. (1963). *Ultrasonics* July–Sept., 135–147.
Thompson, J. H. C. (1933). *Phil. Trans. R. Soc.* **A231**, 339–407.
Viktorov, I. A. (1967). "Rayleigh and Lamb Waves," Plenum Press, New York.

CHAPTER 5

Ultrasonic Holography

E. E. ALDRIDGE

Electronics and Applied Physics Division,
Atomic Energy Research Establishment,
Harwell, Berkshire, England

I. LIST OF SYMBOLS

a — typical dimension of scanning aperture.

b — distance between scanning lines.

c — characteristic dimension of the object taken parallel to the scanning plane.

f — distance between scanning plane and object.

t — time.

v — velocity of scanning transducer.

z — ordinate.

z_1, z_2 — particular values of z.

A — amplitude of ray, real or complex.

B a constant, or complex amplitude of reference beam.
C complex amplitude of illuminating beam.
α, β object ray angles to normal of hologram plane.
γ illuminating light beam angle to normal of hologram plane.
θ reference beam angle to normal of hologram plane.
λ wavelength of sound.
λ_1 wavelength of illuminating light.
$\omega/2\pi$ frequency of ultrasound.
$\omega_1/2\pi$ frequency of illuminating light.
$\omega_r/2\pi$ frequency of electronic signal equivalent to plane reference beam.
*   complex conjugate.

II. INTRODUCTION

Ultrasonic methods for nondestructive testing are widely used and usually employ some form of pulse-echo system, often combined with a scan, for flaw detection (Nondestructive Testing Handbook, 1959). As ultrasound is a wave motion and, like light, can be made to form images, it has long been felt that in a number of applications, an image presentation would yield significant advantages over the more usual systems. The standard method has been to use the ultrasonic analogue of an optical system, that is the scattered ultrasound from the object is collected by an ultrasonic lens and focused into an ultrasonic image which is then converted into an optical image by means of a suitable sensor (Berger and Dickens, 1963; Rozenberg, 1955; cf. chapter by Jacobs, p. 63). Unfortunately the technological difficulties associated with this development of ultrasonic imaging have inhibited its use and consequent evaluation; however it is hoped that by using a newer method many of these difficulties can be removed. The newer method, which is described in this chapter, is to record the ultrasonic image on an ultrasonic hologram, turn this into an optical transparency and then view the image optically in the same way that the image is viewed from an optical hologram (cf. chapter by Ennos, p. 155). This method has an advantage over the other that the difficulties associated with ultrasonic lens systems are circumvented by using the highly developed optical lens systems.

One of the practical limitations of ultrasonic imaging in nondestructive testing is that the objects of interest are usually comparable in size with the limits of resolution, that is they vary in size from fractions of a wavelength to a few tens of wavelengths. (The optical analogue would be the type of image seen in a microscope.) This means that the images are poor by normal standards and, since most objects interact with ultrasound in a rather complex manner, it also means that, in general, the images have to be interpreted in the light of previous knowledge and experience.

A second advantage of the ultrasonic hologram is that it is in a form which lends itself conveniently to optical information processing which might greatly assist this interpretation. From this point of view the ultrasonic hologram is a record of information about the object carried by the ultrasound scattered by the object and the optical system is a computer for deriving some meaningful pattern from the information in the hologram. Since the pattern has to be correlated with previous knowledge and experience to give it meaning, it need not be the image in the usual sense.

Regarding the optical system as a computer raises the question as to whether the hologram could be recorded in a digital computer store and processed digitally, which would be much more convenient and flexible. This will be touched on briefly later.

III. THEORY OF HOLOGRAMS

In order to obtain a simple physical picture, holograms will be treated from the viewpoint of ray theory using the plane reference beam hologram as an example. Later this will be generalised. The formal mathematical theory has been given by Gabor (1949), and Leith and Upatnieks (1963).

A. Plane Reference Beam Hologram

The construction and image reproduction of the plane reference beam hologram is illustrated in Fig. 1. No loss of generality results from considering only two dimensions with the object standing on the X axis and the holographic plane, where the object wavefront overlaps the reference beam wavefront, coinciding with the Z axis as shown in Fig. 1(a). Now consider a ray, defined as a narrow beam which nevertheless is a sufficient number of wavelengths in cross-section to behave like an infinite plane wave over moderate propagation distances, propagating from the top of the object at an angle α to the X axis and meeting the Z axis about z_1 as shown in Fig. 1(a). Along the Z axis, in the region of z_1, this appears as a wave propagating as:

$$A \cos \{\omega t + (2\pi/\lambda) \sin \alpha . z\}, \tag{1}$$

where $\omega/2\pi$ is the frequency, t time, λ wavelength and A the amplitude of the ray. In a similar manner, over the whole of the Z plane, the reference beam propagating at an angle θ to the X axis will appear as a wave propagating as:

$$\cos \{\omega t - (2\pi/\lambda) \sin \theta . z\}, \tag{2}$$

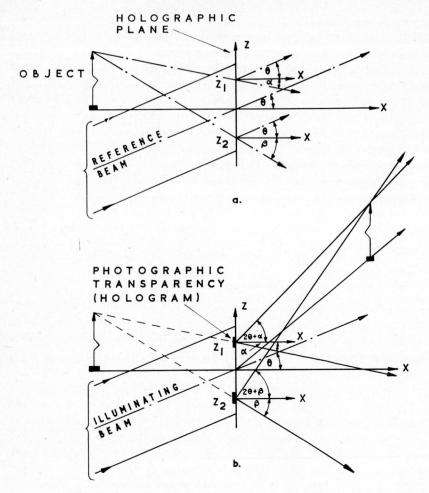

FIG. 1. Formation and image reproduction of a hologram.

where unity amplitude is assumed. Now imagine that in the Z plane these two waves are multiplied together viz.

$$A \cos \{\omega t + (2\pi/\lambda) \sin \alpha . z\} \times \cos \{\omega t - (2\pi/\lambda) \sin \theta . z\}$$

$$= \tfrac{1}{2} A \cos \{(2\pi/\lambda)(\sin \theta + \sin \alpha) z\} + \tfrac{1}{2} A \cos \{2\omega t + (2\pi/\lambda)(\sin \alpha - \sin \theta) z\}, \quad (3)$$

and the second term is filtered off, leaving the first which is independent of time. If a constant B is added to the first term to make the sum always positive, then at a position corresponding to z_1, this sum can be recorded as an intensity pattern on a photographic transparency. Similar intensity patterns can be recorded for all rays from the object and the photographic transparency so covered with these intensity plots, linearly superimposed, is a hologram.

The image reproduction of the hologram is illustrated in Fig. 1(b). The hologram is in the Z plane being illuminated by a plane beam inclined to the X axis at the same angle as the reference beam. For simplicity it is assumed that the transparency is the same size as the original recording plane and that the same wavelength and frequency are used in the image reproduction as in the hologram formation. As with the reference beam, the illuminating beam is described by eqn. (2). After passing through the hologram, the illuminating beam has its amplitude multiplied by the intensity patterns recorded for the original object rays and so for that part of the hologram produced by the ray described by eqn. (1), the beam emerging from the hologram, in the region of z_1, is:

$$[B + \tfrac{1}{2} A \cos \{(2\pi/\lambda)(\sin\theta + \sin\alpha)z\}] \times \cos \{\omega t - (2\pi/\lambda)\sin\theta . z\}$$

$$= B \cos\{\omega t - (2\pi/\lambda)\sin\theta . z\} + (1/4) A \cos\{\omega t + (2\pi/\lambda)\sin\alpha . z\}$$

$$+ (1/4) A \{\omega t - (2\pi/\lambda)(2\sin\theta + \sin\alpha)z\}. \quad (4)$$

The first term of the right-hand side of eqn. (4) is a ray parallel to the illuminating beam, the second term is a ray identical with the original object ray of eqn. (1) and the third term is a ray inclined to the X axis at an approximate angle of $(2\theta + \alpha)$ in the opposite sense. Similar results are produced from all the intensity patterns recorded for all the original object rays. Thus there are three groups of rays produced by the hologram. The first group corresponds to the direct transmission of some of the illuminating beam, the second group is identical with the original object rays and therefore corresponds to a virtual image indistinguishable from the object, and the third group produces a real image in which bumps in the object become dents in the image. This is illustrated in Fig. 1 using two object rays propagating from the same point, one being that described by eqn. (1) and the other being a ray propagating at an angle β to the X axis and meeting the Z plane in the region of z_2.

B. Holograms with General Reference Beams

Let all the rays propagating from the object add up in the holographic plane to a wavefront of $Ae^{i\omega t}$ where A is complex and a function of the co-ordinates in the holographic plane. In a similar manner let the wavefront of the reference beam be $Be^{i\omega t}$ in the same plane. Multiplying these two together yields $(AB^* + A^*B)$ where the asterisk denotes the complex conjugate. If a suitable constant is added, then the sum can be recorded as an intensity plot to make a hologram as before. Illuminating the hologram with a beam having a wavefront $Ce^{i\omega t}$ in the plane of the hologram, gives for the images:

$$(AB^* + A^*B)\, Ce^{i\omega t} = A(B^*C)e^{i\omega t} + A^*BCe^{i\omega t}. \tag{5}$$

Now if (B^*C) is a real constant, then the first term of the right-hand side is obviously a virtual image indistinguishable from the object, whilst the second is a distorted image which, in general, may be virtual or real.

This means that the nature of the reference and illuminating beams can be varied within wide limits. It is not even necessary for (B^*C) to be a real constant; it can have as an argument a function of the co-ordinates of the hologram plane, in which case either a lens can be used to form an image or the illuminating beam may produce, directly, a real image indistinguishable from the object. Generally a skewed reference beam is used to give angular separation of the images so that one may be viewed without interference from the other (Leith and Upatnieks, 1963).

C. Effect of Changing Wavelength

Making a hologram using ultrasound of frequency $\omega/2\pi$ and wavelength λ and then viewing it with coherent light of frequency $\omega_1/2\pi$ and wavelength λ_1, where $\lambda_1 \ll \lambda$, causes image changes the nature of which depend upon the circumstances. Considering the plane reference beam hologram and assuming that the illumination is a plane beam inclined at an angle γ to the X axis, eqn. (4) then becomes

$$[B + \tfrac{1}{2} A \cos \{(2\pi/\lambda)\,(\sin\theta + \sin\alpha)z\} \times \cos \{\omega_1 t - (2\pi/\lambda_1) \sin\gamma . z\}$$

$$= B \cos \{\omega_1 t - (2\pi/\lambda_1) \sin\gamma . z\} + (1/4)\, A \cos \{\omega_1 t + (2\pi/\lambda_1)\,((\lambda_1/\lambda)\sin\theta$$

$$+ (\lambda_1/\lambda)\sin\alpha - \sin\gamma)z\} + (1/4)\, A \cos \{\omega_1 t - (2\pi/\lambda_1)\,((\lambda_1/\lambda)\sin\theta$$

$$+ (\lambda_1/\lambda)\sin\alpha + \sin\gamma)z\}. \tag{6}$$

Choosing the angle γ such that

$$(\lambda_1/\lambda)\sin\theta = \sin\gamma, \tag{7}$$

the second term of eqn. (6) then becomes

$$(1/4)\, A \cos \{\omega_1 t + (2\pi/\lambda_1)\,(\lambda_1/\lambda)\sin\alpha . z\}. \tag{8}$$

This is similar to the original object ray except that its inclination to the X axis is reduced, to a first approximation, in the ratio of the wavelengths. This reduction in inclination applies to all the rays emanating from the hologram and if the rays in Fig. 1(b) are redrawn accordingly, it is seen that the first order effect is to increase the distance that the images lie from the hologram by the ratio of the wavelengths. Additionally there is image distortion.

This can be corrected, either completely or partially, by reducing the hologram photographically, which is equivalent to reducing the wavelength, λ, of the ultrasound, to make the angles of the emergent rays larger. Again by redrawing the rays in Fig. 1(b), it is seen that whilst the size of the images is reduced in the same proportion as the hologram, the distances that the images lie from the hologram are reduced by the square of the reduction.

The effects of wavelength change described above are particular to the plane reference beam hologram illuminated with a plane beam. The effects of wavelength change on other types of hologram varies with the nature of the reference and illuminating beams.

IV. ULTRASONIC HOLOGRAMS

The methods of making an ultrasonic hologram can be divided into two broad classes, those which are the ultrasonic analogues of the optical methods and those which use electronic methods.

In practice the water tanks used and the objects being investigated are not large enough to prevent strong standing wave patterns being formed if continuous tone ultrasound is used. Generally this gives rise to spurious results and as a consequence it is usual to use pulsed ultrasound, where the pulse is about ten to a thousand cycles of carrier, long enough to simulate a continuous tone in the centre of the pulse but short enough to prevent unwanted echoes from interfering with the signal. The unwanted echoes are eliminated with a suitably delayed gate which only passes the signal (range gating).

The optical reproduction requires light of a coherence dependent upon the size of the hologram. Generally the coherence of a laser is not necessary although the intensity of a 1mW helium-neon laser is useful if the images are to be projected onto a screen. For direct viewing the yellow lines of a mercury arc lamp, passed through a pin hole, is suitable. The size of the pin hole will depend upon the degree of coherence required (Taylor and Lipson, 1964).

A. Optical Analogues

With light, the object wavefront, $Ae^{i\omega t}$, is added to the reference beam wavefront, $Be^{i\omega t}$, and the sum is used to illuminate a photographic plate which is so chosen and developed to yield a density in the transparency proportional to the intensity of the light. In effect the plate is used as a square law detector to record the following:

$$(A+B)(A^*+B^*) = BB^* + (AB^* + A^*B) + AA^*. \tag{9}$$

If the reference beam illuminates the plate uniformly, then the first term on the right-hand side is a constant. The next two terms give rise to the holographic action and in conjunction with the first term, if it is large enough, can be regarded as the hologram proper. The fourth term is the object wavefront multiplied by itself. This term introduces a messy central image which is characteristic of all optical holograms.

A number of systems have been proposed for the part of the photographic plate in ultrasound, but the one which appears to have had the most development is that of liquid surface levitation (Worlton, 1968; Mueller and Sheridon, 1966) which, in the past, has been used to form ultrasonic images directly (Schuster, 1959). A travelling ultrasound wave in a liquid transports liquid and if an ultrasonic beam is directed upwards at the surface of the liquid, it will produce a permanent displacement of the surface, across the beam wavefront, usually several orders of magnitude greater than the amplitude of the surface vibrations. The way in which this effect can be used to form a hologram with the images being viewed optically is shown in Fig. 2. An approximate analysis is given in Appendix 1.

Two transducers driven from a signal generator are placed in a water bath and oriented so that their beams overlap at the surface. In addition to the direct displacement of the surface, the interference pattern of the beams produces a static ripple. When an object is placed in the path of one of the beams, this static ripple is modified by the sound scattered by the object and the resultant static surface displacement becomes an ultrasonic hologram of the object. To see the ultrasonic images, a beam of coherent light, reflected from the water surface, is brought to a focus at some convenient plane where the images are viewed by conventional methods. In absence of the ultrasound, the surface is a flat plane and the light beam is focused to a single spot. With the ultrasound present, but without the object, the resulting surface ripple causes the central spot to be diffracted into a number of others lying symmetrically about it. With the object present, these spots become diffused and, as shown in Appendix 1, the two images can be seen by viewing the first order diffraction spots. As shown in Fig. 2, the others are removed by an optical stop. The sound intensities used for this method are of the order of 0.1 W/cm^2, although the threshold sensitivity is about a hundred times less than this (Berger and Dickens, 1963).

Using sound intensities of 0.1 to 1.0 W/cm^2 it is possible to record the surface displacements on a transparent thermoplastic material which is then

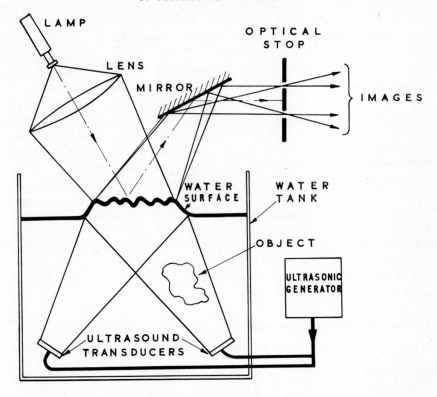

FIG. 2. Ultrasonic analogue of optical holographic system using water surface levitation.

used to phase-modulate coherent light for image reconstruction (Young and Wolfe, 1967).

An alternative to the liquid surface relief is the use of an ultrasonic camera (Smyth *et al.*, 1963; Jacobs, this volume, p. 63). This is essentially a fast electronic scan method in which the hologram is built up in the same way as a television picture. The images are viewed either by means of a photographic transparency taken from the television monitor of the camera, after square law detection of the camera carrier output (Marom *et al.*, 1968), or by feeding the camera output to a Pockels-effect crystal scanned by an electron beam synchronized to the camera electron beam scan (Boutin and Mueller, 1967). The Pockels effect causes modulation of the refractive index of the crystal and so spacial phase modulation of the beam illuminating the crystal.

B. Electronic Methods

In the arrangement of Fig. 2, a small transducer placed just below the surface would have an output proportional to the sum of the reference and object beams. To convert this to a hologram of the optical analogue type described by eqn. (9), all that is needed is to pass the signal through a square law detector and record the result on a photographic transparency at a point corresponding to the position of the transducer. As with the ultrasonic camera, the hologram is built up point by point by synchronized scans of the recording medium and the ultrasonic wavefront. This system is described here because, by analysing that part of the transducer output due to the reference beam, it soon becomes clear that the reference beam can be replaced by an equivalent signal and that, as a result, true multiplication can be used instead of square law detection. Dispensing with the ultrasonic reference beam means that, to produce a hologram, all that is needed is to process, in a suitable manner, the transducer output due to the ultrasound scattered by the object. This method is not possible with light for technological reasons.

A disadvantage of electronic methods is that the hologram has to be built up sequentially by means of a scan.

1. Plane reference beam hologram

Considering a plane reference beam of the form given by equation 2 viz:

$$\cos \{\omega t - (2\pi/\lambda) \sin \theta . z\}, \tag{2}$$

then the contribution to the transducer output by the reference beam will take the form:

$$\cos \{\omega t - (2\pi/\lambda) \sin \theta . vt\} = \cos \{(\omega - \omega_r)t\}, \tag{10}$$

where v is the scanning velocity of the transducer

and $z = vt$

and $\omega_r = (2\pi/\lambda) \sin \theta . v$.

This is simply a sine wave whose frequency differs slightly from the ultrasonic frequency by an amount depending upon the inclination of the reference beam and the velocity of scan, and so the plane ultrasonic reference beam can be replaced with an electrical signal of this form. Whilst this signal can be added to that of the object beam, and one method uses the phosphor of a cathode ray tube to act as a square law detector (MacAnally, 1967), it is more usual to multiply these signals together to produce the hologram directly (for example, see Worlton, 1968).

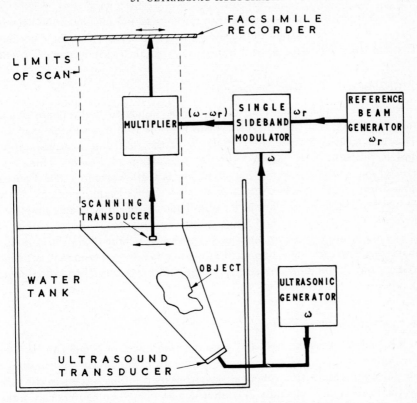

FIG. 3. Ultrasonic hologram using electronic reference beam.

A scheme for producing holograms in this way is shown in Fig. 3, where the object is isonified by a transducer driven by a generator of frequency $\omega/2\pi$, as before, and a small transducer is scanned through the resultant wavefront. The output of the ultrasonic generator is also fed to a single sideband modulator where it is combined with the output of a reference beam generator of frequency $\omega_r/2\pi$ to produce an output of frequency $(\omega-\omega_r)/2\pi$. The output of the scanning transducer is then multiplied by this and the result, with a suitable bias to make it always unipolar, recorded on a facsimile recorder, whose scan is synchronized to that of the transducer. The optical transparency is made from this recording. At the present time the usual practice is to feed the output of the ultrasonic generator directly into the multiplier, omitting the other two items, to give two images which are in line if the isonification is normal, or are angularly displaced if the isonifying beam is at an angle to the scanning plane.

An alternative to the scanning transducer is to use an ultrasonic camera that gives a carrier output and to use its television monitor as the recording medium from which the optical transparency is taken (Mueller *et al.*, 1968).

2. *General reference beam holograms*

Reference beams other than plane can be simulated if the phase of the reference signal is varied in synchronism with the scan in a suitable manner. This will greatly increase the complexity of the electronic instrumentation and is probably only justified for the production of Fourier Transform Holograms (Stroke, 1965). This type of hologram records the Fourier Transform of the images using a spherical reference beam having its centre of curvature in the same plane as the object and a plane object being simulated by using a pulsed system and range gating. The advantage of this hologram is that, by means of the Fast Fourier Transform (Cochran *et al.*, 1967), it may make digital processing economic, leading to an entirely automatic electronic system and eliminating optical processing and the associated photographic procedures.

3. *Synthetic Aperture Scan*

If a pulsed system is used, the scanning transducer of Fig. 3 can be used for both isonifying and receiving the scatter from the object. A transducer producing a beam with a spherical wavefront used in this way gives rise to a synthetic aperture (Appendix 2) from which an image can be obtained in the same way as before. Its resolution is better by a factor of two for the same distance between object and scanning plane. An arrangement for producing a synthetic aperture in conjunction with range gating is shown in Fig. 4.

The output of the ultrasonic generator is fed to a modulator to produce a reference beam, as before, and to a main gate which only allows the generator to drive the transducer for ten to a hundred cycles at a rate governed by the low frequency generator. The echo pulses pass to the multiplier where they are multiplied by that part of the reference beam output passed by the delayed gate, this being synchronized to the main gate and open only for the same duration. The time position of the delayed gate opening, relative to the main gate opening, is chosen so that only echoes from a given plane, of thickness depending upon the pulse duration and at some fixed distance from the scanning plane, give rise to an output of the multiplier. This output is recorded, as before, to produce the hologram of the plane.

A convenient way of simulating a transducer suitable for a synthetic aperture scan, is to use the focal point of a focused transducer.

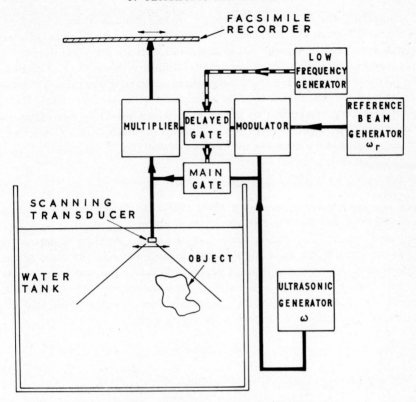

FIG. 4. Ultrasonic hologram using synthetic aperture scan.

4. Synthetic Aperture and C-Scan

If in Fig. 4, the multiplier is replaced by a linear detector, which is operative only during the time the delayed gate is open, then C-Scan results (Non-destructive Testing Handbook, 1959). Assuming the echo signal takes the form $A\cos(\omega t + \phi)$ then the function of the detector is to form the product:

$$A\cos(\omega t + \phi)\cos(\omega t + \phi) \to \tfrac{1}{2}A, \tag{11}$$

that is the reference is replaced by a signal derived from the echo signal and only the amplitude of the echo signal is recorded. A convenient way of obtaining resolution in a C-Scan plot is to use a focused transducer and record only those echoes from the vicinity of the focal plane, which means adjusting the position of the transducer so that the object being scanned lies in the focal plane.

Comparing the C-Scan plot with the image obtained from a synthetic aperture hologram for the same thin object and using the same focused transducer, it can be shown, that to first order, they are the same. Intuitively this is what might be expected since the resolution in both cases is dependent on the focal spot size, a sharp image being achieved in one case by moving the transducer and in the other by adjustment of an optical bench. Holograms are very susceptible to interference which produces time delay or phase variations, and a comparison of a hologram image with a C-Scan plot of the same object is useful in the analysis of this sort of situation.

5. *Images due to scanning raster and resolution*

A hologram made by scanning has a line structure which, in the image reproduction, will function like a diffraction grating and give rise to beams, derived from the illuminating beam, having angular separations dependent upon the ratio of the wavelength to the distance between the lines. In the hologram each of these diffracted beams will act in the same way as the reference beam and produce an image. If these images are not to overlap and interfere with the main ones then geometric considerations show that the following approximate relationship should be obeyed:

$$\lambda/b > 2\theta + c/f,$$

where λ is the wavelength of the ultrasound,

    b is the distance between scanning lines,

    θ is the angle the plane reference beam makes with the normal to the scanning plane,

    c is a characteristic dimension of the object taken parallel to the scanning plane

and f is the distance between the scanning plane and the object.

If "a" is a typical dimension of the scanning aperture, the maximum resolution obtainable is approximately $\lambda f/a$ where λ and f are defined above. For a scanning transducer, with a typical dimension of "b", not to degrade the resolution it follows:

$$b < \lambda f/a$$

or $\lambda/b > a/f.$

V. APPLICATIONS AND LIMITATIONS OF ULTRASONIC HOLOGRAPHY

As ultrasonic and optical holography have much in common, it is perhaps helpful to put the practical relationship in perspective. A striking characteristic of optical holograms is the three-dimensional effect obtained and a typical scene has objects of about two inches high and about two inches apart behind each other. Now two inches corresponds to about a hundred thousand wavelengths and if this is scaled for an ultrasonic situation of 10 MHz in water, the objects become about fifteen yards high and about fifteen yards apart. Typically, an ultrasonic object in nondestructive testing, such as a large flaw, may be thirty wavelengths in size; however, such a size of object is small in optical terms and can only be seen with a microscope. Also for a long ultrasonic pulse of a hundred wavelengths at 10 MHz, corresponding to a range resolution of about a third of an inch in water and over an inch in steel, the equivalent thickness of microscope section is about one

(Holotron Corp.)

FIG. 5.   Photo—Real time liquid surface holographic image.

Object: Milled letters in aluminium plate
Frequency: 10 MHz
Letter size: $\frac{3}{16}$ in. high milled 0·010 in. deep with 0·020 in. line width

thousandth of an inch. From this it is evident that ultrasonic imaging in nondestructive testing has a great deal in common with optical microscopy and that an ultrasonic hologram will tend to produce a rather flat image with the three-dimensional effects very much less apparent than in optical holograms. Thus the image will tend not to differ materially from the images produced by any of the other methods. Examples of the quality of images obtained at the present time are shown in Figs. 5 and 6.

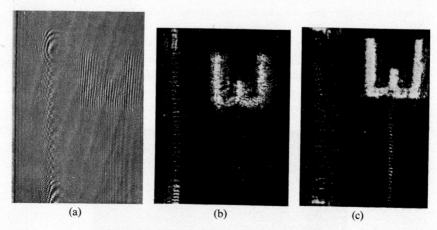

(a)   (b)   (c)

FIG. 6.   (a) Synthetic aperture hologram of two objects. (b) Photo—image of nearer object in focus. (c) Photo—Image of further object in focus.

In Fig. 5 the object is a $\frac{3}{16}$ in. thick aluminium plate in which lettering is milled. The letters are approximately $\frac{1}{4}$ mm deep, $\frac{1}{2}$ mm line wdith and $4\frac{1}{2}$ mm high. The system used was that of liquid surface levitation, similar to that illustrated in Fig. 2, using ultrasound of 10 MHz. With this particular system it has been stated that best results are obtained with pulse lengths of about 80 microseconds at a repetition rate of about 300 per second (Worlton, 1968). An advantage of this type of system is that image reproduction is instantaneous and closed circuit television can be used to view it. Its main disadvantage, i.e. its sensitivity to vibration which by giving rise to long wavelength ripple of the water surface causes the optical images to swing in space and so out of line of sight, appears to have been overcome.

In Fig. 6 the objects are a letter E, 1 cm square with arms $1\frac{1}{2}$ mm to 2 mm wide, cut out of brass sheet $2\frac{1}{2}$ mm thick, and a 2 *BA* cheese head nylon screw. Both objects were held in position with 6 *BA* studding, with the nylon screw about 1 cm and the letter E about 4 cm from the holographic plane. The hologram, made using ultrasound of 10 MHz with a synthetic aperture similar to that illustrated in Fig. 4, is shown in Fig. 6(a). The nylon screw is on the left with the head at the top and the studding, which was tapped into the other end, at the bottom. The letter E is on the right with the supporting studding stretching from the back of the E to the bottom of the hologram. As in this case the reference beam was normal to the hologram, i.e. the electronic reference was the same frequency as that of the isonifying beam, image separation was obtained by tilting the ultrasonic beam direction and letter E by $13\frac{1}{2}$ degrees to the normal and placing the objects off-centre relative to the scan. To isonify both objects at the same time, the pulse length was about 65 microseconds. The repetition rate was about 800 per second and the delayed gate was 5 microseconds wide, situated roughly in the middle of the echo pulse. The hologram was recorded on a facsimile recorder, which was electrically synchronized to the transducer scan before being converted into an optical transparency.

Fig. 6(b) shows the virtual images with the nylon screw in focus. The screw is a poor reflector so its screw thread is rather faint compared to its head and the image of the 6 BA studding supporting it. The letter E is out of focus and its supporting studding indiscernible.

Fig. 6(c) shows the virtual images with the letter E in focus. The studding supporting the E is clearly discernible and the screw is now out of focus. The left-hand thread on the studding is due to an inversion of the optical bench.

The advantages of a mechanical system are flexibility, in that the ultrasonic and recording arrangements can be changed very easily, and high sensitivity resulting from the use of ultrasonic transducers. The disadvantages of the system are the tight tolerances required, about ·01 mm or better at 10 MHz, on the positioning of the scan and recording, and the slow rate at which the recording is made, although this latter is offset to some extent by the ease with which it can be automated. Because of the slow response of mechanical recording systems, the fractional bandwidth of the system can be extremely small, e.g. 50 Hz in 10 MHz, and so thermal noise in the electronics is usually negligible compared to mechanical rumble.

At the present time of writing, ultrasonic holography is still under development and awaiting assessment for nondestructive testing, so much of what follows is tentative. However, owing to the present complexity of instrumentation, ultrasonic holography is not likely to replace the more usual methods except where inspection is an economic necessity and the more usual methods fail. The choice of holographic system will depend upon the circumstances and requirements. A system which yields an instantaneous image enables the N.D.T. inspector to view a component from various angles as it is moved about or, alternatively, to look at moving components if the resulting turbulence in the coupling liquid does not degrade the image unacceptably. A system in which the hologram is made as a recording, such as a transparency, gives a permanent record of the component which can be kept on file and referred to later as if the component which made it were still available for inspection. For example, in the pursuit of quality control, a raw billet might be re-inspected from its hologram if, after the billet had had considerable machining, it was found to have unacceptable flaws.

As mentioned in the introduction, the hologram contains information which is in a convenient form for information processing. An example of where this might be useful is in inspection for small flaws in large grain material, where the effect of the reverberation of the large grains might be removed by spatial filtering leaving the small flaw as a diffraction image. Ultrasonic holography lends itself readily to pulsed systems and this leads to the possibility that, using the range gating facilities of a pulsed system, an object may be divided into sections lying one behind the other, with a hologram being made of each section. This might make for an easier interpretation for inspection purposes rather than a single hologram for the whole object, although there are complications which could arise from unwanted delayed reflections from one section appearing in the hologram of another.

VI. ACKNOWLEDGEMENTS

To Byron Brenden of the Battelle Memorial Institute, Richland, Washington, U.S.A., for very kindly supplying Figure 5, and to my colleagues Arnold Clare and Tony Shepherd for Figure 6.

APPENDIX I

Liquid Surface Levitation Holograms

This is a semi-qualitative treatment of the nature of a light beam reflected from a water surface disturbed by ultrasound as in ultrasonic holography. For simplicity it is convenient to consider only plane beams and to choose axes as shown in Fig. 7, where the Z axis is in the water surface and the X axis is perpendicular to the surface and the inclination of the ultrasonic reference beam to the X axis is θ and the inclination of the incident light beam to the X axis is β. The inclination of the isonifying beam will also be assumed to be at angle θ to the X axis in the opposite sense.

Let the ultrasonic reference beam wavefront in the surface be

$$\exp\{i\omega t + i(2\pi/\lambda)\sin\theta.z\},$$

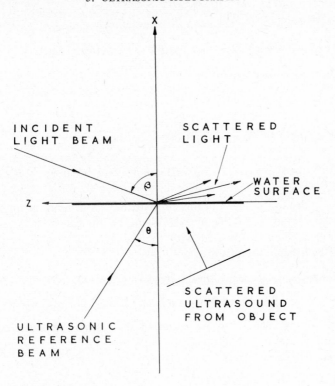

FIG. 7. Appendix 1—co-ordinate system.

and the wavefront in the surface due to the ultrasound scattered by the object be

$$A \exp\{i\omega t - i(2\pi/\lambda)\sin\theta.z + i\phi\},$$

where A and ϕ are due to the object, are both functions of z and are small. The resultant surface displacement will take the form

$$aA \cos\{(4\pi/\lambda)\sin\theta.z - \phi\} + b,$$

where a and b are constants.

The incident light beam will have a wavefront in the plane of the surface of the form

$$\exp\{i\omega_1 t + i(2\pi/\lambda_1)\sin\beta.z\}.$$

The reflected light from the surface will suffer a spatial phase modulation due to the surface displacement so that the wavefront of the reflected light beam in the plane of the surface will take the form

$$\exp\left[i\omega_1 t + i(2\pi/\lambda_1)\sin\beta.z + ib + iaA\cos\{(4\pi/\lambda)\sin\theta.z - \phi\}\right].$$

Using the well-known expansion

$$\exp\{i\rho\cos\delta\} = \sum_{n=-\infty}^{\infty} i^n J_n(\rho)\exp\{in\delta\},$$

the reflected light beam wavefront takes the form

$$\sum_{n=-\infty}^{\infty} i^n J_n(aA)\exp\left[i\omega_1 t + i(2\pi/\lambda_1)\{\sin\beta + 2n(\lambda_1/\lambda)\sin\theta\}z + ib - in\,\phi\right].$$

This is a collection of light beams propagating, in general, at angles $\sin^{-1}\{\sin\beta + 2n(\lambda_1/\lambda)\sin\theta\}$ to the X axis. Using an optical stop to remove all beams except those associated with the values of $n = \pm 1$ gives for the reflected light beam wavefront:

$$iJ_1(aA)\exp\left[i\omega_1 t + i(2\pi/\lambda_1)\{\sin\beta + 2(\lambda_1/\lambda)\sin\theta\}z + ib - i\phi\right]$$

$$+iJ_1(aA)\exp\left[i\omega_1 t + i(2\pi/\lambda_1)\{\sin\beta - 2(\lambda_1/\lambda)\sin\theta\}z + ib + i\phi\right].$$

Comparing these two terms with that of the ultrasound scattered by the object and assuming (aA) is small so that

$$J_1(aA) \cong \tfrac{1}{2}aA,$$

it is seen that the first corresponds to a real reversed image and the second to a virtual image of the object, as in the situation shown in Fig. 1. The images have an angular displacement from each other of $\dfrac{4(\lambda_1/\lambda)\sin\theta}{\cos\beta}$ approximately.

Appendix II

Synthetic Aperture

In what follows, the Synthetic Aperture will be treated from what is hoped is the simple point of view of geometric optics. The more formal mathematical treatment has been given elsewhere (Cutrona et al., 1961).

In a synthetic aperture, the image of the object is formed by rays reflected from the object back to the isonifying source. For this purpose it can be considered that the object is made up of a myriad of reflecting facets, each one reflecting a ray, the orientation of which depends upon the orientation of the facet relative to the isonifying rays. The image of the object is then obtained by suitably processing the reflected rays which pass through the image forming or receiving aperture.

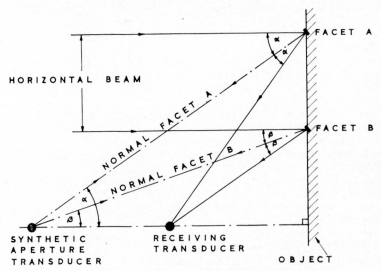

FIG. 8. Appendix II—synthetic aperture.

In Fig 8 is shown a vertical object being isonified by a horizontal beam. Consider, typically, two reflecting facets on the object at A and B with their normals inclined to the horizontal at angles α and β respectively. From the laws of reflection, the reflected rays from these facets will be at angles 2α and 2β to the horizontal and as drawn in the figure it is assumed that both of these rays are received by a point transducer which will also pick up rays from other facets as well. All the facets on the object will send out rays, and by moving the transducer parallel to the object, all these rays can be received and an image of the object ultimately obtained.

Provided the angles are small, the normals of the facets at A and B will intersect at twice the distance that the receiving transducer is from the object, and as is shown in the figure, a point transducer placed at this intersection can isonify the object and receive reflections back from the facets at A and B. This point transducer is the synthetic aperture transducer and simple geometric argument shows that all the facets reflecting rays to the receiving transducer at a given position, under the action of the horizontal beam, will reflect rays back to the synthetic aperture transducer in the corresponding position. Thus, as for the receiving transducer, an image of the object can be obtained by suitably processing the output of the synthetic aperture transducer as it scans the object.

If the angles are not small, the synthetic aperture scan will give rise to image distortion as can be seen by comparing the phase shift of the ray paths of facets at A and B in the two cases.

REFERENCES

Berger, H. and Dickens, R. E. (1963). ANL–6680, Argonne National Laboratory, Illinois.

Boutin, H. and Mueller, R. K. (1967). *J. acoust. Soc. Am.* **42** (5), 1169.

Cochran, W. T., Cooley, J. W., Farin, D. L., Helms, H. D., Kaenel, R. A., Lang, W. W., Maling, Jr., G. C., Nelson, D. E., Rader, C. M. and Welch, P. D. (1967). *Proc. IEEE* **55** (10), 1664–1674.

Cutrona, L. J., Vivian, W. E., Leith, E. N. and Hall, G. O. (1961). *Trans. IRE* **MIL-5**, 127–131.

Gabor, D. (1949). *Proc. R. Soc.* **A197**, 454–487.

Leith, E. N. and Upatnieks, J. (1963). *J. opt. Soc. Am.* **53**, 1377–1381.

MacAnally, R. B. (1967). *Appl. Phys. Lett.* **11** (8), 266–268.

Marom, E., Fritzler, D. and Mueller, R. K. (1968). *Appl. Phys. Lett.* **12** (2), 26–28.

Mueller, R. K. and Sheridon, M. K. (1966). *Appl. Phys. Lett.* **9** (9), 328–329.

Mueller, R. K., Marom, E. and Fritzler, D. (1968). *Appl. Phys. Lett.* **12** (11), 394–395.

Nondestructive Testing Handbook. (1959). (R. C. McMaster, ed.) Vol. II. The Ronald Press, New York.

Rozenberg, L. D. (1955). *Soviet Phys. Acoust.* **1**, 105.

Schuster, K. (1959). *Jenaer Jb.*, 217.

Smyth, C. N., Poynton, F. Y. and Sayers, J. F. (1963). *Proc. Instn elect. Engrs* **110** (1), 16.

Stroke, G. W. (1965). *Appl. Phys. Lett.* **6**, 201–3.

Taylor, C. A. and Lipson, H. (1964). "Optical Transforms". G. Bell and Sons Ltd., pp. 32–37 and 44–45.

Worlton, D. C. (1968). BNWL–SA–1772, Battelle Memorial Inst., Pacific N.W. Lab., Richland, Washington.

Young, J. D. and Wolfe, J. E. (1967). *Appl. Phys. Lett.* **11** (9), 294.

CHAPTER 6

Optical Holography and Coherent Light Techniques

A. E. Ennos

*Division of Optical Metrology, National Physical Laboratory,
Teddington, Middlesex, England*

I. LIST OF SYMBOLS

a_x amplitude of object wave at point x on hologram plate.

A_x total amplitude at point x on hologram plate.

$\phi(x)$ phase of object wave at point x on hologram plate.

θ angle between reference and object wave.

k phase factor of object wave.

λ wavelength.

I_x intensity of light at point x on hologram plate.

T_a amplitude transmission.

A_r reconstructed amplitude.

δ fringe spacing of hologram detail.

Δ path-length change.

α_1 angle between surface motion vector and incident ray.

α_2 angle between surface motion vector and scattered ray.

α   angle of bisection between incident and scattered ray.

ψ   angle between surface motion vector and bisector of incident and scattered rays.

β   angle of incidence of illuminating ray.

d   amplitude of surface vibration.

T   periodic time of vibration.

t   instantaneous time.

$\bar{A}$ time-averaged amplitude.

I   intensity of image reconstructed from time-averaged hologram.

I_0 intensity of image reconstructed from stationary parts of surface.

Φ   argument of Bessel-function distribution of intensity I.

II. Introduction

Optical methods have been used for a longer period than any other as a means of nondestructive testing. The human eye and brain, aided perhaps by a simple lens or microscope, are the most powerful combination for detecting surface flaws and errors of shape without contact having to be made with the object, and a great many inspection processes still depend heavily upon them. Unfortunately, these methods of inspection are very subjective, with all the disadvantages of training skilled personnel, the setting of a suitable standard of perfection, etc. Where the inspection methods are confined to certain specific measurements of dimensions, these can of course be accurately carried out, but they are at best only a partial solution to the problem of complete testing. However, it is not the intention of this chapter to dwell upon the well-known methods of geometrical optics, or even of classical interferometry, which have for many years been used successfully to make test measurements, but rather to give an account of some of the new methods which are currently being developed as a result of the invention of the laser. It will be shown that these new methods are capable of approaching closer to the ultimate goal of attaching a quantitative figure to some of the assessments which are at present made by the eye, coupled with operator experience.

The importance of the laser for testing purposes lies in the fact that the regular train of coherent light waves that it emits can be used as a kind of finely-divided scale for making precise measurements over long distances and in three dimensions. The unit of measurement is the wavelength of light (about one half a micro metre, or twenty micro-inches) so that the measuring sensitivity can be very high. It is of course not novel to use light waves as the yardstick for precise measurement, as can be instanced by the tool-room interferometers used to calibrate slip-gauges, but, as will be seen in the course of this chapter, the highly-coherent laser radiation allows one to extend measurements to a far wider class of subject than previously. In many cases this is achieved via the medium of holography.

Holography is a new science which is having remarkably wide application (Chapter 5 on acoustic holographic methods). Basically, it deals with a storage method for use with any type of wave-motion, in which both the amplitude and phase of the wave-motion can be captured and released at any convenient period later on. This contrasts with the normal storage mechanisms, as embodied in photographic plates, photosensitive electric detectors, etc., which can only react to the intensity, or (amplitude)^2, of the wave-motion. Because all the information about the shape of an object is embodied in the complex wave of light which emanates from it when it is illuminated, the holographic process allows one to record this shape in permanent form, in the hologram. Thus at any later time the shape can be regenerated and used as a sort of three-dimensional template, against which any slight deviations in the form of the object can be compared.

Stemming from this general idea, it will be shown how holographic methods in optics can be used not only for comparison of similar shapes, but also to measure accurately how structures deform under the effect of mechanical stress or thermal gradients. The way in which the surface of an object vibrates can also be analysed with great precision. A further application is to the early detection of minute structural changes in a material, consequent upon its suffering fatigue processes.

Since holography plays such a large part in many of the new techniques it is appropriate that a more detailed description of the process should now be given.

III. THE HOLOGRAPHIC PROCESS

Optics is largely concerned with the process of forming images. The different parts of an object, whether it be self-luminous or merely scattering light which is directed upon it, send out waves which have a certain amplitude and a certain relative phase to one another. The function of the imaging system is to collect the waves travelling in different directions from the object and somehow bring them together again in another region to reproduce the same, or nearly the same, distribution of amplitude and phase. In this respect most lens systems fail to do the job perfectly, sometimes deliberately so. For example, the photographic lens, and even the human eye, deliberately distort the optical image so that its depth is much smaller than the depth of the object field. This allows an intensity pattern which is not too badly out of focus to be recorded or detected on a plane surface–the photographic film or retina. Holography works on a different principle altogether from imagery with lenses. It records the amplitude and phase of the total wave leaving the object surface at some distance away from it in such a way that the same amplitude and phase may be 'played out' again at a later time. It can in this way reproduce the optical effect of the object very much more perfectly than can any lens system, and without the use of lenses.

Since, as has been stated already, photographic plates or other detectors are sensitive only to intensity, and not to phase, some special way of phase-recording must be used. This is done by interfering together the wave from the object with a simple plane or spherical wave, which acts something like the carrier wave in radio transmission. In Fig. 1(a) is drawn a simple arrangement for recording the hologram of an illuminated object, in this case a four-

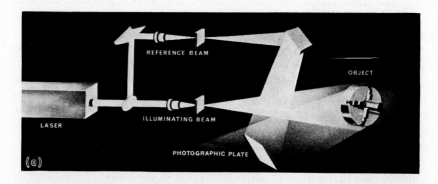

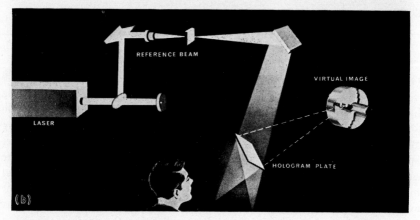

FIG. 1. (a) Recording a hologram. (b) Reconstruction of image from hologram. (Crown copyright reserved).

jaw chuck holding a metal tube. The light scattered from it is allowed to fall onto a photographic plate placed some arbitrary distance away. By itself, this scattered light would produce only a uniform fogging of the emulsion, but before being recorded it is 'mixed' with some of the light direct from the illuminating source, the laser. The two waves, one having a complex distribution of phase and amplitude and the other a uniform distribution, interfere together to form a pattern of fine light and dark fringes, which are recorded by the photographic emulsion as a hologram. The appearance of a typical hologram is shown in Fig. 2, and it can be seen that in no way does it bear any relation to the 'normal' optical image.

The image is reconstructed from the hologram by illuminating it with the direct light from the laser, the 'reference' beam alone, as shown in Fig. 1(b). The fine fringe detail on the hologram plate now acts like a complex diffraction grating, so that while some of the light passes straight through it (the 'zero order diffraction'), other parts of the beam will be diffracted on either

FIG. 2. Hologram plate in kinematic mount (for accurate re-positioning of plate).
(Crown copyright reserved).

side of this direction (the 'first order diffracted waves'). One of these waves is identical to the wave originating from the object, and so if an observer looks through the hologram plate he will appear to see the original object in place, although the object has subsequently been removed. The image is a virtual one in that no actual rays of light emanate from it. However, since a wave identical to that originally leaving the object has been generated, the observer will be able to see the full effects of perspective and depth in it, and, by moving his head from side to side, change the apparent viewpoint, with accompanying parallax effects. The image is a truly 'three-dimensional' one.

The mechanism of image formation in holography can be very simply explained in mathematical terms by reference to Fig. 3 which shows a cross-section of the photographic plate and impinging waves. At a particular point x on the hologram plate, let the resultant wave from all points on the object have amplitude a_x and phase $\phi(x)$. It can thus be represented by $a_x \exp[-i\phi(x)]$. The reference beam, considered as a plane wave of unit amplitude falling on the plate at angle of incidence θ, can be represented by $\exp(-ikx)$ where k is related to θ and the wavelength λ by $k = (2\pi/\lambda)x \sin\theta$.

The total amplitude reaching the photographic plate at point x will then be

$$A_x = a_x \exp[-i\phi(x)] + \exp(-ikx). \tag{1}$$

The intensity I_x is obtained by multiplying by its complex conjugate

$$I_x = (1 + a_x^2) + \exp(ikx)\{a_x \exp[-i\phi(x)] + a_x^* \exp[i\phi(x) - 2ikx]\}. \tag{2}$$

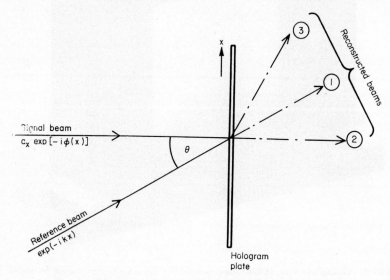

FIG. 3. Diagrammatic representation of holographic process.

The photographic plate darkens according to how much light falls on it and it simplifies matters if it can be assumed that the amplitude transmission T_a of the processed plate is proportional to the intensity falling on it,

$$\text{Thus} \quad T_a = BI_x \quad \text{where} \quad B \quad \text{is a constant.} \tag{3}$$

When the reference beam of amplitude $\exp(-ikx)$ illuminates the hologram, the transmitted amplitude will then be

$$A_r = T_a \exp(-ikx) = B I_x \exp(-ikx) \tag{4}$$

or

$$A_r = B\{(1+a_x)^2 \exp(-ikx) + a_x \exp[-i\phi(x)] + a_x{}^* \exp[i\phi(x) - 2ikx)]\}. \tag{5}$$

The first term represents the light that passes straight through the hologram, the second term is the reconstructed wave, identical to the object wave in all but absolute amplitude, and the third term corresponds to the first order diffracted wave on the other side to the straight through direction. This is termed the conjugate wave and in some cases can give rise to a real image of the object. If the response of the photographic emulsion is not linear according to eqn. (3), extra terms corresponding to higher orders of diffraction result. These do not materially affect the faithfulness of the reconstructed image.

There are a number of important points about the holographic process which must be emphasized if one is to make use of it practically. Firstly, it bears strong resemblances to interferometry and, as such, very stable conditions must be ensured during the exposure time. A relative motion of as little as a quarter of a wavelength between the photographic plate and the object during this time can blur out the fine fringe detail such that no image can be reconstructed. Secondly, highly coherent light is necessary, since the interference process takes place between waves which may travel along considerably different distances. It is a matter of geometry to assess what are the path length differences, but for an object of any size it may be several centimetres. The laser of course provides the necessary temporal and spatial coherence for this. Thirdly, the photographic emulsion must have a high resolution if there is to be a useful field of view. This comes about because, if the angle between a 'reference' beam ray and any one particulars cattered ray coming from the object is θ then the fringe spacing δ is given by

$$\delta = \frac{\lambda}{\sin\theta}. \tag{6}$$

Thus for $\theta = 30°$, $\delta = 2\lambda$, or about one micrometre. The photographic emulsion must in this case resolve 1000 lines/mm, which can only be achieved by using high resolution emulsions which are inherently insensitive; thus high intensity illumination must be employed if the exposure time is not to become too large.

IV. INTERFEROMETRY WITH HOLOGRAMS

Fig. 1(b) shows how the image of an object can be reconstructed from a hologram by illuminating it with the reference beam. If this reconstruction process is carried out with the object still resting in its original position, then the three-dimensional image will superimpose exactly over the object, and, by looking through the hologram plate, object and image will appear as one. Suppose that the object is moved very slightly, or becomes distorted to some extent. In the case of the object illustrated, this can be done by tightening one of the jaws of the chuck. The observer will then see something like Fig. 4. Bands of interference fringes will cover the surface, the number and spacing of which will depend on how much the object has been distorted. Optically, the effect is very similar to that occurring in a conventional interferometer used to test lenses and mirrors, only here one is getting interference effects with surfaces which are by no means optically smooth. The interference effects are in fact entirely independent of the roughness of the surface, for

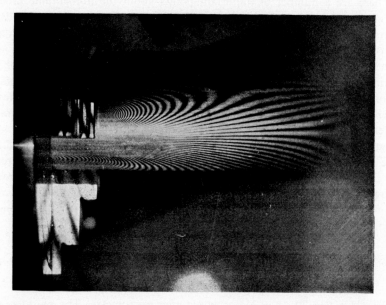

FIG. 4.   Interference fringes on brass tube held in 4-jaw chuck, caused by tightening jaws.
(Crown copyright reserved).

in hologram interferometry one is effectively 'subtracting' the detailed shape of a surface from itself, or something very similar to it, leaving only the effect of the change in position or shape. One can regard it, in fact, as being a form of differential interferometry.

The technique described above is a 'live' method of measuring small changes in shape of an object when subjected to some process, mechanical or otherwise, and allows continual observation of the effect. A complementary technique, which records only the change in shape that has taken place in a fixed interval of time, is to record a double exposure hologram. To illustrate this in the case of the lathe chuck, one exposure (of half the normal duration) is made before screwing up the jaw, and a second one, of similar duration, after the distortion has taken place. On processing the hologram, the reconstructed image will be found to be covered with interference bands similar to those obtained in the 'live' fringe experiment. These are often referred to as 'frozen' fringes, since the information regarding the change in shape that has taken place between exposures has been 'frozen' into the hologram.

Before describing some of the measurement problems to which hologram interferometry has been successfully applied, brief mention must be made of the way in which the interference patterns are interpreted, for this indicates the scope and limitations of the method. Referring to Fig. 5, which is the cross-section of part of a typical rough surface, let us assume that the interference pattern results from movement of this elementary area by an amount

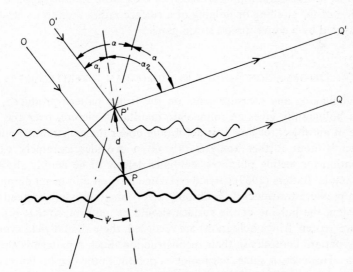

Fig. 5. Geometry of interference fringe formation.

d in direction PP'. The surface is assumed to be illuminated by parallel light rays in a direction OP, and viewed along direction QP. These directions make angles α_1 and α_2 with PP'. If the surface moves by only a small amount, it can be assumed that the interference effect is caused by the combining together of waves scattered from corresponding points on the surface, when in its original and displaced positions. The path-length change Δ can then be obtained geometrically as

$$\Delta = d(\cos \alpha_1 + \cos \alpha_2). \tag{7}$$

This can be expressed in another way by measuring the angles from the bisector of angle OPQ, as in the figure. Then

$$\Delta = 2d \cos \left(\tfrac{1}{2}\alpha_1 + \alpha_2\right) \cos \tfrac{1}{2}(\alpha_1 - \alpha_2) = 2d \cos \alpha \cos \psi \tag{8}$$

where α is the angle of bisection, and ψ the angle between bisector and displacement direction.

Constructive interference (giving a bright fringe) occurs when $\Delta = n\lambda$, and dark fringes are obtained when $\Delta = (n + \tfrac{1}{2})\lambda$ (n being an integer, λ the wavelength of the light).

From eqn. (8), the path difference, and thus the fringe order number, is seen to be proportional to the resolved part of the motion in the direction of the bisector. Interference patterns, then, will only give measurements of motion resolved into one particular direction; when the directions of illumination and viewing are nearly normal to the surface, this direction is also nearly normal to it. Hologram interferometry is thus more suitably applied to the detection of the swelling or bulging of a surface rather than to its stretching, as measured by a conventional strain gauge.

V. Deformation Studies by Hologram Interferometry

Without making any measurements on the fringe pattern produced, interference holography gives an immediate qualitative picture, over the whole surface of an object, of how it is deforming under the constraints applied to it. Even without further analysis this often provides extremely valuable information for testing purposes, enabling defects to be readily shown up. For example, Butters (1969) carried out some tests on thin metal diaphragms used in pressure transducers. With only a very small pressure applied to the diaphragm, the bulging of the surface should be uniform, so that if double-exposure 'frozen' fringe holograms are recorded, these should yield symmetrical ring-pattern contours in their reconstruction. Such a pattern is shown in Fig. 6(a). However, if some 'hard spot' is present in the diaphragm material, it will show up as a distortion in the contour pattern, as illustrated in Fig. 6(b).

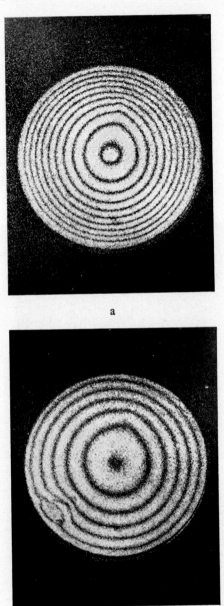

a

b

FIG. 6. Interference fringe due to distortion of a metal pressure-transducing diaphragm under pressure (a) good diaphragm (b) diaphragm having a 'hard spot' at one edge. (Courtesy of Dr. J. N. Butters, Loughborough University).

'Live' fringe experiments carried out on the same diaphragm also showed that considerable wander in the 'zero' had taken place in use, due to creep of the metal. Since each contour represents a displacement of half a wavelength, or approximately 0·3 microns, relative to its neighbour, the test is extremely sensitive. Another example in which interference holography has been successfully applied to measuring mechanical deformations is in the study of the buckling of thin-walled cylinders under compressive loads. Figure 7 shows the type of complex pattern obtained by Leadbetter and Allan (1969) when a steel cylinder was compressed to near its buckling point. A wealth of information on the local deformations round the edges can be obtained in this way, and, from the stress analyst's point of view, enable him to provide exact quantitative data for testing his theoretical models.

FIG. 7. Interference fringes on steel cylindrical shell subjected to axial compression. 'Frozen' fringes obtained by slight increase in force between double exposure, prior to visual buckling of cylinder. (Courtesy of I. Leadbetter, National Engineering Laboratory). (Crown copyright reserved).

Deformations can also be produced by thermal effects, and these may readily be observed by interference holography. As an example, Magill and Wilson (1967) demonstrated the way in which a power transistor deformed as it heated up. Wolfe and Doherty (1966) examined the behaviour of thermoelectric cooling modules in operation. In both of these examples the information so obtained could help in evaluating the design as well as in testing for faulty components. A word of caution should be mentioned here, however. Excessive heating, besides giving rise to an unnecessarily large number of fringes in the interference pattern, also causes refractive index variations in the surrounding air which can distort the fringes and even prevent a hologram from being recorded if there is air turbulence during the exposure time.

Small-scale creep is another quantity well suited to measurement by interference holography. Bradford (1969) has studied the 'recovery' of shape of a number of plastic materials after they have been indented, for example by dropping a steel ball on to the surface, or in the case of a plastic tile, by treading on it with a stiletto heel. It is surprising over what a long period the recovery continues to take place. Similarly, pieces of broken wood gradually change their overall shape due to drying out the fibres and stress relief. This can be readily detected and measured. The effect of dampening the surface of a rough-sawn piece of wood was shown in the same way by Ennos (1966) whose double-exposure holograms yielded a ring-contour pattern of frozen fringes, indicating an overall swelling of the surface.

Being a non-contacting means of testing, hologram interferometry can be used for measuring deformations within a transparent enclosure, e.g. a vacuum vessel, or an encapsulated component. Provided that there is some kind of 'window' to let the light in, testing is possible however poor the optical quality of the window. For example, Haines and Hildebrand (1966) demonstrated the movement of a wire-mesh grid within an image intensifier assembly, caused by electrostatic forces when a potential was applied. In the same way changes in the gas density distribution within a transparent vessel may be measured. Heflinger et al. (1966) illustrated this by showing 'frozen' fringes formed when the filament of an ordinary electric lamp was switched on.

Nondestructive testing by interference holography is primarily a laboratory technique, since the stability conditions necessary to record a hologram call for a vibration-free environment. With experimental development, however, this limitation is gradually being removed, such that testing facilities are already being marketed in the USA for use on the factory floor. By a combination of high-powered laser sources and the use of test benches well isolated from vibration, tests can be carried out in a routine way on large areas of laminated aircraft panels to inspect the epoxy resin bonding between the laminates for poor adhesion. To do this the panel is either strained by application of a light pneumatic pressure, or alternatively by a gentle uniform heating of one side. The 'live', or 'frozen' fringe techniques can both be used, and then any defect of bonding shows up as a discontinuity in the contours. Grant and Brown (1969) of the G.C.-Optronics Corporation, which specializes in holographic testing methods, also describe a tyre-testing apparatus based on the same lines, which is capable of detecting sub-surface defects such as a separation between the sidewall and outer ply of the rubber tyre. Here again, a slight excess pressure is used to obtain the fringe pattern.

There is no doubt that these new techniques are still in the early stages of development and that further progress in interference holographic testing will be made and new applications found. For example, development of the pulsed solid-state laser, which is very much brighter than the continuous wave gas lasers more usually used in holography, may pave the way for testing methods which can be used directly in the factory. The relatively low-power gas lasers require that photographic exposure times of the order of a second or more are necessary, which calls for extreme stability of the apparatus. Pulsed lasers, being powerful enough to expose the photographic materials in their pulse time of a fraction of a microsecond, will need no such special stability. The present drawback to their widespread use for this purpose is their relatively short coherence length.

In addition to the qualitative testing described above, by which flaws in materials may be detected with great sensitivity, interference holography can be used to give a full quantitative solution to the problems of deformation. The theory for this has been briefly indicated in the previous section, which showed that from one fringe pattern only a single component of the surface motion resolved in one direction is obtainable. However, by recording holograms from more than one direction (perhaps by recording a large area hologram) the complete solution can be obtained. Details of the methods for doing this (and, for example, of separating the translation motion of the object from the deformation) will not be considered here. It may however be mentioned that work is in progress towards finding a simple way of measuring the in-plane component of strain, similar to that recorded by a strain gauge or by photo-elastic techniques (see Ennos, 1968).

VI. ENGINEERING INSPECTION OF SHAPE BY INTERFERENCE HOLOGRAPHY

It is an attractive idea to try to use a holographically-recorded image as an optical template. For example, the image of a 'master' shape could then be compared in shape with production-line components. If the comparison between the 'master' shape and the component yielded an interference fringe contour map, this could then be used to measure the difference between them.

Unfortunately, however, unless the surfaces of the components are very smooth, this technique will not work because the individual machining marks caused by turning, grinding, etc. are large compared to the wavelength of light and do not correspond exactly in the two surfaces under consideration. The conditions for meaningful optical interference cannot then be met. However, certain types of engineering surfaces are very smooth (e.g. precision cylinder bores) and are required to be manufactured with great precision.

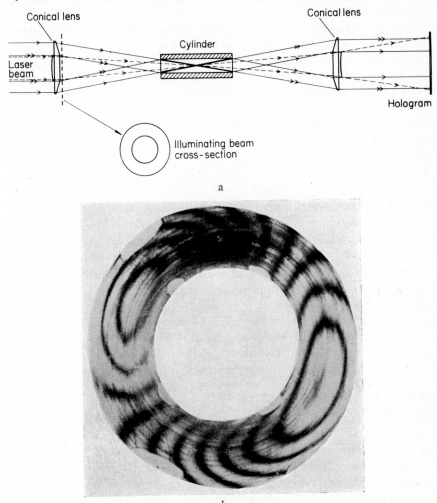

FIG. 8. (a) Optical arrangement for holographic comparison of cylinder bores. (b) Fringes on image of cylinder wall due to comparison of 'master' cylinder with one having a curved axis. (Crown copyright reserved).

Apparatus to measure the shape of such cylinder bores holographically has been developed (Archbold *et al.*, 1967; Archbold and Ennos, 1969) which yields, as its output, a contour map of the difference in shape between two of the cylinders. To obtain the hologram, the internal walls of one cylinder are illuminated with collimated laser light which falls on them at an oblique angle of incidence, such that the light is reflected only once (Fig. 8(a)). Special conical lenses are used to do this, and the central part of the beam is used as the 'reference' for recording the hologram. If a 'master' cylinder is used to record the hologram, then when this is replaced by any other cylinder, interference fringes of the sort shown in Fig. 8(b) will apparently be formed on the surface of the cylinder. The annular nature of the figure corresponds to a view of the cylinder when looking axially into it from one end. From the shape of the fringe pattern, the error in form of the second cylinder may be deduced.

A testing facility of this type relies very largely on exact mechanical location of the object under test, for a tilt or translation of a perfect cylinder would also give rise to fringes which might be interpreted as a difference in shape. For this reason a kinematic mounting method is used. The apparatus described in the second reference above was designed for inspecting diesèl injector cylinder bores 10 mm dia., 58 mm length, and these were hung from four ball-ended supports, two at each end, with a stop to locate the cylinder axially. This constitutes a kinematic location such that the same fringe pattern appears whenever the cylinder is replaced with the same orientation.

Interpretation of the fringe patterns caused by a mismatch between two different cylinders is not so straightforward as when a surface is illuminated and viewed normally. However, it is evident that each point on the annular image corresponds to a particular point of the cylinder wall. Without going into the method of interpretation in detail, the salient points to remember are that each successive fringe represents an increase in radial distance at the corresponding point on the cylinder wall of $\lambda/(2 \cos \beta)$ (where β is the angle of incidence), and also that the fringes of zero order pass through the location points. For a typical case, where $\beta = 81°$, the $\cos \beta$ factor 'densitises' the measurement such that one fringe corresponds to about seven half wavelengths, 2 micrometres. This is of the same order as the tolerances required in the finished product.

There are two drawbacks to using hologram interferometry directly as a means of inspection for errors in shape. The principal one has already been mentioned–the need for having very smooth surfaces. The other is the oversensitivity of the method for many inspection requirements, where the tolerances are measured in thousandths of an inch. There are two general lines of attack on this probelm which will only be outlined briefly here, as they have been by no means fully developed. The first is to generate on the holographic image of the object a set of contour lines corresponding to the variations of height of its surface. This can be done by recording a double exposure holo-gram of the stationary object in which either the reference beam direction or the wavelength is altered between exposures (see Hildebrand and Haines, 1967). By so doing, contours of any desired height separation may be generated. Such contours are useful for measuring the shape of a surface, but less useful when used for checking the correctness of a large number of similar shaped surfaces. An alternative approach by Burch and Cook (1969) combines holography with the moiré fringe technique of measurement. Their aim is to compare any one of a number of rough-surfaced components with a holo-graphically-recorded master. The output would then be in the form of a fringe contour map having contour intervals of perhaps one or two thousandths of an inch, representing the error of form. It is beyond the scope of this chapter to go into details of the method.

VII. HOLOGRAPHY APPLIED TO VIBRATION ANALYSIS

One of the most direct applications of holography is to the study of the mode patterns of a regularly vibrating surface. The simplest method of doing this is to record a hologram of the surface while it is vibrating (Powell and Stetson, 1965). The reconstructed image is then found to be crossed by a pattern of interference fringes. Figure 9 shows a typical example of this, which was obtained by 'holographing' a turbine blade that was vibrated by cementing on to the back a small piezo electric crystal. It is noticeable that one of the fringes is considerably brighter than the rest, and it will be shown below that this fringe in fact delineates the nodal or stationary regions of the blade. By taking holograms at different frequencies of vibration the various modes of vibration can thus be found, leading to a greater under-standing of the elastic properties of the material from which the vibrating object was made.

To explain the formation of the fringe pattern, we must refer back to the mathematical outline given in Section III describing the holographic process. If the surface of the object is vibrating sinusoidally in a normal direction, with amplitude d and periodic time T, this will cause a variation in phase of the 'signal' wave of $(4\pi d/\lambda) \sin (2\pi t/T)$, where t is the instantaneous time measured relative to some zero. The amplitude received at the point x on the hologram plate will now be represented by

$$a_x \exp \left[-i\left(\phi(x) + \frac{4\pi}{\lambda} d \sin \frac{2\pi t}{T} \right) \right]. \tag{9}$$

Fig. 9. 'Time-averaged' fringe pattern on turbine blade vibrating at 12·3 kHz. This is a reconstruction from a hologram recorded while the blade was vibrating. (Crown copyright reserved).

If the time of exposure is long compared with the periodic time (which is usually the case), then the time average of this term must be taken to get the mean amplitude $\bar{A}$ which will be recorded and subsequently reconstructed.

$$\bar{A} = \frac{2a_x}{T} \exp\left[-i\phi(x)\right] \int\limits_0^{T/2} \exp\left[-\frac{4\pi i}{\lambda} d \sin \frac{2\pi t}{T}\right] dt$$

$$= \frac{2a_x}{\pi} \exp\left[-i\phi(x)\right] \int\limits_0^{\pi} \exp\left[-\frac{4\pi i}{\lambda} d \sin \gamma\right] d\gamma.$$

Therefore $$\bar{A} = a_x \exp\left[-i\phi(x)\right].J_0\left(\frac{4\pi d}{\lambda}\right). \tag{10}$$

The reconstructed intensity $I = |A|^2$

or

$$I = J_0^2\left(\frac{4\pi d}{\lambda}\right)I_0 \qquad (11)$$

where I_0 is the reconstructed intensity of the stationary object. The function $J_0^2(\Phi)$ is an oscillatory one, having unit value at the origin, with decreasing values at the succeeding maxima. The minima have zero value. From eqn. (11), when $d = 0$, $T = I_0$, and so the full brightness of reconstruction is obtained for the nodal points. The subsequent maxima have relative brightness values $0.163, .080, .061 ...$ and occur very approximately at values of Φ of $\pi, 2\pi, 3\pi ...$ Thus bright reconstructions are obtained not only for the nodal regions, but also for those whose amplitude of oscillation d takes on approximate values $\frac{1}{4}\lambda, \frac{1}{2}\lambda, \frac{3}{4}\pi ...$ etc. Therefore an apparent fringe pattern will be formed on the reconstructed image surface, contouring equal amplitudes of vibration. The intensity of each succeeding fringe will diminish, and the spacing of fringes will correspond approximately to one-quarter wavelength of vibration amplitude.

Vibration analysis carried out in this way by recording holograms of the vibrating surface have obvious application to the component parts of rotating machinery, where a knowledge of the resonant frequencies and modes of vibration are very important. By recording a single hologram a pictorial display of the modes is immediately obtained, which is a great improvement on probe methods where the amplitude of vibration is measured point by point on the surface. Vibration analysis of this type can also reveal defects in a structure, since the fringe pattern may become distorted when there is some inhomogeneity in the material under examination, at least at one particular frequency.

Although this simple technique of 'time-averaging' the vibration is one which can be readily carried out on any vibrating surface and at any chosen frequency, it has the drawback that the method is essentially a 'blind' one. Until after the holograms are processed the presence or absence of an interesting resonance pattern is not known. An alternative if somewhat more elaborate technique is to apply the principle of the stroboscope so that fringe patterns may be viewed 'live'. This can be done, as in a 'live' fringe experiment, by first recording the surface when stationary, replacing the hologram plate in its original position so that the reconstructed image is superimposed on the object, and then observing the moving fringes which are formed when the surface vibrates. To arrest their motion, the illuminating beams used in the reconstruction stage are pulsed on at the same frequency as that of the vibration. When a short pulse duration relative to the periodic time is used, fringes of high contrast will be seen even at high orders of interference, since the method is essentially one of two-beam interferometry. The spacing of the fringes will correspond to uniform increments of amplitude of vibration, of one half wavelength in this case.

Using a mechanical chopping arrangement to pulse on the light and an optical means for synchronizing the pulse and vibration frequency Archbold and Ennos (1968) showed that the stroboscopic method of examination could detect sharp resonance conditions of a vibrating surface when a range of frequencies were continuously scanned through. Observation of a single cycle of vibration could also be observed by slowly varying the phase between the pulse and the vibrator drive signal. (A similar technique is of course used in conventional stroboscopic analysis.) It appears that with the development of electro-optic shutters the stroboscopic technique may be further developed so that a wider frequency range could be covered than when mechanical chopping is used.

VIII. Fatigue Detection by Holography

The methods of hologram interferometry described so far do not depend on the nature of the surface of the object being studied. In fact, they depend upon the surface structure not changing at all. For example, if in the course of an experiment to detect the creep motion of a metallic specimen its surface becomes corroded in any way, optical interference can no longer take place, because the micro-structure of the surface has changed, and it now scatters the light in an entirely different way. In the inverse sense, this phenomenon can be made use of to detect the early onset of structural changes in a surface before they can be observed directly. It is thus possible to carry out fatigue tests on metal specimens subjected to mechanical working, in a shorter time than needed using conventional techniques.

According to the description of holography given in Section III, the hologram acts as a diffraction grating which, when illuminated by the reference beam alone, diffracts a part of the incident light to regenerate the object wave. It can be shown that this process is strictly reversible; illuminating the hologram with the object wave alone will in fact regenerate the reference beam. Marom (1969) made use of this fact in designing an apparatus to detect fatigue in metal bars subjected to mechanical vibration. A hologram of part of the surface was first recorded in the usual way, using a collimated beam as the reference wave. The processed hologram plate, which can be regarded as a filter 'matched' to the surface in question, was then replaced in its holder and the specimen alone illuminated by the laser source. The hologram reconstructs a parallel beam of light travelling in the same direction as the reference beam. This was focused by means of a lens on to a photodetector, and the output of the detector measured as a function of the time during which the specimen was subjected to the vibration. Measurements were taken only when the specimen was in the rest position. Figure 10 shows the results obtained for a bar of aluminium vibrated at its resonant frequency.

From the optical point of view, this experiment is one of correlating the wave scattered by the original surface with the wave scattered by the surface modified by the mechanical treatment. It is beyond the scope of this chapter to go into correlation theory (see Chapter 7 by Cox, p. 181), and only the principal predictions that it makes for this application will be given. Firstly, it can be shown that for an elastic deformation of the surface, the degree of correlation of the two waves is affected only by second-order terms. This means that the intensity of the reconstructed beam will not alter appreciably for small amounts of strain. On the other hand, for a non-linear deformation, e.g. a dislocation of the surface, the correlation decays almost linearly with the extent of the dislocation. This predicts that it should be possible to detect the formation of cracks in the presence of elastic deformation by the falling off in intensity of the reconstructed reference beam.

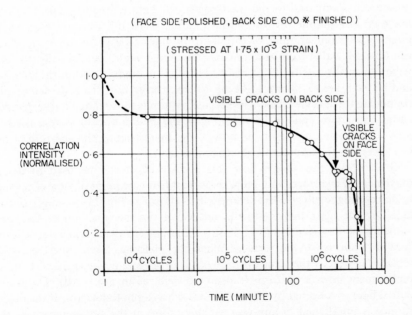

FIG. 10. Decay of correlation intensity with time for vibrating aluminium bar (after Marom). Drop in intensity indicates change in surface structure.

The results shown in Fig. 10 bear this out. The correlation intensity starts to drop well before any visible cracks appear on either the front or rear surface of the specimen. There is almost an order of magnitude of time between the onset of the decay in correlation intensity and the appearance of visible cracks on the face which is being examined. Similar tests on other materials give the same type of results, showing that it is always possible to detect incipient failure by this method earlier than by conventional techniques. Although much work needs to be done to develop the method, it appears promising as a time-saving diagnostic tool.

IX. LASER SPECKLE EFFECTS AND THEIR USES IN TESTING

The properties of the coherent radiation emitted by a laser may in some circumstances be used directly as a means of testing, without going to the complication of recording a hologram. One of the characteristics of laser light is that when it is scattered from an object, the surface appears to be covered in 'speckles'–fine light and dark areas which move when the viewpoint is changed (e.g. Fig. 12(a)). A simple qualitative explanation of this effect is that each element of 'speckle' represents the smallest area which the eye, or optical system, can just resolve, and, since this area may be quite large and 'unsmooth' compared to the wavelength, the light scattered from it will be made up of a number of waves with random phase differences. These waves interfere with one another vectorially to produce a resultant intensity, which can take any value from zero to some maximum value. Statistically, the distribution of the intensities of resolvable areas is random within these limits, and so each area is likely to be of different brightness from its neighbour, producing the speckle effect. The speckle pattern is thus related to the detailed surface structure and to the resolving power of the optical system used to view it. Reducing the resolving power, e.g. by restricting the aperture of the eye, increases the apparent size of the speckle.

Groh (1969) has shown how this speckle effect can be used to detect movement of a surface, and, more important from the point of view of testing, to detect fatigue effects. His arrangement, shown in Fig. 11, is simply one of directing the laser light onto the surface of the specimen under test, and allowing the scattered light to fall on a photographic plate, which records the speckle 'imprint' of the surface. When the plate has been processed, it is replaced in exactly the same position, so as to act as a negative mask. Correct replacement gives a uniform transmitted field, as in Fig. 12(d). The transmitted light is collected by a lens and focused on a photodetector. If the object surface is misaligned in any way, or the nature of the surface changes, the matching of negative and speckle pattern will not be perfect, and a greater

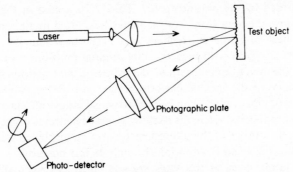

FIG. 11. Experimental arrangement for comparing recorded and actual speckle pattern from a surface (after Groh).

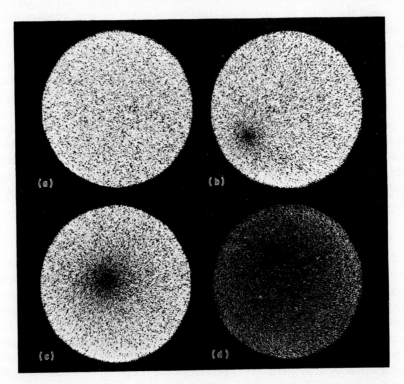

FIG. 12. Patterns caused by super-position of recorded and actual speckle patterns. (a) Misaligned by a large amount; (b) with some translational and rotational misalignment; (c) with less misalignment; (d) with perfect alignment. (Courtesy of Dr. G. Groh, Phillips Zentrallaboratorium, Hamburg.).

signal is recorded by the photodetector. This is illustrated in Fig. 12(b) and (c), which shows the appearance of the field transmitted by the photographic plate when there are different degrees of translational and rotational misalignment of the object with respect to its original position.

Just as in the holographic method described in the previous section, the correlation between the speckle pattern, before and after a subtle change in surface properties takes place, can be used to detect fatigue effects or corrosion. The simplicity of the method lends itself to dynamic methods of measurement. For example, the output of the photodetector can be displayed on an oscilloscope while the specimen is vibrating through its rest position.

A further application of the laser speckle effect has been described by Archbold *et al.* (1969) for the detection of the vibration nodal patterns on a surface. The method complements those of holography described in Section VII. The surface under examination is illuminated by a laser beam and viewed with a telescope having a small entrance pupil. This reduces the resolving power of the system so that the speckle pattern appears fairly coarse. At the same time, some of the light direct from the laser is fed into the telescope to give a uniform field of view, covering the same area as the surface under examination. This light and the 'speckle' light mutually interferes to give a speckle pattern of modified appearance. If the object surface now moves towards or away from the observer, the speckles will go bright and dark cyclically just as would interference fringes. The resultant 'twinkling' of the speckle pattern enables one to detect movement with great sensitivity. Furthermore, if the surface is vibrating the speckles will be blurred out to a uniform brightness, but the nodal regions, being stationary, will still be covered in high-contrast speckles. The method, being visual, is a quick way of surveying a vibrating surface for its nodes without the complication of photographic recording.

X. PRACTICAL LIMITATIONS OF COHERENT LIGHT METHODS OF TESTING

Techniques using the wave-properties of coherent light are essentially interferometric, and have both the sensitivity and need for stability that interferometry entails. As briefly mentioned in Section III, holography requires a stable environment, or, alternatively, high intensity, short duration illumination. At the time of writing, the majority of holographic work has been done with the relatively low power helium–neon laser, emitting red light of wavelength 632·8 nanometres, since this is the most commercially available source of continuous wave laser light. Its power is limited to about 100 mW, but its coherent length is typically 30 cm, and so objects of considerable depth can be studied. Improved methods of increasing the coherence length beyond this are already becoming commercially available. With the recently developed photographic plates of resolving power 2000–3000 line pairs/mm (e.g. Agfa-Gaevert 8E75) and using a 10 mW laser, it is possible to record an object of about 30 cm cube in less than 1 sec exposure. The higher-powered argon lasers, emitting a range of wavelengths in the blue and green, can deliver, typically, up to 2W continuous, but, without additional optical elements, they have only a short coherence length. When this limitation is removed (and the necessary accessories are becoming commercially available) the size of object which can be recorded goes up by a linear factor of five or more. Argon lasers are still, however, extremely expensive and, being inefficient convertors of energy, require large power supplies and water cooling. The ultimate factor limiting the size of object which can be studied is the natural turbulence of the air. Although this can be reduced by shielding etc., it will, in the end, prevent time-lapse work being done, e.g. a study of creep motion over an extended period.

Pulsed lasers, e.g. ruby (emitting red light of 694·3 nanometres) or frequency-doubled neodymium glass (emitting in the green at 530 nanometres), overcome the stability problem by their short pulse length, typically 30 nanoseconds for a Q-switched laser. Until recently, however, it has been difficult to obtain coherence lengths of longer than a few cm. It is also not possible to carry out 'live-fringe' experiments with a source giving only short duration pulses, and reconstruction has to be carried out with a laser of a different wavelength. However, there is a wide field of application for pulsed laser holography which so far has only been touched upon.

One of the possible future developments in coherent light testing is to use the powerful carbon dioxide laser, which emits in the infra-red at 10·6 microns wavelength. This is attractive in such applications as inspection for shape because the longer wavelength is closer to normal machine tolerances. However, holography is, so far, not possible at this wavelength, for materials sensitive to it are not available. The carbon dioxide laser may however be used in direct interferometry for machine-tool control.

REFERENCES

Archbold, E. and Ennos, A. E. (1963). *Nature. Lond.* **217**, 942–943.
Archbold, E. and Ennos, A. E. (1969). Proceedings of " Symposium on the Engineering Uses of Holography ", University of Strathclyde, September 1968. Cambridge University Press, London.
Archbold, E., Burch, J. M. and Ennos, A. E. (1967). *J. scient. Instrum.* **44**, 489–494.
Archbold, E., Burch, J. M., Ennos, A. E. and Taylor, P. A. (1969). *Nature, Lond.* **222**, 263–265.
Bradford, W. R. (1969). Proceedings of " Symposium on the Engineering Uses of Holography ", University of Strathclyde, September 1968. Cambridge University Press, London.

A. E. ENNOS

Burch, J. M. and Cook, R. W. E. (1969). Proceedings of " Symposium on Engineering Uses of Holography ", University of Strathclyde, September 1968. Cambridge University Press, London.

Butters, J. N. (1969). Proceedings of " Symposium on Engineering Uses of Holography ", University of Strathclyde, September 1968. Cambridge University Press, London.

Ennos, A. E. (1966). *Perspective, Lond.* **8,** 276–286.

Ennos, A. E. (1968). *J. scient. Instrum.* (*J. Phys. E.*) Series 2, **1,** 731–734.

Grant, R. M. and Brown, G. M. (1969). *Mater. Evaluation* **27,** 79–84.

Groh, G. (1969). Proceedings of " Symposium on Engineering Uses of Holography ", University of Strathclyde, September 1968. Cambridge University Press, London.

Haines, K. A. and Hildebrand, B. P. (1966). *Trans. IEEE* **I.M. 15,** 149–161.

Heflinger, L., Wuerker, R. F. and Brooks, R. E. (1966). *J. appl. Phys.* **37,** 642–649

Hildebrand, B. P. and Haines, K. A. (1967). *J. opt. Soc. Am.* **57,** 155–162.

Leadbetter, I. K. and Allan, T. (1969). Proceedings of " Symposium on Engineering Uses of Holography ", University of Strathclyde, September 1968. Cambridge University Press, London.

Magill, P. J. and Wilson, A. D. (1967). *Proc. IEEE* (Letter) **55,** 2032–2033.

Marom, E. (1969). Proceedings of " Symposium on Engineering Uses of Holography ", University of Strathclyde, September 1968. Cambridge University Press, London.

Powell, R. L. and Stetson, K. A. (1965). *J. opt. Soc. Am.* **55,** 1593–1598.

Wolfe, R. and Doherty, E. T. (1966). *J. appl. Phys.* **37,** 5008–5009.

CHAPTER 7

Data Handling Techniques

C. W. COX

Dept. of Electrical Engineering, South Dakota School of Mines and Technology, Rapid City, S. Dakota, U.S.A.

I. List of Symbols

a, b	test parameters
a_0, a_1	transmitted data
a_0, b_0	nominal values of test parameters a, b
a_{jk}	change in the jth observation per unit of change in test parameter p_k
c_j	pattern category j
C	capacitance
C_{jk}	normalized linear product $\mathbf{x}_j$ and $\mathbf{y}_k$ or cost of decision favoring H_j when H_k is transmitted
d	distance between two point faults
d_k	kth fault location
$d_{\min}$	minimum resolving distance between point faults
D_j	decision in favor of H_j
E_x	energy of signal x
f	frequency of sinusoidal signal variation
f_1	filter corner frequency
F	arbitrarily defined bandwidth
F_0	single sided signal bandwidth containing all non-zero sinusoidal responses
F_n	noise bandwidth
$g(f^2)$	defining function for a class of low-pass filters
$g_j(x)$	the jth discriminant function
G	a defined gaussian test statistic
$h(t)$	impulse response (weighting function) of transducer
h_{jk}	the jth component of the transducer response to the kth basis function
$h_k(t)$	transducer response to the kth basis function
$H(f)$	transducer frequency response (Fourier transform of $h(t)$)
$\overline{H}(s)$	transducer transfer function (Laplace transform of $h(t)$)
H_j	jth test hypothesis
$\mathbf{i}$	unit vector of Gibb's notation
j	$\sqrt{-1}$
$\mathbf{j}$	unit vector of Gibb's notation
$\mathbf{k}$	unit vector of Gibb's notation
k_0	constant multiplier on filter response
K	decision threshold
K_0	low frequency gain of a low-pass filter
$K_{ky}(\tau)$	covariance function of variables $x(t)$ and $y(t)$

$\overline{m}$ mean value

$n(t)$ noise in $x(t)$

$N(f)$ noise spectrum

N_0 single-sided white noise power density

$p(x)$ probability density function of X

$p(x, y)$ joint probability density function of X and Y

p_j the jth of a set of test parameters *or* the negative of the location of the jth pole on the complex plane

$P(x)$ probability distribution function of X

$P(x, y)$ joint probability distribution function

P_d detection probability

p_E probability of error

P_f false alarm probability

P_x power of signal x

r_j expected loss if H_j is true

r_s scanning rate

R resistance *or* total average risk

$R_{xx}(\tau)$ autocorrelation function of $x(t)$

$R_{xy}(\tau)$ cross-correlation function of variables $x(t)$ and $y(t)$

$s(t)$ useful signal component of $x(t)$

S/N signal-to-noise ratio

t continuous time variable

t_d time required for probe to traverse d

t_0 a particular instant or interval of time

T arbitrarily defined time duration

T_0 time duration of signal including all non-zero values

T_s sampling interval

$u(t)$ unit step function

V_0 a one-dimensional data output

$\mathbf{V}_r$ phasor value of a test reference voltage

$\mathbf{V}_s$ the phasor voltage value of a test response

w_{jk} the kth weight in the jth linear discriminant

$\mathbf{x}$ a data vector whose components are the observations $(x_1, x_2, ..., x_j, ...)$

$\overline{x}$ mean value

x_B decision boundary

x_j the jth of a set of data observations *or* the jth component of a real or abstract data vector

$x_j(t)$ a time varying observable

x_p projection of vector **x** onto vector **y**

X random variable

$X(f)$ Fourier transform of $x(t)$

$\overline{X}$ mean value

$\overline{X^2}$ mean-squared value

z a defined random test variable *or*

$z_k(t)$ output of the kth terminal of an orthonormal filter

Z   complex impedance

$\mathbf{Z}(a,b)$ complex impedance depending on test parameters a, b

$\mathbf{Z}_\Delta$ variation of **Z** from nominal value

$\langle x, y \rangle$ inner product of data vectors **x** and **y**

$|\mathbf{x}|$ magnitude (length) of data vector **x**

x^* complex conjugate of x

$P[A]$ probability that A occurs

$P[A|B]$ probability of A given B

$\mathscr{F}[\,]$ Fourier transform of [ ]

$\mathscr{F}^{-1}[\,]$ inverse Fourier transform of [ ]

E [ ]   expectation of [ ]

$\widetilde{(\ )}$ time average of ( )

$\overline{(\ )}$ ensemble average of ( )

$\Delta(\ )$ an increment of ( )

$\Delta_k(\ )$ the kth increment in ( )

α, β variations of test parameters a, b from nominal values a_0, b_0

Γ_j the jth set of observations

δ_{jk} Kronecker delta

$\delta(x)$ unit impulse (Dirac delta)

θ angle between vectors **x** and **y**

$\boldsymbol{\theta}$ parameter vector

θ_j jth test parameter

μ_n nth central moment

$\Lambda(x)$ likelihood ratio

σ standard deviation

σ^2 variance

τ time interval

$\mathbf{\phi}_j$ the jth of a set of orthonormal basis vectors

$\phi_j(t)$ the jth of a set of orthonormal functions

$\Phi_{xx}(f)$ power density function of $x(t)$

$\Phi_{xy}(f)$ cross-power density of $x(t)$ and $y(t)$

$\Psi_{xx}(f)$ energy density function of $x(t)$

$\Psi_{xy}(f)$ cross-energy density function of $x(t)$ and $y(t)$

II. BACKGROUND

It would be difficult to argue with the observation that the field of non-destructive testing is entering an era of greatly increased emphasis on data handling techniques and information processing. The present state of the art of gleaning useful information from tests is simply inconsistent with the sophistication of the tests themselves, with the demands of current technology for the diametrically opposed requirements of speed and of accuracy in results, and with the availability of new devices, techniques, and knowledge which would appear to be adaptable to the problem. Tests are being developed which yield signals that are rich in information, but much of it comes as by a foreign language for which a translating key is not yet available. The great amount of work being done, with different motives, on information processing and the new availability of low-cost computing and logic modules in myriad forms brings new promises to the development of more effective translators and interpreters of test data.

In this chapter, emphasis will be on techniques which have not yet found wide application in the field of nondestructive testing but which must be considered as more sophisticated data processing schemes are sought. Attention will be directed primarily to those tests often performed at production line speeds, usually scanning tests applied to each unit in a run. Batch sampling techniques form an important part of the total testing picture, but they will not be specifically considered here, though some techniques discussed may find applications to this problem.

III. STATE OF THE ART

A dominant characteristic of data readout from nondestructive tests has been its heavy dependence on visual interpretation of graphically displayed data with considerable reliance on the human operator to provide whatever dimensionality the data vector possesses. Typically, the operator has before him a single trace, a small number of traces, or a pattern or figure on an oscillograph or other recording device showing the variation in test response as the specimen is scanned. He relies on his experience as an observer to tell him when a given signal change represents an unacceptable deviation from a normal specimen. The experienced human operator is, in most cases, an essential component of the testing system, and the best is subject to all the fallibility and variability generally to be expected of humans.

Data vectors at present tend to be at most two-dimensional. For example, the two components of such a vector might be, and often are, the magnitude and phase of a sinusoidal signal. The test itself, on the other hand, is apt to be of rather high dimensionality, having a number, larger than two, of parameters which can exhibit anomalous behavior. Much effort has been expended, with some success, to perform transformations on raw data vectors which will make each component sensitive to changes in a particular test parameter and relatively insensitive to all others.

Up to the present time, efforts in these directions have been especially directed to just one type of test — for example, to an eddy current test, or to an ultrasonic test. There is a great need for techniques which are more general, with applicability to a large class of scanning tests.

IV. The Scanning Test

Before the discussion of signal processing can proceed very far, the nature of the class of tests yielding the signals must be sufficiently specified to place bounds on the problem. To this end, a general test model should be established. It is also helpful to review very briefly current methods which may shed light on the problem of developing new methods.

A. The Test Model

The model which will be assumed is illustrated in Fig. 1. It includes the test sample and the test instrument along with its scanning probe. The excitation unit and response unit may exist in a single scanning head, as implied in Fig. 1, or they may be in different locations, though they will normally be fixed in position with respect to each other, as in Fig. 2. The scanning head is assumed to move with a velocity r_s with respect to the sample.

The excitation energy must be kept below levels which would be destructive to the sample. The probe forms the *excitation signal* into the *response signal* in a manner which is partially determined by the condition of that part of the test sample which lies within the field of the probe. The *raw data vector* could, in the simplest case, be nothing more than the response signal amplified. More often it will be a multidimensional vector whose components are continuous or discrete signals representing parameters of the test signal. The

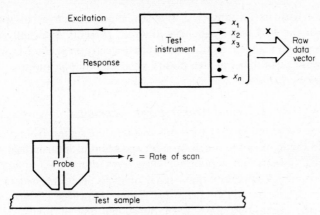

FIG. 1. Model of the scanning test.

term *vector*, as used here, is quite general. The *n*-dimensional vector is simply an ordered set of observations, $(x_1, x_2, ..., x_n)$, which will be designated **x**.

The many examples from practice of instruments which yield data vectors of more than one component are excited in a variety of ways. Some typical probing signals are single-frequency continuous wave (CW) signals, multifrequency CW signals, periodically pulsed signals, or wave packets formed of gated CW signals.

A particular signal-processing scheme would generally be as applicable to one of these tests as to another. For example, a double-frequency eddy

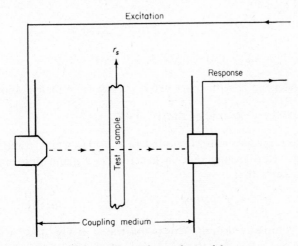

FIG. 2. Alternative probe model.

current test with its response fed into two highly-selective bandpass filters centered on the two test frequencies filtered, presents essentially the same signal processing problem as does the pulsed electromagnetic test when each response pulse is sampled at two points, then the sampler outputs smoothed (cf. chapter by Libby).

B. Some Archetypical Techniques

When one views the existing methods of analyzing signals from nondestructive tests in the light of the much broader spectrum of information and communications systems developing for objectives other than nondestructive testing, it becomes apparent that signal analysis in nondestructive testing remains in a comparatively primitive state. However, it also becomes apparent that techniques have been developed for particular transducer types which may be considered to be first steps in the directions of the most modern developments in such broad areas as information theory, decision processes, and pattern recognition. It seems appropriate to discuss a few examples of such systems.

1. Impedance Plane Methods

Here we discuss briefly in rather general terms a technique which has been extensively applied in well-documented cases (McMaster, 1959; Hochschild, 1958).

Suppose a probe can be devised having its complex impedance primarily dependent on the state of the two parameters (a, b) which during scan show small variations (α, β) about some nominal point (a_0, b_0). That is, the probe impedance is

$$\mathbf{Z} = \mathbf{Z}(a, b) = R(a, b) + jX(a, b), \tag{1}$$

where

$$a = a_0 + \alpha, \quad b = b_0 + \beta, \quad j = \sqrt{-1}. \tag{2}$$

If $\mathbf{Z}_\Delta$ is the change in $\mathbf{Z}$ as the parameters change from their nominal values,

$$\mathbf{Z}(a, b) = \mathbf{Z}(a_0, b_0) + \mathbf{Z}_\Delta(\alpha, \beta). \tag{3}$$

Now suppose the probe is connected in a bridge which compares to its nominal value. Recognizing that in general $\mathbf{Z} \gg \mathbf{Z}_\Delta$ and utilizing eqn. (2) one can easily show that

$$\mathbf{V}_s = \mathbf{A}\mathbf{Z}_\Delta (\alpha, \beta) = \mathbf{A}[R_\Delta (\alpha, \beta) + jX_\Delta(\alpha, \beta)], \tag{4}$$

where $\mathbf{A}$ is a complex constant. This is illustrated in Fig. 3(a), where $\mathbf{V}_s = 0$ for $(a, b) = (a_0, b_0)$, or $(\alpha, \beta) = (0, 0)$.

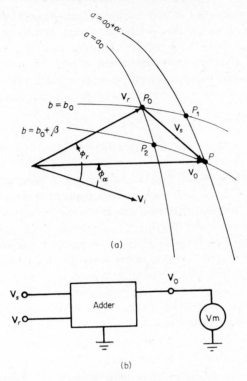

FIG. 3. The impedance-plane technique.

Amplifiers and phase shifters provide the reference voltage $\mathbf{V}_r$ of adjustable magnitude and adjustable phase with respect to $\mathbf{V}_s$. In Fig. 3(a), its magnitude and phase have been adjusted to place its origin at the center of curvature of the $a = a_0$ characteristic, which has been established by holding a constant at its nominal value while b has been allowed to range on both sides of b_0. Now when the amplified bridge output, $\mathbf{V}_s$, is added to $\mathbf{V}_r$, as in Fig. 3(b), the resulting voltage, $\mathbf{V}_0$, has a *magnitude* relatively insensitive to changes in b, but quite sensitive to changes in a. For example, a change in both b and a to the value of point P results in a magnitude $|\mathbf{V}_0| = V_0$, essentially the same as that resulting in a change in a only to P_1. On the other hand, a change in b only to point P_2 produces practically no change in V_0. In this case, $|\mathbf{V}_0| = V_0$ represents a one-dimensional signal vector particularly sensitive to the change α in parameter a; that is approximately,

$$V_0 = x = x(a). \tag{5}$$

By adding another adjustable amplifier and phase-shifter combination,
one *may* be able to set another reference voltage with its origin at the center
of the curvature of the $b=b_0$ curve. Then he will have established a two-
dimensional data vector, each component of which is relatively sensitive to
one parameter while being relatively insensitive to the other. Then, approximately, the data vector is characterized by the column matrix

$$\mathbf{x} = \begin{bmatrix} x_1(\alpha) \\ x_2(\beta) \end{bmatrix} . \tag{6}$$

Had the contours of constant a and constant b been orthogonal, then a
two-dimensional vector with each element depending on a single parameter
could have been accomplished by a simple phase rotation on $\mathbf{V}_s$ followed by a
resolution into its real and imaginary components. Such orthogonality is not
the good fortune of most test instrument designers, however.

Here we have a system whose utility is based on a transformation of the
raw data vector which yields new data vectors more or less independent of
each other and sensitive to one test parameter each. This idea has much in
common with the more general ideas advanced for *estimators* in modern
communication and control theory.

2. *A Test Requiring Pattern Recognition*

Tests are presently in use which provide two or more channels, for example,
the multifrequency electromagnetic tester (Renken *et al.*, 1958) or the multichannel sampled-pulse test (Renken, 1968). A two-channel sample recording
from such a test, applied to thin-wall tubing, is shown in Fig. 4.

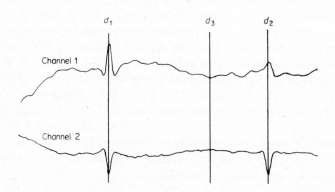

FIG. 4. A two-dimensional pattern.

It is known that Channel 1 is sensitive to a particular class of inside fault but relatively insensitive to similar faults on the outside. Channel 2, on the other hand, reacts to either kind of fault. With this prior knowledge of the test instrument's behavior, one is able to establish quickly that there is an outside fault at point d_1 and an inside fault at point d_2. In other words, he *recognizes* that the *two-dimensional pattern* formed by the two observations at d_1 falls into one category and that at d_2 falls into another. He may also question whether there is a fault at point d_3, coming to a *decision*, based on what he has *learned* about the test, that there is no fault at that point. The test can, of course, be extended by increasing the number of channels, thereby increasing the dimensionality of the pattern and possibly the number of fault categories that can be recognized.

This simple test contains the primordial elements of a number of techniques which are receiving attention today and which are rapidly advancing toward a high degree of sophistication. For example, the problem of deciding whether at point d_3 there is actually a fault-indicating signal or just noise may be automated by application of the techniques of *decision theory*. The classification of the fault for a given set of observations lies in the area of *pattern recognition*, which further attempts to replace the human operator by a machine. Finally, attempts to give the machine the benefits of experience that the human operator can provide leads to the study of *learning*, or *adaptive*, systems.

3. A "Trainable" Test

A recently developed multichannel test (Libby and Wandling, 1968) utilizes raw data whose components do not necessarily exhibit the parameter selectivity of the previous example. It is assumed, however, that the tester can be "tuned" so that each channel responds significantly only to a set of parameters whose number does not exceed the number of channels.

In a three-channel tester, for example, each component of the data vector is assumed to be dependent on three parameters only, say p_1, p_2, and p_3, or

$$x = (x_1, x_2, x_3), \tag{7}$$

where

$$x_1 = x_1(p_1, p_2, p_3)$$

$$x_2 = x_2(p_1, p_2, p_3) \tag{8}$$

$$x_3 = x_3(p_1, p_2, p_3).$$

When the parameters are small enough changes from a set of nominal quantities, the above becomes

$$x_1 = a_{11}p_1 + a_{12}p_2 + a_{13}p_3$$

$$x_2 = a_{21}p_1 + a_{22}p_2 + a_{23}p_3 \qquad (9)$$

$$x_3 = a_{31}p_1 + a_{32}p_2 + a_{33}p_3.$$

The test is concerned with detecting two of the parameters, say p_2 and p_3, and it is implemented as shown in Fig. 5, where the circles with arrows represent adjustable gains and Σ indicates a summing device. The technique is equivalent to solving for p_2 and p_3 by successive elimination of variables.

This system is interesting in the light of modern signal processing techniques because the adjusting of the gains involves *training* the unit by introducing a *training set* of parameter patterns and *teaching* the machine (by adjustment of the gains) to recognize the values of p_2 and p_3.

V. Characterization of Signals

When a discussion of signal analysis techniques is attempted, it soon becomes clear that means to describe or specify the properties of the signal itself are essential. Many of the most stringent constraints on a processing scheme come from the peculiar nature of the signal to be processed, and if it cannot be properly assessed and specified, solutions to processing problems will tend to have a hit-and-miss character.

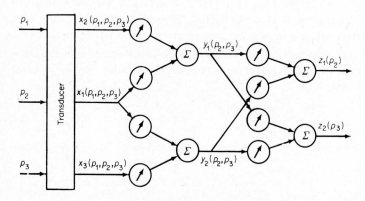

Fig. 5. A "training" scheme.

A. Data Vectors

The discussion of signals in mathematical terms lends itself beautifully to the development of those abstractions which are the delight of the mathematician. In the long run such abstractions will work to the advantage of solutions to signal processing problems, provided that basic hypotheses find counterparts in the real world. In this section, however, an effort is made to provide the minimum abstraction necessary to discuss modern techniques but, at the same time, to tie them to concepts which are familiar to most engineers. It is fortunate that many of the abstractions on observable signals can be expressed in vector, or vector-like forms, thereby utilizing the familiar geometric properties of two- and three-dimension vectors to illustrate the n-dimensional properties of *signal space*.

1. Discrete Observations

When the data consists of a set of simultaneous observations, $x_1, x_2, ..., x_n$, it will generally be convenient to consider the observations to be the components of a *data vector* imbedded in signal space. The vector may be expressed as the set of ordered numbers

$$\mathbf{x} = (x_1, x_2, ..., x_n). \tag{10}$$

The *inner product* (or dot product) of two n-dimensional data vectors, x and y, is defined by

$$\langle x, y \rangle = \sum_{k=1}^{n} x_k y_k^*, \tag{11}$$

where, for generality, we admit the possibility of complex components whose conjugates are indicated by the asterisk superscript. The length $|\mathbf{x}|$ of the data vector is defined by

$$|\mathbf{x}|^2 = \langle x, x \rangle. \tag{12}$$

When the components are real and are considered to be mutually perpendicular, the above is nothing more than a statement of the Pythagorus Theorem.

Suppose one forms the set of n unit vectors, similar to the familiar $\mathbf{i}, \mathbf{j}$, and $\mathbf{k}$ vectors of Gibbs notation in three-dimensions,

$$\boldsymbol{\phi}_1 = (1, 0, 0, ..., 0)$$

$$\boldsymbol{\phi}_2 = (0, 1, 0, ..., 0)$$

$$\tag{13}$$

$$\boldsymbol{\phi}_n = (0, 0, 0, ..., 1).$$

Obviously,

$$\langle \phi_j, \phi_k \rangle = \delta_{jk} = \begin{cases} 1, & j = k \\ 0, & j \neq k. \end{cases} \tag{14}$$

Such vectors are called *orthonormal*, which, geometrically, means that they are of unit length and mutually perpendicular. They may be considered as *basis vectors* along which the components of **x** lie, a role geometrically identical to that of the **i**, **j**, and **k** vectors of Gibbs. In the manner of Gibbs, then, one may write

$$\mathbf{x} = \sum_{k=1}^{n} x_k \boldsymbol{\phi}_k. \tag{15}$$

It follows immediately from eqns. (14) and (15) that the jth component of **x** is the inner product

$$x_j = \langle x, \phi_j \rangle. \tag{16}$$

2. Energy Signals†

The *energy* of a signal, $x(t)$, is defined as

$$E_x = \int_{-\infty}^{\infty} |x(t)|^2 dt = \int_{-\infty}^{\infty} x(t)x^*(t)dt. \tag{17}$$

A physical analogy to this expression is provided in the form of heat generated in a 1-ohm resistor with $x(t)$ volts applied. Whenever E_x is finite, the signal $x(t)$ is called an *energy signal*.

A set of energy signals, $\phi_1(t), \phi_2(t), ..., \phi_n(t)$, are orthonormal if

$$\int_{-\infty}^{\infty} \phi_j(t)\phi_k^*(t)dt = \delta_{jk}. \tag{18}$$

Any energy signal may be expressed as a series of such *orthonormal* functions, forming the *orthogonal* series

$$x(t) = \sum_{k=1}^{n} x_k \phi_k(t), \tag{19}$$

---

† See Burdic (1968), Cooper and McGillem (1967), and Lathi (1968).

where the equality becomes more and more exact as $n \to \infty$, and

$$x_j = \int_{-\infty}^{\infty} x(t)\phi_j^*(t)dt. \tag{20}$$

These two equations define what is often called a *generalized Fourier series*. The integral of the product of two such signals (possibly complex) is

$$\int_{-\infty}^{\infty} x(t)y^*(t)dt = \sum_{k=1}^{n} x_k y_k^*. \tag{21}$$

Clearly, then, this integral is the inner product, as defined in eqn. (11), between two data vectors formed from the coefficients in the orthogonal series approximating the signal. As a matter of fact, for data vectors formed in this way eqns. (10)–(16), for discrete observations, all hold. Therefore, it suits our purposes admirably to define the integration in eqn. (18) as the inner product of $x(t)$ and $y(t)$, i.e.

$$\langle x, y \rangle = \int_{-\infty}^{\infty} s_a(t)s_b^*(t)dt. \tag{22}$$

It must be emphasized that the inner product may be evaluated by using the vector form of the signal or it may be evaluated from the integral without even selecting a set of orthogonal components.

Now we may collect eqns. (18), (19), and (20) into energy-signal counterparts of eqns. (14), (15) and (16)

$$x(t) = \sum_{k=1}^{n} x_k \phi_k(t) \qquad (\text{or } \mathbf{x} = \sum_{k=1}^{n} x_k \phi_k),$$

$$\text{where } x_j = \langle x, \phi_j \rangle \text{ and } \langle \phi_j, \phi_k \rangle = \delta_{jk}. \tag{23}$$

When data vectors are formed from coefficients in the orthogonal series, the relationships shown for discrete observations apply exactly.

The length of the discrete data vector becomes the energy in the energy signal. That is, eqn. (17) becomes

$$E = |\mathbf{x}|^2 = \langle x, x \rangle. \tag{24}$$

3. Power Signals†

When the energies are infinite, as in periodic signals and random signals, equations such as (17) and (22) cannot be evaluated. However, the average power over some interval, T, is finite. We shall define the *signal power* of such *power signals* as

$$P_x = \frac{1}{T} \int_{t_0}^{t_0+T} |x(t)|^2 dt = \frac{1}{T} \int_{t_0}^{t_0+T} x(t)x^*(t)dt$$

$$= \overline{|x(t)|^2}, \tag{25}$$

where the wavy overbar is used to indicate time averages. This has the physical analogy of the average rate of heat generation in a 1-ohm resistor with an applied voltage $x(t)$. For *stationary* signals, which are the only ones considered here, t_0 may be any convenient time.

The power signals $\phi_1(t), \phi_2(t) \ldots, \phi_n(t)$ are *orthonormal over T* if

$$\frac{1}{T} \int_{t_0}^{t_0+T} \phi_j(t)\phi_k^*(t)dt = \overline{\phi_j(t)\phi_k(t)} = \delta_{jk}. \tag{26}$$

The power signal $x(t)$ can be expressed, in a manner similar to that for energy signals, as

$$x(t) = \sum_{k=1}^{n} x_k\phi_k(t), \tag{27}$$

where

$$x_k = \frac{1}{T} \int_{t_0}^{t_0+T} x(t)\phi_k^*(t)dt. \tag{28}$$

This form includes the conventional Fourier series, as well as numerous other possible expansions.

Here, again, the coefficients can be considered to be the components of a data vector, **x**, with the same correspondences to the discrete case that were noted for energy signal, provided the inner product is now defined

$$\langle x, y \rangle = \frac{1}{T} \int_{t_0}^{t_0+T} x(t)y^*(t)dt = \overline{x(t)y^*(t)}. \tag{29}$$

With this definition of the inner product, eqn. (23) holds for power signals as well as for energy signals.

---

† See Burdic (1968), Cooper and McGillem (1967), and Lathi (1968).

4. *Random Signals*†

In this section, a knowledge of basic probability is assumed. However, it seems advisable to review briefly some aspects of stationary random processes before proceeding with the discussion.

In the discussion we shall use the empirical, or relative-frequency definition of probability. That is, if n is the number of occurences of A in N experiments, then

$$P\,[A] = \text{Probability of event } A$$

$$= \lim_{N \to \infty} \frac{n}{N}$$

$$= \frac{n}{N} \text{ for large } N. \tag{30}$$

The response signal records of N identical experiments performed simultaneously (or of one experiment performed N times) are shown in Fig. 6. In this experiment the random variable X is the set of signals, $x_1(t), x_2(t), \ldots, x_n(t)$. The *probability density function* is

$$p(x_k) = \frac{P[x_k < X \leqslant x_k + \Delta_k x]}{k}$$

$$= \frac{\Delta_k N}{N \Delta_k}, \tag{31}$$

where $\Delta_k N$ is the number of records taking on a value within the slit of height $\Delta_k x$ above x_k.

The *probability distribution function* is

$$P(x_k) = P[X \leqslant x_k] = \sum_{j=1}^{k} p(x_j)\, \Delta_j x. \tag{32}$$

When the number of records becomes very large, this becomes

$$P(x) = \int_{-\infty}^{\infty} p(\zeta)d\zeta, \tag{33}$$

---

† See Burdic (1968), Cooper and McGillem (1967), Lathi (1968), Papoulis (1965), and Jenkens and Watts (1968).

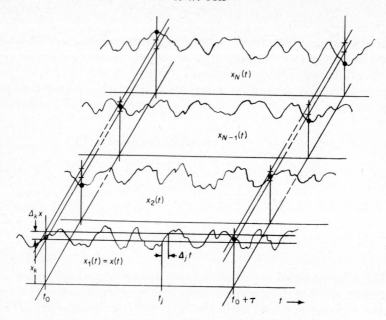

Fig. 6. Ensemble of random signals from the same random process.

carrying with it the implication that

$$p(x) = \frac{dP(x)}{dx}.$$ (34)

The probability density function could have been evaluated at some other time than t_0, say $t_0 + \tau$. The random process is stationary if

$$p(x)|_{t=t_0} = p(x)|_{t=t_0+\tau}$$ (35)

for any t_0 and τ.

The *expectation* of X at t_0 is the average of the N values it has at $t = t_0$,

$$E[X] = x_{\mathrm{avg}} = \sum_{k=1}^{K} \frac{x_k \Delta_k N}{N}$$

$$= \sum_{k=1}^{K} x_k p(x_k)\, \Delta_k x.$$ (36)

When x can take on all values, this becomes

$$E[X] = \int\limits_{-\infty}^{\infty} xp(x)dx. \tag{37}$$

Similarly, the expectation of a function of the random variable X is

$$E[f(X)] = [f(x)]_{\text{avg}} = \int\limits_{-\infty}^{\infty} f(x)p(x)dx = \overline{f(x)}. \tag{38}$$

Here the overbar is used to indicate the average over the entire ensemble of signals at $t = t_0$.

Table I lists some of the most commonly used statistical expectations and the standard symbols and terminology used in references to them.

TABLE I

Some Common Statistical Expectations

$f(x)$	Parameter defined by $E[f(x)]$	Symbol
X^n	nth moment of X	α_n
X	mean (average of X)	$\overline{X}$ or m
X^2	mean-squared value of X	$\overline{X^2}$
$(X-m)^n$	nth central moment of X	μ_n
$(X-m)^2$	variance of X = (standard deviation of X)^2	$\sigma^2 = \overline{X^2} - m^2$

The last entry in the table follows easily from the second and third when it is noted that the expectations operator, $E[A]$, is distributive.

When there are two random variables, X and Y, their *joint probability density* function is

$$p(x_j, y_k) = \frac{P[(x_j < X \leqslant x_j + \Delta_j x) \text{ and } (y_k < Y \leqslant y_k + \Delta_k y)]}{\Delta_j x \Delta_k y}, \tag{39}$$

and the *joint probability distribution* function becomes

$$P(x_j, y_k) = p[(X \leqslant x_k) \text{ and } (Y \leqslant y_k)]. \tag{40}$$

In the continuous case

$$P(x, y) = P[(X \leqslant x) \text{ and } (Y \leqslant y)] \tag{41}$$

and

$$p(x, y) = \frac{\partial^2 P(x, y)}{\partial x \partial y}, \tag{42}$$

or

$$P(x, y) = \int_{-\infty}^{\infty} \int_{-\infty}^{\infty} p(x, y) dx dy. \tag{43}$$

The joint expectation of a function of X and Y is

$$E[f(X, Y)] = \int_{-\infty}^{\infty} \int_{-\infty}^{\infty} f(x, y) p(x, y) dx dy = \overline{f(x, y)}. \tag{44}$$

Let us now return to Fig. 6 and look at the content of just the first signal, $x_1(t) = x(t)$. It can be shown that the expectation of this single trace is

$$E[X(t)] = \lim_{T \to \infty} \frac{1}{T} \int_{-T/2}^{T/2} x(t) dt = \overset{\sim}{x(t)}. \tag{45}$$

When the expectation computed by eqn. (45) from one record of the experiment is the same as that computed across a large ensemble of experiments, by use of eqn. (37), the process is said to be *ergodic*. Ergodic processes are also stationary, though the inverse is not necessarily true.

Equation (38) has the extension, for *ergodic processes*, to

$$E[f(x)] = \lim_{T \to \infty} \frac{1}{T} \int_{-T/2}^{T/2} f(x, t) dt = \overset{\sim}{f(x, t)}$$

$$= \int_{-\infty}^{\infty} f(x) p(x) dx = \overline{f(x)}, \tag{46}$$

where $f(x, t)$ implies a function of a single signal recording and $f(x)$ implies a function of the values taken by an ensemble of recordings at a given time. Another way of expressing the same data is to equate ensemble averages, indicated by the straight overbar, to time averages, indicated by the wavy overbar.

For a two-variable ergodic process, eqn. (44) may be extended to

$$E[f(X, Y)] = \overline{f(x, y)} = \widetilde{f(x, y, t)}. \tag{47}$$

We are now in a position to talk of random data vectors in the same geometric sense that deterministic signals have been described. If the randomly varying signals $\phi_1(t), \phi_2(t), ..., \phi_n(t)$ are selected so that

$$E[\Phi_j(t)\Phi_k{}^*(t)] = \delta_{jk}, \tag{48}$$

then one may express the randomly varying signal $x(t)$ as

$$x(t) = \sum_{k=1}^{n} x_k \phi_k(t), \tag{49a}$$

where

$$x_k = E[X(t)\Phi_j{}^*(t)]. \tag{49b}$$

Thus, if the inner product of a pair of random signals is expressed as

$$\langle x, y \rangle = E[X(t)Y^*(t)] = \widetilde{x(t)y^*(t)}, \tag{50}$$

then all the geometric interpretations given to power signals may be applied to random signals, provided they are ergodic.

B. Correlation†

In Fig. 7 is shown a pair of data vectors, $\mathbf{x}$ and $\mathbf{y}$, in the three-dimensional signal space characterized by the basis vectors ϕ_1, ϕ_2, and ϕ_3. From conventional vector analysis, we have the well-known relationship

$$\langle x, y \rangle = |\mathbf{x}| \, |\mathbf{y}| \cos \theta,$$

or

$$\cos \theta = \frac{\langle x, y \rangle}{|\mathbf{x}| \, |\mathbf{y}|}. \tag{51}$$

The angle θ is a measure of the *correlation* between the two data vectors. It is zero when the vectors are orthogonal, and it reaches a maximum of unity when they are colinear.

---

† See Burdic (1968), Cooper and McGillem (1967), Lathi (1968), Papoulis (1965), Jenkens and Watts (1968), Van Trees (1968), and Bendat and Piersol (1966).

C. W. COX

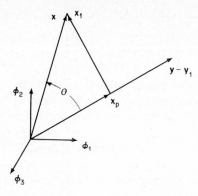

Fig. 7. Illustrating correlation and orthogonalization.

The *projection* of **x** onto **y** is

$$\mathbf{x}_p = |\mathbf{x}| \, (\cos \theta) \, \frac{\mathbf{y}}{|\mathbf{y}|}$$

$$= \frac{\langle x, y \rangle}{|\mathbf{x}| \, |\mathbf{y}|} \, \mathbf{y} = C_{xy} \mathbf{y}. \tag{52}$$

The coefficient

$$C_{xy} = \frac{\langle x, y \rangle}{|\mathbf{x}| \, |\mathbf{y}|} \tag{53}$$

can be used in the formation of two uncorrelated data vectors, $\mathbf{x}_1$ and $\mathbf{y}_1$, from the given correlated pair, since on Fig. 7

$$\mathbf{x}_1 = \mathbf{x} - \mathbf{x}_p.$$

Thus, a new set of orthogonal vectors

$$\mathbf{y}_1 = \mathbf{y}$$
$$\mathbf{x}_1 = \mathbf{x} - C_{xy} \mathbf{y} \tag{54}$$

has been formed in which

$$\cos \theta_1 = \frac{\langle x_1, y_1 \rangle}{|\mathbf{x}_1| \, |\mathbf{y}_1|} = 0.$$

This has been the *Gram–Schmidt orthogonalization* of a set of two vectors. This may easily be extended to n vectors in an n-dimensional space. For example, the orthogonal vectors, $\mathbf{y}_1$, $\mathbf{y}_2$, ..., $\mathbf{y}_n$ may be formed from the non-orthogonal vectors $\mathbf{x}_1$, $\mathbf{x}_2$, ..., $\mathbf{x}_n$ by the generalization

$$\mathbf{y}_k = \mathbf{x}_k - \sum_{j=1}^{k-1} C_{jk}\,\mathbf{y}_j, \quad k = 1, 2, ..., n, \tag{55}$$

where

$$C_{jk} = \frac{\langle y_j \cdot x_k \rangle}{|y_j|^2}. \tag{56}$$

Practical applications to nondestructive testing of this type of orthogonalization, provided it can be instrumented, become apparent when compared to their archetype in the form of the impedance plane tester discussed earlier.

Now suppose the vectors $\mathbf{x}$ and $\mathbf{y}$ are observables having some vector variation $\mathbf{x}_v$, $\mathbf{y}_v$ about a vector mean $\bar{\mathbf{x}}$, $\bar{\mathbf{y}}$. Then the vector variations about the means are

$$\mathbf{x}_v = \mathbf{x} - \bar{\mathbf{x}} \tag{57}$$

$$\mathbf{y}_v = \mathbf{y} - \bar{\mathbf{y}}.$$

The cosine of the angle between these two vectors is called the *correlation coefficient*

$$\rho_{xy} = \cos\theta_v = \frac{\langle x_v, y_v \rangle}{|\mathbf{x}_v|\,|\mathbf{y}_v|}$$

$$= \frac{\langle (x-\bar{x}),\, (y-\bar{y}) \rangle}{\langle (x-\bar{x}),\, (x-\bar{x}) \rangle \langle y-\bar{y}),\, (y-\bar{y}) \rangle}. \tag{58}$$

The numerator of the correlation coefficient,

$$K_{xy} = \langle (x-\bar{x}),\, (y-\bar{y}) \rangle \tag{59}$$

is called the *covariance* of x and y. The *variances* of x and y are

$$\sigma_x^2 = |\mathbf{x}|^2 = \langle x, x \rangle \tag{60}$$

$$\sigma_y^2 = |\mathbf{y}|^2 = \langle y, y \rangle. \tag{61}$$

These three items of terminology are borrowed from statistics, but they are becoming useful in the discussion of deterministic or mixed systems. Clearly,

C. W. COX

the square root of the variance is the familiar root-mean-square value of a power signal. The more usual form of eqn. (58) is

$$\rho_{xy} = \frac{K_{xy}}{\sigma_x \sigma_y}. \tag{62}$$

Suppose the variables $x(t)$ and $y(t+\tau)$ are stationary and expressed as the orthogonal series

$$x(t) = \sum_{k=1}^{n} x_k \phi_k(t)$$

$$y(t+\tau) = \sum_{k=1}^{n} y_{k,\tau} \phi_k(t) = y_\tau(t).$$

The covariance now becomes

$$K_{xy}(\tau) = \langle (x-\bar{x}), (y_\tau - \bar{y}) \rangle$$

$$= \langle x, y \rangle - \langle \bar{x}, \bar{y} \rangle. \tag{63}$$

If

$$R_{xy}(\tau) = \langle x, y \rangle = \int_{-\infty}^{\infty} x(t) y^*(t+\tau) dt, \tag{64}$$

then (63) becomes

$$K_{xy}(\tau) = R_{xy}(\tau) - \langle \bar{x}, \bar{y} \rangle. \tag{65}$$

In the terminology standard for random variables

$K_{xx}(\tau), K_{yy}(\tau) = $ *autocovariance functions*

$K_{xy}(\tau), K_{yx}(\tau) = $ *cross-covariance functions*

$R_{xx}(\tau), R_{yy}(\tau) = $ *autocorrelation functions*

$R_{xy}(\tau), R_{yx}(\tau) = $ *cross-correlation functions.*

For stationary ergodic random variables, eqns. (63) and (64) become

$$K_{xy}(\tau) = E[\{X(t) - \bar{x}\}\{Y(t+\tau) - \bar{y}\}^*]$$

$$= \overline{[X(t) - \bar{x}][Y(t+\tau) - \bar{y}]^*}$$

$$= [x(t) - \bar{x}][Y(t+\tau) - \bar{y}]^* \tag{66}$$

and

$$R_{xy}(\tau) = E[X(t)Y^*(t+\tau)]$$

$$= \overline{X(t)Y^*(t+\tau)} = x(t)y^*(t+\tau). \tag{67}$$

Physical significance can be attached to the correlation function by references to eqns. (17) and (25), i.e.

$$R_{xx}(0) = E_x = \text{signal energy of } x(t) \qquad (68a)$$
$$\text{for energy signals}$$

$$R_{xx}(0) = P_x = \text{average signal power of } x(t) \qquad (68b)$$
$$\text{for power signals.}$$

For methods of measuring the quantities discussed in this section, the reader is referred to the rather complete coverage by Bendat and Piersol (1966).

C. Characteristic Frequency Spectra†

The familiar signal frequency spectrum, based on the Fourier transform of the signal, does not quite suffice to characterize the frequency-dependent nature of all signals. Other spectra are needed to complete the picture, particularly when random power signals are involved.

The standard signal frequency spectrum is simply the Fourier transform of the signal. For the signal $x(t)$ it is

$$X(f) = \mathscr{F}[x(t)] = \int_{-\infty}^{\infty} x(t)\exp\left(-j2\pi ft\right) dt, \qquad (69a)$$

and the signal can be generated from its spectrum by the inverse Fourier transform,

$$x(t) = \mathscr{F}^{-1}[x(f)] = \int_{-\infty}^{\infty} X(f)\exp\left(j2\pi ft\right) df. \qquad (69b)$$

Parceval's Theorem allows the energy of $x(t)$ to be expressed in the alternative ways

$$E_x = \int_{-\infty}^{\infty} |x(t)|^2 dt = \int_{-\infty}^{\infty} |X(f)|^2 df. \qquad (70)$$

The quantity

$$\Psi_{xx}(f) = |X(f)|^2 = X(f)X^*(f) = X(f)X(-f) \qquad (71)$$

† See Burdic (1968), Cooper and McGillem (1967), Lathi (1968), Papoulis (1965), Jenkens and Watts (1968), Van Trees (1968), and Bendat and Piersol (1966).

is called the *energy density spectrum*, since it has the units of (signal energy)/Hertz. Thus, eqn. (70) becomes

$$E_x = \int_{-\infty}^{\infty} \Psi_{xx}(f)df = R_{xx}(0), \quad \text{when } E_x < \infty. \tag{72}$$

It follows from eqns. (69) and (64) that for energy signals

$$\Psi_{xx}(f) = \mathscr{F}[R_{xx}(\tau)] = \int_{-\infty}^{\infty} R_{xx}(\tau) \exp(-j2\pi f\tau) \, d\tau. \tag{73}$$

This can be generalized to give the *cross-energy density spectrum*

$$\Psi_{xy}(f) = \mathscr{F}[R_{xy}(\tau)] = \int_{-\infty}^{\infty} R_{xy}(\tau) \exp(-j2\pi f\tau) \, dr. \tag{74}$$

A power signal may be considered the limiting case of an energy signal of duration T, $x_T(t)$, identical with the power signal on the interval T, centered at $t = 0$. The signal power is

$$P_x = \lim_{T \to \infty} \frac{E_{Tx}}{T} = \lim_{T \to \infty} \int_{-\infty}^{\infty} \frac{\Psi_{xx}(f)}{T} \, df.$$

The quantity

$$\Phi_{xx}(f) = \lim_{T \to \infty} \frac{\Psi_{Txx}(f)}{T} \tag{75}$$

is called the *power density spectrum* for obvious reasons. Again using eqns. (69), one finds that the *cross-power density* spectrum is

$$\Phi_{xy}(f) = \lim_{T \to \infty} \frac{1}{T} X_T(f) Y_T^*(f)$$

$$= \lim_{T \to \infty} \mathscr{F}\left[\frac{1}{T} \int_{-\infty}^{\infty} x_T(t) y_T^*(t+\tau) dt \right].$$

But $x_T(t)$ and $y(t)$ are zero except for $-(T/2) < t < (T/2)$, in which interval they are equal to $x(t)$ and $y(t)$. Therefore

$$\Phi_{xy}(f) = \mathscr{F}\left[\lim_{T\to\infty} \frac{1}{T} \int_{-T/2}^{T/2} x(t)y^*(t+\tau)dt\right] = \mathscr{F}[x(t)y^*(t+\tau)]$$

$$= \mathscr{F}[R_{xy}(\tau)] = \int_{-\infty}^{\infty} R_{xy}(\tau)\exp(-2\pi f\tau)\,d\tau. \tag{76}$$

We see, then, that the Fourier transform of the correlation function of an energy signal is its energy-density spectrum, but that the Fourier transform of the correlation function of a power signal is its power density spectrum.

Since for power signals,

$$R_{xy}(\tau) = \mathscr{F}^{-1}[\Phi_{xy}(f)] = \int_{-\infty}^{\infty} \Phi_{xy}(f)\exp(2\pi f\tau)\,df, \tag{77}$$

the counterpart of eqn. (72) for power signals is

$$P_x = \int_{-\infty}^{\infty} \Phi_{xx}(f)df = R_{xx}(0), \text{ when } E_x \to \infty. \tag{78}$$

VI. TIME–BANDWIDTH CONSIDERATIONS†

Two recurring items of consideration in any signal processing study are the *time duration* and the *bandwidth* of the signal. The specification of these interrelated quantities is not always an easy matter. They will be approximations whose basis depends quite heavily on which phenomenon they are considered to effect and on the type of signal. Power signals, for example, cannot have a finite time duration specified, as can energy signals. Nevertheless, processing times available are finite, even for power signals.

A. The Sampling Theorem and Bandwidth

One indication of bandwidth comes from the familiar *sampling theorem* of Shannon and Weaver (1949):

If the frequency spectrum of a continuous function of time is limited to

---

† See Burdic (1968), Lathi (1968), Wainstein and Zubakov (1962), Woodward (1953), Cook and Bernfeld (1967), Schwartz (1959), and Lerner (1961).

frequencies within the interval $-F_0 < f < F_0$, then the function is completely defined by time domain samples taken at intervals

$$T_s = \frac{1}{2F_0}. \tag{79}$$

If the signal is non-zero only in the time interval $t_0 < t < t_0 + T_0$, then the *time duration* is

$$T_0 = NT_s = \frac{n}{2F_0}, \tag{80}$$

where n is the number of samples. Signals sampled in this way are said to be sampled at the *Nyquist rate*.

Practically speaking, this simply means that if the signal can be adequately described by n equally-spaced samples in the interval T_0, then the signal is *bandlimited* to the one-sided frequency limit

$$F_0 = \frac{n}{2T_0}, \tag{81}$$

and any processing device with a bandwidth such that all frequencies below F_0 are unaffected will not modify the signal. The implication is that a signal can be completely specified by a set of

$$n = 2F_0 T_0 \tag{82}$$

numbers.

From another point of view, the quantity $n/2 = F_0 T_0$ may be considered the *time–bandwidth product*, provided that *bandwidth* is defined as the least extent of the frequency spectrum covering *all* non-zero frequency components in the amplitude spectrum and that *time duration* is the interval of time containing all non-zero portions of the signal. It should be pointed out, however, (see Lerner, 1961) that these terms enjoy a variety of definitions, often such that the time–bandwidth product is near unity for a large class of signals.

The most widely-used definition of bandwidth is that it is the continuous range of frequency within which the amplitude exceeds a value 3 dB below the maximum value within the frequency range. This is the familiar *half-power* or -3-dB point. The most familiar concept of time duration is undoubtedly the *time-constant*. For this kind of signal, the time–bandwith product becomes

$$FT = \frac{1}{2\pi}, \tag{83}$$

where F is the one-sided -3-dB bandwidth and T is the time constant. When the bandwidth is expressed in radians/sec, this becomes

$$WT = 2\pi FT = 1. \qquad (84)$$

Many definitions of bandwidth and time duration are such that the above relationship is approximated for a wide class of signals (Burdic, 1968; Wainstein and Zubakov, 1962; Lerner, 1961).

When the signal under study is a power signal, time duration must be associated with the correlation function. Applying the sampling theorem to the autocorrelation function of a random signal simply chooses the time interval, T_0, to be such that samples of the random signal taken at t and $t + T_0$ are uncorrelated. The signal represented in Fig. 8 should, for example, be sampled at the Nyquist rate as indicated for its correlation function.

On the other hand, a time–bandwidth concept more in line with that of eqn. (83) may be useful in some applications. One possibility for a bandwidth

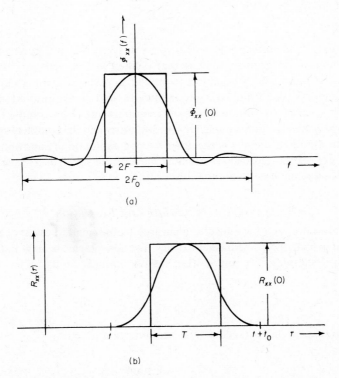

FIG. 8. Time-duration and bandwidth criteria based on signal power.

definition is that of a rectangular power density spectrum having the same average power as the one under study. In Fig. 8(a), then

$$E_x = 2F\Psi_{xx}(0) \qquad \text{for energy signals}$$

$$P_x = 2F\Phi_{xx}(0) \qquad \text{for power signals.}$$

(85)

In a similar manner, the time duration can be defined as that of a rectangular correlation function under which the area is the same as that of the given signals, i.e.

$$\int_{-\infty}^{\infty} R_{xx}(\tau)\, d\tau = T R_{xx}(0). \tag{86}$$

By eqns. (78) and (76), these two equations become

$$R_{xx}(0) = 2F\Phi_{xx}(0)$$

$$\Phi_{xx}(0) = T R_{xx}(0),$$

and

$$FT = \tfrac{1}{2}. \tag{87}$$

The time–bandwidth product defined in this way for any power signal is of the same order of magnitude as that defined in eqn. (83) for the simple exponential pulse, and it is subject to similar limitations and inaccuracies.

In summary, the definitions of bandwidth and time duration must be carefully selected to suit the problem requirements. For complete reconstruction of signals, for example, bandwidth must span the complete spectrum of the band-limited signal. On the other hand, a definition of bandwidth based on that part of the spectrum containing *most* of the signal energy, or power, may be quite adequate in another application.

B. Transducer Bandwidth and Resolution

The resolution of a transducer is measured by the smallest distance between flaws for which the transducer provides distinctly separate flaw indication. If z is the location of a point in scan, the two impulse faults,

$$f_1(z) = \delta(z - z_1)$$

$$f_2(z) = \delta(z - z_1 - d),$$

(88)

are separated by the distance d. The scanning rate is

$$r_s = \frac{dz}{dt} = \text{constant},$$

so if at $t = 0$, $z = 0$, then

$$z = r_s t$$

and the time variations of fault signals into the transducer will be of the forms

$$x_1(t) = \delta(t - t_1)$$

$$x_2(t) = \delta(t - t_1 - t_d),$$

(89)

where

$$t_d = \frac{d}{r_s}.$$

(90)

The smallest value for d for which the transducer detects two separate pulses is a measure of the transducer resolution.

A simple example will illustrate the effect of transducer bandwidth on resolution.

In a linear transducer, H, the input $x(t)$ is related to the output $y(t)$ by the convolution of the input with the system weighting function, or impulse response. So

$$y(t) = \int_{-\infty}^{\infty} x(t)h(t - \tau)\, d\tau.$$

(91)

In the frequency domain this is

$$Y(f) = H(f)X(f).$$

(92)

One rather simple low-pass system is that characterized by

$$H(f) = \frac{f_1^2}{(f_1 + jf)^2}.$$

(93)

This is the simple second-order system whose pole configuration (of the Laplace transform) consists of a single second-order pole at $s = -2\pi f_1$. Its impulse response, or response when $x(t) = \delta(t)$, is

$$y(t) = h(t) = 4\pi^2 f_1^2 t \exp\left(-2\pi f_1 t\right) u(t),$$

(94)

where

$$u(t) = \begin{cases} 1, & t > 0 \\ 0, & t < 0. \end{cases}$$

(95)

In Fig. 9(a) is shown the frequency spectrum of the transducer and its response to a pair of impulses separated so that $t_0 = 1/2\pi f_1$. Clearly, this transducer does not resolve points of this separation.

Now triple the transducer bandwidth, as in Fig. 9(b), and excite it again with the same pair of impulses. This time the response, also shown in the figure, strongly suggests the presence of two input pulses, and we can say with a high probability of being correct that the two points are resolved. If this case represents the least uncertainty considered admissible, then

$$t_d = \frac{d_{\min}}{r_s} = \frac{3}{2\pi f_1}$$

and

$$d_{\min} = \frac{3r_s}{2\pi f_1}. \tag{96}$$

Since f_1 is proportional to bandwidth, by any bandwidth criterion, the resolving distance is inversely proportional to bandwidth and directly proportional to scanning rate. Thus

$$d_{\min} = \frac{kr_s}{F}, \tag{97}$$

where k is a constant depending upon the criteria established for bandwidth and "separateness" of responses to pulse pairs. For example, when the criterion for separateness is that used above and the bandwidth is defined by eqn. (85),

$$F = \frac{E_\chi}{2\Psi_{xx}(0)} = \frac{\int_{-\infty}^{\infty} |H(f)|^2 \, df}{2|H(0)|^2} = \frac{\pi}{4} f_1 \tag{98}$$

and

$$d_{\min} = \frac{3r_s}{8F}. \tag{99}$$

Although this example may represent a number of transducers with fair accuracy, it must be noted that in a given problem the mathematical model must be the best available representation of the real system for it to yield meaningful conclusions.

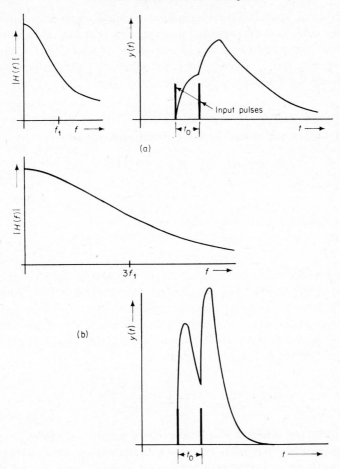

FIG. 9. Effect of bandwidth of resolution.

VII. TRANSDUCER OUTPUT NOISE†

Since the function of the transducer in nondestructive testing is to yield flaw-related signals, any signals which are not flaw-related must be considered transducer output noise. The noise may originate either in the test sample, in such forms as swaging ripples or grain size variations, or in the transducer itself because of, for example, component drift and time jitter in sample

---

† See Burdic (1968), Cooper and McGillem (1967), Lathi (1968), Papoulis (1965), Jenkens and Watts (1968), Van Trees (1968), Bendat and Piersol (1966), Rosie (1960), Wainstein and Zubakov (1962), Woodward (1953), Cook and Bernfeld (1967), and Schwartz (1959).

pulses. Noise in transducers will generally be either random in nature or it will be periodic, as is the perennial nuisance of power line feed-through.

Most noise of interest here can be considered additive. That is, the transducer response, $x(t)$, is to consist of useful signal, $s(t)$, plus noise, $n(t)$, which is statistically independent of the useful signal. Thus,

$$x(t) = s(t) + n(t). \tag{100}$$

When $x(t)$ is ergodic, it has the autocorrelation function

$$R_{xx}(\tau) = R_{ss}(\tau) + R_{sn}(\tau) + R_{ns}(\tau) + R_{nn}(\tau)$$

$$= R_{ss}(\tau) + R_{nn}(\tau) + 2m_n m_s, \tag{101}$$

where m_s and m_n are the mean values.

A. Gaussian White Noise

In the absence of complete information, the best assumption about the noise amplitudes are that they are *gaussian*. This is because of the *central limit theorem* that states, roughly, that the aggregate effect of a large number of separate random processes is a random variable whose amplitudes tend to be *gaussian*, or *normally distributed*. When the effect is the noise component, $n(t)$, this means that the distribution of values is

$$p(n) = \frac{1}{(2\pi\sigma_n)^{\frac{1}{2}}} \exp\left(-\frac{(n-m_n)^2}{2\sigma_n^2}\right). \tag{102}$$

The gaussian (or normal) distribution has a number of useful properties, including the facts that sums of gaussian variables are also gaussian, that linear operations on gaussian signals yield gaussian signals, and that all derivatives are continuous.

Suppose all values of noise at separate times are uncorrelated, a condition that can only be approximated in nature. Then

$$R_{nn}(\tau) = E[n(t)n(t+\tau)] = P_n \delta(t) \tag{103}$$

and the power density spectrum is

$$\Phi_{nn}(f) = \mathscr{F}[R_{nn}(\tau)] = P_n = \frac{N_0}{2}. \tag{104}$$

Thus the noise contains all frequencies in equal amounts. This is called *white noise*, by analogy to white light. Passed through an ideal filter of one-sided bandwidth, F, it yields a total average power

$$P = \frac{N_0}{2}\,(2F) = N_0\,F,$$

so N_0 is the *power per unit of bandwidth*.

Gaussian white noise is white noise with normally distributed amplitudes.

B. Band-limited Noise

Band-limited white noise has the power density spectrum

$$\Phi_{nn}(f) = \begin{cases} \dfrac{N_0}{2}, & |f| < F \\[2mm] 0, & |f| > F. \end{cases} \tag{105}$$

This may be considered the result of passing white noise through an ideal low-pass filter of bandwidth F. The autocorrelatoin function is

$$R_{nn}(\tau) = \mathscr{F}^{-1}[\Phi_{nn}(t)] = \frac{N_0}{2\pi}\,\frac{\sin F\tau}{F\tau} \tag{106}$$

and approaches $(N_0/2)\,\delta(\tau)$ as $F \to \infty$, so the nearby (in time) values of noise become less and less correlated as the frequency span of the noise expands.

Non-white band-limited noise satisfies the condition to an arbitrary closeness, that

$$\Phi(f) = \begin{cases} \Phi(f), & |f| < F \\ 0, & |f| > F. \end{cases} \tag{107}$$

C. Transducer Noise Bandwidth

If the tranducer impulse response is $h(t)$, then the Fourier transform of the response to noise alone is

$$Y_n(f) = H(f)N(f).$$

The power density spectrum of the output can be shown to be

$$\Phi_{yy(n)}(f) = \mathscr{F}[R_{yy}(\tau)]$$

$$= \int_{-\infty}^{\infty} R_{yy}(\tau)\exp\left(-j2\pi f\tau\right)\,d\tau$$

$$= |H(f)|^2\Phi_{nn}(f). \tag{108}$$

For white noise inputs,

$$\Phi_{yy(n)} = |H(f)|^2 \frac{N_0}{2} \tag{109}$$

and the total noise power out is

$$P_{y(n)} = \int_{-\infty}^{\infty} \Phi_{yy(n)}(f) \, df = \frac{N_0}{2} \int_{-\infty}^{\infty} |H(f)|^2 \, df. \tag{110}$$

The *noise bandwidth* of the transducer is the bandwidth of a rectangular noise power output spectrum having a constant power density equal to the zero-frequency value of the actual power density output with white noise inputs and having the same total power as the actual spectrum. Thus

$$F_n = \frac{P_{y(n)}}{2\Phi_{yy(n)}(0)} = \frac{\displaystyle\int_{-\infty}^{\infty} |H(f)|^2 \, df}{2|H(0)|^2} . \tag{111}$$

Note that the noise equivalent bandwidth is the same as the bandwidth defined by (98).

It should be pointed out that when the noise output of the transducer originates largely in the test sample its bandwidth can be adjusted, because signal frequencies originating in the sample are directly proportional to the scanning rate. That is,

$$\frac{F_n}{r_s} = \text{constant, for input to probe.} \tag{112}$$

D. Signal-to-Noise Ratio

The usual definition of signal-to-noise ratio is the ratio of peak signal power to average noise power,

$$\frac{S}{N} = \frac{[s(t)]^2}{P_n} = \frac{\{\mathscr{F}^{-1}[S(f)]\}^2}{\displaystyle\int_{-\infty}^{\infty} \Phi_{nn}(f) \, df} = \frac{\left[\displaystyle\int_{-\infty}^{\infty} S(f) \exp(+j2\pi ft) \, dt\right]^2}{\displaystyle\int_{-\infty}^{\infty} \Phi_{nn}(f) \, df} . \tag{113}$$

Transducer response to signal alone has the Fourier transform

$$S'(f) = H(f)S(f)$$

and the response to noise alone has the power density spectrum

$$\Phi_{yy(n)}(f) = |H(f)|^2 \Phi_{nn}(f).$$

The signal-to-noise ratio of the transducer output is then

$$\frac{S}{N} = \frac{\left[\displaystyle\int_{-\infty}^{\infty} H(f)S(f) \exp(j2\pi ft)\, dt \right]^2}{\displaystyle\int_{-\infty}^{\infty} |H(f)|^2 \Phi_{nn}(f)\, df}. \tag{114}$$

For white noise input, this becomes

$$\frac{S}{N} = \frac{2\left[\displaystyle\int_{-\infty}^{\infty} H(f)S(f) \exp(j2\pi ft)\, dt \right]^2}{N_0 \displaystyle\int_{-\infty}^{\infty} |H(f)|^2\, df}. \tag{115}$$

VIII. SPECTRAL PROCESSING

The oldest form of signal processing is very likely to be the frequency-spectrum modifications implemented by the familiar passive low-pass, high-pass, and band-pass filters. The demands of the communications industries and the insights provided by the Fourier transform have resulted in a well-developed technology, supported by a vast store of literature, of both theoretical and design orientation. The new availability of low-cost active elements to be applied to spectral filters has renewed interest in their application under more stringent constraints than was previously practical.

A. Filter Synthesis

Such a vast store of literature has accumulated on the subject of passive filter synthesis that it would be presumptuous to attempt to survey the field here. The interested reader may find Van Valkenburg (1960) and Kuo (1960) to provide good introductions to the subject.

Active filters are now receiving considerable attention and probably hold the most promise in extending spectral filter applications to nondestructive testing. The basic active elements used in active filter synthesis are the following:

(a) The operational amplifier

(b) The negative-immittance converter (NIC)

(c) The controlled source

(d) The gyrator.

Of these components, the most generally useful to the filter designer is the operational amplifier, since it can be used in the construction of the other three. Excellent introductions to the applications of operational amplifiers appear in certain manufacturers' publications (Burr-Brown Corp., 1963; Philbrick Researches, Inc., 1966).

The active filter is synthesized by connecting to the terminals of the above active elements passive two-terminal or four-terminal networks selected to provide the proper system impulse response (or its Fourier transform, the frequency response). An advantage of active filter synthesis is that most filter functions can be realized using resistance–capacitance (*RC*) networks, avoiding the troublesome problem of inductances.

The techniques of realizing filters using active components are too varied to allow their coverage in this brief treatment. A very readable treatment of the basic principles in the utilization of all four basic active elements appears in Burr–Brown Corp. (1966), though emphasis is on realization using operational amplifiers. More general treatments of active filters occur in Kuo (1960), Huelsman (1968), Su (1965), and Ghausi (1965), which include practical realizations not requiring operational amplifiers.

B. Some Widely Applicable Filters

There are a few filters which find such wide applicability in signal processing that they deserve special mention.

An essential part of practically any signal processing scheme is some sort of wide-band filtering, either low-pass, high-pass, or band-pass. Since it is possible to derive high-pass and band-pass functions from low-pass functions, we may limit our discussion to the latter. The filters we shall now discuss will often require active-network realizations.

1. The RC Filter

The *RC filter* (Schwartz, 1959) is the simplest low-pass filter whose impulse response has a Laplace transform with a simple pole on the negative real axis. It is implemented by a simple series resistance and shunt capacitor to give the impulse response whose Fourier transform is

$$H(f) = \frac{K_0 f_1}{f_1 + jf}, \quad \text{where} \quad f_1 = \frac{1}{2\pi RC}. \tag{116}$$

The -3-dB bandwidth is $F = f_1$, and the *fall-off* (or *roll-off*) rate well past $f = F$ is -20 dB per decade of frequency increase. Transient response to step inputs is sluggish.

2. The Gaussian Filter

The *gaussian filter* is important because it approximates a variety of practical filters. For example, consider n RC filters cascaded with isolating amplifiers separating them. The result is

$$H(f) = \frac{K_0 f_1^n}{(f_1 + jf)^n} \tag{117a}$$

so that the impulse response is

$$h(t) = K_0 (2\pi f_1)^n t^n \exp(-2\pi f_1 t) u(t). \tag{117b}$$

It can be shown (Schwartz, 1959) that for large n this is approximated by

$$h(t) = \left\{ H_0 \exp\left[-\tfrac{1}{2}\left(\frac{t - t_0}{T} \right)^2 \right] \right\} u(t),$$

where $t_0 = 1/f_1$. This tendency can be observed even for $n = 2$, as in the example of eqn. (93) and Fig. 9. The Fourier transform of the above approximation is

$$H(f) = \sqrt{(2\pi)} T H_0 \exp\left(-\tfrac{1}{2}[2\pi fT]^2 \exp\left(-j2\pi t_0 f\right)\right).$$

But $H(0) = K_0$, and the -3-dB point $f = F$ is determined by setting $H(F) = 1/\sqrt{2}$, so that

$$H(f) = \left\{ K_0 \exp\left[-\tfrac{1}{2}\left(\frac{0 \cdot 59 f}{F} \right)^2 \right] \right\} \exp\left(-2\pi t_0 f\right) \tag{118a}$$

and

$$h(t) = \{2.98 K_0 \exp[-\tfrac{1}{2}(7 \cdot 5 Ft)^2]\} u(t). \tag{118b}$$

3. Butterworth Filters

A large class of low-pass filters are of the form

$$H(f) = \frac{K_0}{[1+g(f^2)]^{1/2}}, \tag{119}$$

where K_0 is the dc gain and $g(f^2)$ is a polynomial selected to give the desired amplitude response. When the Laplace transform of the impulse response is of the form

$$\bar{H}(s) = H\left(\frac{s}{2\pi}\right) = \frac{k_0}{a_n s^n + a_{n-1} s^{n-1} + \ldots + a_1 s + 1}$$

$$= \frac{k_0/a_n}{(s+p_1)(s+p_2)\ldots(s+p_n)}, \tag{120}$$

the filter is said to be of nth order with poles at $s = -p_1$, $s = -p_2$, ..., $s = -p_n$ and all zeros at infinity. The poles may be either real or complex. The filter characteristics depend on the selection of the function $f(g)^2$.

In *Butterworth* filters the n poles lie on a circle whose center is at the origin of the complex s-plane. These filters are *maximally flat*; that is, the $g(f^2)$ has been selected to maximize the flatness in the pass band in the sense that the first n derivatives of $H(f)$ are zero at $f = 0$. Their *fall-off* (or *roll-off*) rate outside the pass-band is $-20n$ decibels per decade of frequency change, and the response is down 3 decibels at the cut-off frequency for any order, n. These filters have a tendency to overshoot and ring on sudden input changes, a tendency which increases as the order of the filter increases.

4. Chebyshev Filters

Chebyshev filters are filters whose poles lie on an ellipse, rather than on a circle as did those of the Butterworth filter. It is an *equal-ripple* filter whose frequency response ripples a prescribed small amount about the nominal value to give the same response at the cut-off frequency as at zero frequency. Chebyshev filters have overshooting and ringing tendencies similar to Butterworth filters.

5. Bessel Filters

Bessel filters get their name from the fact that the denominator of $\bar{H}(s)$ is the nth order Bessel polynomial. They approximate the ideal delay

$$\bar{H}(s) = K_0 \exp(-sT), \quad \text{or} \quad H(f) = K_0 \exp(-j2\pi fT) \tag{121}$$

in an optimum manner to yield a *maximally flat delay* or *linear phase* function. These filters possess very desirable transient response characteristics with very little overshoot or ringing. They suffer somewhat in comparison with Butterworth or Chebyshev filters in that their roll-off rate after cut-off is somewhat retarded.

For lucid and compact introductory treatments of these, and other filters of the class satisfying (119), the reader is referred especially to Chapter 13 of Kuo (1960) and to Chapter 4 of Ghausi (1965). Filters of these types are now readily available as off-the-shelf modules for a considerable range of n and cut-off frequencies.

C. Matched Filters

Matched filters have been extensively applied in radar and sonar signal analysis, but have as yet found little application in nondestructive testing. Nevertheless, many of the recurring filtering problems in nondestructive testing can be profitably attacked from this point of view.

The matched filter is deduced from an attempt to maximize the signal-to-noise ratio of eqn. (115) using a procedure depending on Schwartz's inequality,

$$\left(\int fg^* \, dx \right)^2 \leqslant \left(\int ff^* \, dx \right) \left(\int gg^* \, dx \right). \tag{122}$$

Associating f above with $H(f)$ in eqn. (115) and g with $[S(f)\exp(i2\pi ft)]^*$ Schwartz's inequality becomes equivalent to

$$\frac{S}{N} = \frac{2}{N_0} \int_{-\infty}^{\infty} S(f)^2 \, df = \frac{2}{N_0} \int_{-\infty}^{\infty} s^2(t) \, dt = \frac{2E_s}{N_0}. \tag{123}$$

If in eqn. (122) $g = kf$, corresponding to $H(f) = kS^*(f)\exp(-j2\pi ft)$, then the left side of Schwartz's inequality is

$$k^2 \left(\int ff^* \, dx \right) = k^2 \left(\int ff^* \, dx \right) \left(\int gg^* \, dx \right),$$

satisfying the equal sign in eqn. (122). Since this is the extreme value in the range of the inequality, eqn. (123) becomes, for *white noise contamination*,

$$\left(\frac{S}{N} \right)_{\max} = \frac{2E_s}{N_0}, \tag{124}$$

and this maximum signal-to-noise-ratio occurs when the filter function is

$$H(f) = kS^*(f)\exp(-j2\pi fT). \tag{125}$$

This is the *matched filter* matched to the input signal of known form, $s(t)$. The reason for replacing t by the *time delay* T becomes evident when it is noted that

$$\text{if} \quad \mathcal{F}[x(t)] = X(f), \quad \text{then} \quad \mathcal{F}[x(-t)] = X^*(f) = X(-f) \tag{126}$$

and the impulse response of the matched filter is therefore

$$h(t) = \mathcal{F}^{-1}[H(f)] = k\mathcal{F}^{-1}[S(f)\exp(j2\pi fT)]^*. \tag{127}$$

Thus, the filter impulse response is the expected signal delayed by T and reversed in time. For realizability, $h(t)$ must be zero for $t < 0$, so to realize a matched filter exactly it must be possible to select a finite time delay exceeding the time duration of the signal to which it is matched.

The output of a matched filter when the input is signal only is

$$y(\tau) = \int_{-\infty}^{\infty} s(t)h(\tau-t)\,dt$$

$$= \int_{-\infty}^{\infty} s(t)s(T-\tau+t)\,dt$$

$$= R_{ss}(\tau-T), \tag{128}$$

or the delayed autocorrelation function of the input signal. For this reason the matched filter is often called a *correlation detector*.

When the noise is colored, rather than white, so that

$$\Phi_{nn}(f) = |N(f)|^2 \tag{129}$$

then

$$\left(\frac{S}{N}\right)_{max} = \int_{-\infty}^{\infty} \frac{|S(f)|^2}{|N(f)|^2}\,df, \tag{130}$$

occurring when the filter function is

$$H(f) = k\,\frac{S^*(f)}{|N(f)|^2}\exp(-j2\pi f)$$

$$= \left(\frac{k_1}{N(f)}\right)\left(\frac{k_2\,S^*(f)\exp(-j2\pi f)}{N^*(f)}\right). \tag{131}$$

TABLE II

Matched Filter Approximations

Filter Type	FT	Relative (S/N) in dB
One-stage RC	0·2	−1·0
Ideal Low-pass	0·7	−0·8
Gaussian	0·2 to 0·7	

This result demonstrates that if the colored noise spectrum is known, the signal-to-noise ratio may be optimized by the use of a *prewhitening filter* followed by a filter matched to the signal component of its output.

Although the matched filter is often not exactly realizable, it can often be approximated by rather simple filter function. It is pointed out by Schwartz (1959) that the filter matched to the rectangular pulse can be closely approximated by the filters of Table 2 when their time–bandwidth (-3-dB) products are as indicated.

The prewhitening requirement for coloured noise often leads to band-pass filtering when the matched filter is approximated. For example, suppose a first-order approximation to the noise spectrum is $N(f) = K_n/jf$, so

$$N(f) = \frac{K_n}{f} \ . \tag{132}$$

Then from eqn. (131)

$$H(f) = \frac{k}{K_n^2} f^2 |S(f)| \ . \tag{133}$$

Since $S(f)$ is low-pass, this result is a band-pass filter.

Approximating the whitening process in this way has the advantage of placing a zero at the origin which eliminates the very low frequencies due to component drift and gradual changes in sample character not resulting from fault conditions. The effect is to cause an output signal of zero mean with an approximate double differentiation of the signal that would have resulted with matching on the assumption of white noise. The result will be similar to that shown in Fig. 10, where the peaking effect on signal and the band-limited whitening effect on noise are evident.†

---

† This technique has been successfully used by C. J. Renken in work, largely unpublished, to enhance the signal-to-noise ratio from scan-type electromagnetic tests.

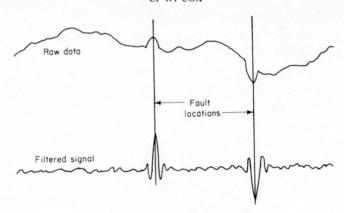

Fig. 10. The effects of band-pass filtering

IX. Decision Processes†

The *decision process* as applied to nondestructive testing will be the process of observing a data vector, then arriving at a decision as to just which one of a set of discrete test conditions gave rise to it, with a specified probability of a correct decision.

The most widely applied decision process at present is probably that of deciding between two alternatives: (1) the data vector is the result of noise plus a known signal, or (2) the data vector is the result of noise alone. These alternatives are of obvious importance when one is concerned with the reception of radar or sonar signals or with the reception of pulse-code-modulated (PCM) communication signals. However, the choice in such a binary process does not have to be between "noise plus signal" and "noise alone." The choice may be similarly made between a pair of mutually exclusive test conditions. This binary decision process is often studied under the heading of *detection theory* or *hypothesis testing*.

A second, and more general, heading under which decision theory is studied is that of *estimation theory*, having to do with the estimation of the parameters which determine the signal structure.

We shall consider briefly the binary process where $\mathbf{x}$ is a data vector whose form is known except for a set of parameters $(\theta_1, \theta_2, ..., \theta_n)$ which must be estimated at each observation. That is, the vector

$$\boldsymbol{\theta} = (\theta_1, \theta_2, ..., \theta_n) \tag{134}$$

---

† See Jenkens and Watts (1968), Wainstein and Zubakov (1962), Woodward (1953), Selin (1965), Helstrom (1960), McMullen (1968), Berkowitz (1965), Hancock and Wintz (1966), DeFranco and Rubin (1968), and Malinka (1966).

must be estimated whenever a decision must be made as to the nature of X. Although X can be any dimension, for simplicity we shall discuss the one-dimension case of the sort embodied by, say, a single oscillograph trace. Then

$$\mathbf{x} = x_1 = x. \tag{135}$$

An example of a one-dimensional signal with a two-dimensional parameter vector is the gaussian process where

$$p(x) = \frac{1}{2\pi\sigma} \exp\left(\frac{-(x-m)^2}{2\sigma^2}\right) \tag{136}$$

in which the parameter vector may be

$$\boldsymbol{\theta} = (\theta_1, \theta_2) = (m, \sigma). \tag{137}$$

Let us turn to a binary decision problem where there are two possible parameter vectors,

$\boldsymbol{\theta}_0$ = parameters of observed signal when a_0 is transmitted $\tag{138a}$

$\boldsymbol{\theta}_1$ = parameters of observed signal when a_1 is transmitted. $\tag{138b}$

The hypotheses H_0 and H_1 are applied to the random data signal X as follows:

H_0: X is decided by $\boldsymbol{\theta}_0$; or, a_0 was transmitted $\tag{139a}$

H_1: X is decided by $\boldsymbol{\theta}_1$; or, a_1 was transmitted. $\tag{139b}$

In this case, the signal is known to contain a transmission of either a_0 or a_1; the problem is to decide which. The actual observed signal is corrupted by noise and takes a range of values. The probability density function of the output based on the assumption that a_0 was transmitted is

$$p(x|\boldsymbol{\theta}_0) = P[(x < X \leqslant x + \Delta x)|H_0], \tag{140a}$$

where $P[A|B]$ is read "the probability of A, given B." Similarly,

$$p(x|\boldsymbol{\theta}_1) = P[(x < X \leqslant x + \Delta x)|H_1]. \tag{140b}$$

The observations will generally cluster about the transmitted values. For example, they may be normally distributed, as shown in Fig. 11. Somehow or other a *decision boundary* x_B, must be determined; then all observations

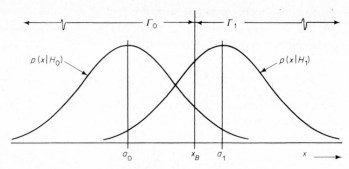

FIG. 11. Probability densities of observations resulting from just two separate causes.

in Γ_0 to the left of x_B are presumed to result from transmission of a_0 and all observations in Γ_1 to the right of x_B are presumed to result from transmission of a_1.

Let us now define some statistical parameters of the test:

$$D_j = \text{decision in favor of } H_j \tag{141}$$

$$\Gamma_j = \text{range of } x \text{ such that } \begin{cases} -\infty < x < x_B, & j = 0 \\ x_B < x < \infty, & j = 1 \end{cases} \tag{142}$$

$$P[\theta_j] = P[\theta = \theta_j] \tag{143}$$

$$C_{jk} = \text{cost of deciding } H_j \text{ when } H_k \text{ is true} \tag{144}$$

$$r_i = \text{expected loss if } H_j \text{ is true.} \tag{145}$$

Then

$$P[D_j|H_k] = \int_{\Gamma_j} p(x|\theta_k)\, dx, \tag{146}$$

$$P[D_0|H_k] + P[D_1|H_k] = 1 \tag{147}$$

and

$$P[\theta_1] + P[\theta_1] = 1. \tag{148}$$

In addition, we note that

$$r_0 = C_{00}\, P[D_0|H_0] + C_{10}\, P[D_1|H_0] \tag{149a}$$

$$r_1 = C_{01}\, P[D_0|H_1] + C_{11}\, P[D_1|H_1]. \tag{149b}$$

and define

$$R = \text{total cost of the result}$$

$$= \text{total average risk}$$

$$= r_0\, P[\theta_0] + r_1\, P[\theta_1]. \tag{150}$$

We shall now discuss briefly the commonly used criteria for establishing the decision boundaries.

1. Bayes Solution

When the *a priori* statistics, $P[\theta_0]$ and $P[\theta_1]$, are known and the costs are known, the *Bayes Solution*, which minimizes the total average risk, can be applied. This results in defining a *likelihood ratio*,

$$\Lambda(x) = \frac{p(x|\theta_0)}{p(x|\theta_1)}, \tag{151}$$

and a *threshold*,

$$K = \frac{p(x_B|\theta_0)}{p(x_B|\theta_1)} = \frac{P[\theta_0]}{P[\theta_1]} \frac{C_{10} - C_{00}}{C_{01} - C_{11}}. \tag{152}$$

The decision rule is

$$\begin{Bmatrix} \Lambda(x) < K, \text{ assign } X \text{ to } \Gamma_0 \\ \\ \Lambda(x) > K, \text{ assign } X \text{ to } \Gamma_1 \end{Bmatrix}. \tag{153}$$

The minimum value of R is called the *Bayes risk*.

2. The Ideal Observer Strategy

This strategy minimizes the average probability of error. That is, it minimizes

$$P[\text{error}] = P_B = P[\theta_0] P[D_1|H_0] + P[\theta_1] P[D_0|H_1]. \tag{154}$$

It turns out to be equivalent to the Bayes solution when

$$C_{10} - C_{00} = C_{01} - C_{11}.$$

The threshold is then

$$K = \frac{P[\theta_0]}{P[\theta_1]}. \tag{155}$$

3. The Minimax Criterion

This is neatly described by Helstrom (1960) and by Selin (1965) as a contest between Nature and the observer. No matter what strategy the observer chooses, Nature will choose $P[\theta_0]$ in such a way as to minimize her losses. The observer, on the other hand, tries to choose a strategy based on the most unfavorable choice of $P[\theta_0)]$ by Nature.

The *minimax criterion* directs the observer to apply the Bayes strategy against the worst possible $P[\theta_0]$ Nature can direct against him. The minimum risk may be considered a function of $P[\theta_0]$ and, since $P[\theta_0] = 0$ and $(P[\theta_0] - 1)$ represent certainties,

$$(R_{\min} \quad \text{when} \quad P[\theta_0] = 0) = (R_{\min} \quad \text{when} \quad P[\theta_0] = 1) = 0.$$

All other values of minimum risk are positive and the plot of $R_{\min}$ must peak somewhere between $P[\theta_0] = 0$ and $P[\theta_1] = 1$, as shown in Fig. 12. The value of $P[\theta_0]$ at the peak is the one that Nature, in her perversity, will pick, and we shall call Nature's choice of probabilities $P_B[\ ]$. The threshold then becomes

$$K = \frac{P_B[\theta_0]}{P_B[\theta_1]} \frac{C_{10} - C_{00}}{C_{01} - C_{11}}. \tag{156}$$

4. *The Neyman–Pearson Criterion*

The *Neyman–Pearson criterion* is applicable when neither the *a priori* statistics $P[\theta_0]$ and $P[\theta_1]$ nor the costs are known. The strategy is to select the *detection probability*,

$$P_d = P[D_1|H_1], \tag{157}$$

then set the decision boundary to minimize *false alarm probability*

$$P_f = P[D_1|H_0]. \tag{158}$$

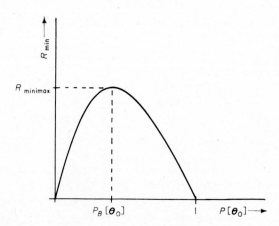

FIG. 12. The minimax solution.

The reasons for adopting such a strategy become apparent when applied to such problems as radar detection, where H_0 hypothesizes noise alone and H_1 hypothesizes noise plus transmitted signal. Since high detection probabilities and low false alarm probabilities are desirable, the Neyman–Pearson criterion is a logical one.

Since both P_d and P_f depend on x_B, it is possible to plot P_d as a function of P_f as in Fig. 13. Such a plot is called the *receiver operating characteristic* in communication works. The false alarm probability is zero only when the probability of accepting H_1 is zero, which also makes the probability of a detection zero, so all operating curves pass through the point $(P_d, P_f) = (0, 0)$. Also, if all decisions are in favor of H_1, then the false alarm and detection probabilities are both unity, so all operating characteristics terminate on $(P_d, P_f) = (1, 1)$. The operating characteristics will depend, in the detection example, on the signal-to-noise ratio. When it is infinite, for example, there is certain detection so the lines $P_d = 1$, $0 < P_f < 1$ is the operating characteristic. When the ratio is zero, false alarms and correct elections occur with equal probability, resulting in the straight line shown. When the signal-to-noise ratio falls between zero and infinity, the characteristic falls within the envelope formed by zero and infinity ratios.

Once the operating values P_{d0} and P_{f0} are determined, the threshold is found by

$$K = \frac{dP_d}{dP_f} (P_{d0}, P_{f0}), \tag{159}$$

which is the slope of the tangent line at the selected point.

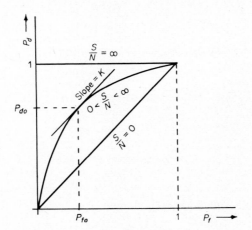

FIG. 13. A receiver operating characteristic.

When the noise is white and gaussian, it turns out that if we define a new variable

$$z = \log \Lambda(x), \tag{160}$$

then

$$P_f = \int_{z_B}^{\infty} p(z|\boldsymbol{\theta}_0) \, dz, \tag{161}$$

where

$$p(z|\boldsymbol{\theta}_0) = \frac{1}{(4\pi E_s/N_0)^{\frac{1}{2}}} \exp\left[-\frac{(z+E_s/N_0)^2}{4(E_s/N_0)} \right], \tag{161a}$$

and

$$P_d = \int_{z_B}^{\infty} p(z|\theta_1) \, dz, \tag{162}$$

where

$$p(z|\theta_1) = \frac{1}{(4\pi E_s/N_0)^{\frac{1}{2}}} \exp\left[-\frac{(z-E_s/N_0)^2}{4(E_s/N_0)} \right]. \tag{163}$$

These two expressions determine the operating characteristic. The result is shown graphically in Fig. 14. The dependence of the operating characteristic on the signal-to-noise ratio, $(S/N) = (2E_s/N_0)$, is evident.

Another gaussian quantity in this process is the statistic

$$G = \tfrac{1}{2}(N_0 z + E) = \int_{0}^{T} x(t)s(t) \, dt = G(T). \tag{164}$$

The output of a realizable matched filter with input $x(t)$ is

$$y(t) = \int_{-\infty}^{\infty} x(\tau)h(t-\tau) \, d\tau$$

$$= k \int_{0}^{t} x(\tau)s(\tau) \, d\tau = kG(t). \tag{165}$$

The matched filter with time delay T, that is with

$$h(t) = ks(T-t), \tag{166}$$

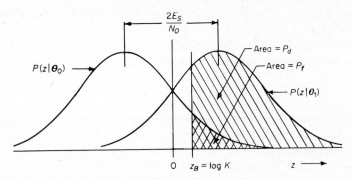

FIG. 14. The gaussian decision process.

has its peak output at $t = T$ when the input contains the design signal. This suggests the implementation of the decision process as shown in Fig. 15, where the two hypotheses are

H_0: noise only in $x(t)$

H_1: signal plus noise in $x(t)$.

This is equivalent to setting threshold as in Fig. 14 according to

$$G_B = \tfrac{1}{2}(N_0 z_B + E). \tag{167}$$

The possibilities decision techniques hold for nondestructive testing are obvious, and they will increase in number as the test activity becomes more and more an integral part of the manufacturing process. An application to ultrasonic flaw detection is suggested by Malinka (1966).

X. PATTERN RECOGNITION†

The term *pattern recognition* has come to possess a meaning much more general than the original analogy to the visual recognition of displayed patterns, such as printed characters or oscillograph traces. The original idea

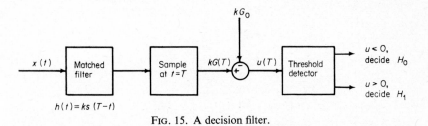

FIG. 15. A decision filter.

† See Okajima *et al.* (1963), Nilsson (1965), and Sklansky (1966).

is still very important as, for example, in the classifying of the many recurring types of curve segments in electro-cardiographs (Okajima *et al.*, 1963) (and possibly in nondestructive tests). More generally, a *pattern* is any set of simultaneous observations, and it can therefore be assigned the vector value

$$\mathbf{x} = (x_1, x_2, ..., x_n) \qquad (168)$$

where the components are discrete observations, all taken simultaneously. In a scan-type test, such a pattern results from simultaneous readings of an n-channel transducer output.

It is desired that the patterns indicate to the observer which one of a set of, say, m categories of test condition gave rise to it. It is convenient to assign each category a number, so that category j is

$$c_j = j, \qquad j = 1, 2, 3, ..., m. \qquad (169)$$

Patterns are classified by designing *discriminant functions* $g_1(\mathbf{x})$, $g_2(\mathbf{x})$, ..., $gm(\mathbf{x})$, defined such that

$$\left\{ \begin{array}{l} \text{When } g_i(\mathbf{x}) > g_j(\mathbf{x}), \text{ for all } j \neq i, \text{ the pattern} \\ \mathbf{x} \text{ is said to belong in category } i \end{array} \right\}. \qquad (170)$$

Categorization is thus accomplished by finding the maximum discriminant function, and the pattern sorting criterion is fixed by the choice of discriminant function. Fig. 16 illustrates the process.

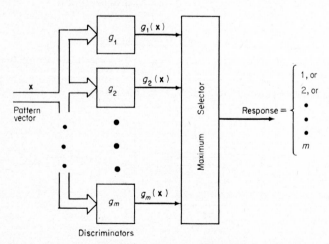

FIG. 16. A pattern recognizing machine.

The simplest classifier is the *pattern* dichotomizer, defined by $m = 2$. Then (170) may be written

$$g_1(\mathbf{x}) - g_2(\mathbf{x}) \begin{cases} > 0 & \text{for category 1} \\ < 0 & \text{for category 2.} \end{cases} \qquad (171)$$

The left side of this inequality may be replaced by the single discriminant function

$$g(\mathbf{x}) = g_1(\mathbf{x}) - g_2(\mathbf{x}) \qquad (172)$$

and the classifier becomes the single discriminator and simple threshold detector illustrated in Fig. 17.

Discriminant function can take on a great variety of forms, but we shall consider here just the *linear machine* having the *linear discriminant*,

$$g_j(\mathbf{x}) = w_{j1} x_1 + w_{j2} x_2 + \ldots + w_{jn} x_n + w_{j(n+1)}$$

$$= x_j w_j + w_{j(n+1)}. \qquad (173)$$

Among these is the *minimum-distance classifier* (Nilsson, 1965) which assigns to each category a point in the pattern vector space. Then each pattern is placed in that category associated with the assigned point closest to the pattern to be classified.

A linear machine with $m = 2$ is realized, as shown in Fig. 18, and is called a *threshold logic unit*. This is a special case of the dichotomizer of Fig. 17 in which

$$g(x) = w_1 x_1 + w_2 x_2 + \ldots + w_n n_n + w_{n+1}$$

$$= \langle x, w \rangle + w_{n+1}, \qquad (174)$$

where $\mathbf{w} = (w_1, w_2, \ldots, w_n)$ is the weight vector of the threshold logic unit. The discriminant divides the pattern vector space into two parts separated by the hyperplane (or plane if $n = 3$, or line if $n = 2$)

$$g(\mathbf{x}) = 0. \qquad (175)$$

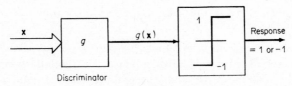

FIG. 17. The threshold logic unit.

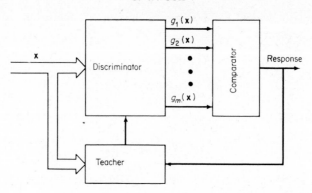

FIG. 18. A pattern recognizer with training.

The effectiveness of a pattern classifier is greatly dependent on a fortunate choice of discriminant function, and much work is now being done on attempts to make this choice optimum. The techniques of decision theory, discussed in the preceding section, play a large part in such work.

The potential in the use of pattern classifiers will probably be greatest when *training* is incorporated. A trainable classifier *learns* by having its discriminator modified as a *teacher*, either human or machine, updates its responses to input patterns, as indicated in Fig. 18. The machine of Libby and Wandling (1968) in Fig. 5 incorporates "training" as the "teacher" adjusts the potentiometers in reactions to responses from a "training set" of patterns into the machine. In this case the teacher is human.

The literature abounds in techniques training algorithms for pattern classifiers, though relatively few of them have reached the point of application to real problems (Sklansky, 1966). A class of techniques which is developing rapidly involves the use of the threshold logic unit as in Fig. 19. Here, for convenience, $w_{n+1} = 0$ and

$$g(\mathbf{x}) = \langle x, w \rangle.$$

The teacher thus operates during training to reduce the inner product of $\mathbf{x}$ with $\mathbf{w}$. This is accomplished by adjusting the weights w_1, w_2, ..., w_n. The primary problem here is the selection of an algorithm for adjusting the weights.

Space does not permit more than this cursory discussion of trainable pattern classifiers. The interested reader is referred especially to Nilsson (1965) and Sklansky (1966) and the extensive bibliographies therein for his introduction to these techniques.

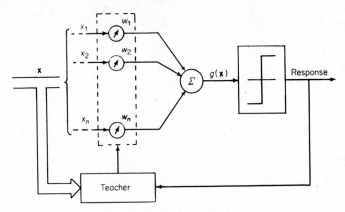

FIG. 19. A trainable threshold logic unit.

XI. GENERALIZED FILTERS

Conventional spectrum analyzers are filters which determine the *harmonics* of periodic power signals. That is, they provide the amplitudes of the basis functions in the orthogonal series expansion of the signal, the basis function in this case being the imaginary exponentials forming the sines and cosines. In this section we shall discuss briefly a more general spectrum analyzer in which the harmonics are the amplitudes of the basis functions for the orthogonal expansion of energy signals. In other words, they are the x_k in eqns. (23).

Typical orthogonal functions are Bessel functions, Laguerre polynomials, and properly selected linear combinations of exponentials. The latter are generally of most interest in signal processing, because they provide a great variety of waveforms and are relatively simple to generate in the laboratory (Huggins, 1956; Haddad and Braun, 1961).

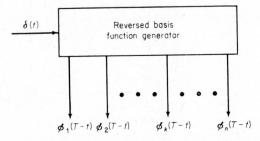

FIG. 20. The orthogonal basis function generator.

In one method of implementing such a filter, the first step is to construct an *n*-terminal device which upon excitation by the unit impulse will generate the *n* basis functions delayed by *T* and reversed in time, as indicated in Fig. 20, where *T* is the maximum time duration of any energy signal to be filtered. If the device is now excited by the energy signal $x(t)$, the output of the *k*th terminal is the convolution of the signal and the *k*th basis function, or

$$z_k(t) = \int_{-\infty}^{\infty} \phi_k(T-\tau)x(t-\tau)\,d\tau. \tag{176}$$

Sampled at $t = T$, this yields

$$z_k(T) = \int_{-\infty}^{\infty} \phi_k(T-\tau)x(T-\tau)\,d\tau$$

$$= \int_{-\infty}^{\infty} x(t)\phi_k(t)\,dt$$

or, from (20) for real-valued functions,

$$z_k(T) = x_k. \tag{177}$$

This is the *k*th generalized harmonic, and the generalized filter is implemented as in Fig. 21.

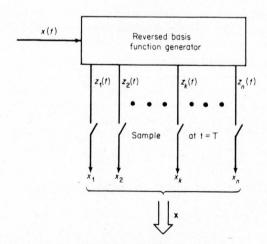

FIG. 21. The orthonormal filter.

Suppose the transducer output in response to a unit impulse input is $h(t)$. Then the transducer response to the kth basis signal is

$$h_k(t) = \int_{-\infty}^{\infty} h(\tau - t)\phi_k(\tau)\, d$$

$$= \sum_{j=1}^{n} h_{jk}\, \phi_j(t). \tag{178}$$

The response to the energy signal $x(t)$ is

$$y(t) = \int_{-\infty}^{\infty} h(\tau - t)y(\tau)\, d$$

$$= \int_{-\infty}^{\infty} h(\tau - t)\left(\sum_{k=1}^{n} x_k\, \phi_k(\tau)\right) d\tau$$

$$= \sum_{j=1}^{n} \sum_{k=1}^{n} h_{jk}\, x_k\, \phi_j(t)$$

$$= \sum_{j=1}^{n} y_j\, \phi_j(t) \tag{179}$$

where

$$y_j = \sum_{k=1}^{n} h_{jk}\, x_k. \tag{180}$$

The above equation demonstrates that the h_{jk} transform the input vector, **x**, into the output vector, **y**, and they consequently completely characterize the system, which consists of transducer *plus* test sample. Since the components of (178) can be generated by exactly the same methods as those of $x(t)$ in Fig. 21, the h_{jk} can be obtained directly by exciting the probe with a basis function then filtering the result, as shown in Fig. 22. Litman and Huggins (1963) have suggested this technique, along with an optimum selection of basis functions, in this case probing signals, to detect small parameter changes.

These techniques hold promise for the analysis of the individual pulses in pulse systems. A major problem in their application has been the difficulty of synthesizing filters fast enough to process the short pulses that result from

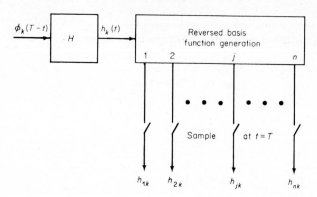

FIG. 22. System parameters obtained by use of the orthonormal filter.

most pulsed tests. Perhaps a solution to this problem may be the adoption of a pulse stretching technique similar to that used in oscilloscopes of extremely wide bandwidth.

XII. DIGITAL PROCESSING

Many of the techniques that have been discussed will require digital processing for their practical implementation. Such processing, if done on-line, might be accomplished in a time-sharing arrangement with a large remote general purpose computer, or it might be accomplished with a small computer especially adapted for process control and on-line data reduction. Computers of the latter type, with memories upward from 4000 words and a great variety of input–output devices, are being supplied by an increasing number of manufacturers. They are almost certain to play a large role in the next generation of nondestructive testers.

It is impossible, in the space allowed here, to attempt to set forth the essentials of digital computer techniques. However, the designer who is contemplating their use should be aware of a couple of pitfalls that may await him.

Signals to be processed digitally must first be converted to digital form through the use of analog-to-digital (A–D) converters. Such converters employ sampling techniques without sampling intervals long enough to allow conversion and storage of data plus any real-time computing that is required. This places definite lower limits on the duration of pulses that can be accurately converted. Present equipment can convert and store at a rate in the order of 1000 samples per second, but this may drop to less than 100 samples per second when real-time computing is required.

The sampling requirement also makes it absolutely essential that high frequency noise is absent, since sampling on the peaks and valleys of such noise can greatly bias the results. Therefore, signals should be pre-filtered with low-pass filters before conversion to digital form.

REFERENCES

Bendat, J. S. and Piersol, A. G. (1966). "Measurement and Analysis of Random Data." John Wiley, New York.

Berkowitz, R. S. (1965). "Modern Radar". John Wiley, New York.

Burdic, W. S. (1968). "Radar Signal Analysis". Prentice-Hall, Englewood Cliffs, N.J.

Burr–Brown Corp. (1963). "Handbook of Operational Amplifier Applications". Burr-Brown Research Corp., Tuscon.

Burr–Brown Corp. (1966). "Handbook of Operational Amplifier Active RC Net-Works". Burr–Brown Research Corp., Tucson.

Cook, C. E. and Bernfeld, M. (1967). "Radar Signals". Academic Press, New York.

Cooper, G. R. and McGillem, C. D. (1967). "Methods of Signal and System Analysis". Holt, Rinehart and Winston, New York.

DeFranco, J. V. and Rubin, W. L. (1968). "Radar Detection". Prentice-Hall, Englewood Cliffs, N.J.

Ghausi, M. S. (1965). "Principles and Design of Linear Active Networks". McGraw-Hill, New York.

Haddad, R. A. and Braun, L., Jr. (1961). In "Adaptive Control Systems" (E. Mishkin and L. Braun, Jr. eds.), pp. 311–322. McGraw-Hill, New York.

Hancock, J. C. and Wintz, P. A. (1966). "Signal Detection Theory". McGraw-Hill, New York.

Helstrom, C. W. (1960). "Statistical Theory of Signal Detection". Pergamon Press, Oxford.

Hochschild, R. (1958). In "Progress in Nondestructive Testing", (E. Stanford and J. Fearon, eds.), pp. 76–99. Macmillan, New York.

Huelsman, L. P. (1968). "Theory and Design of Active RC Networks", McGraw-Hill, New York.

Huggins, W. H. (1956). IRE Trans. Circuit Theory, 44, 210–216.

Jenkens, G. M. and Watts, D. G. (1968). "Spectral Analysis and Its Applications". Holden-Day, San Francisco.

Kuo, F. F. (1960). "Network Analysis and Synthesis", Second Edn. John Wiley, New York.

Lathi, B. B. (1968). "An Introduction to Random Signals and Communication Theory". International Textbook Co., Scranton, Pennsylvania.

Lerner, R. M. (1961). In "Lectures on Communication System Theory" (E. Baghdad, ed.). McGraw-Hill, New York.

Libby, H. L. and Wandling, C. R. (1968). AEC Res. and Dev. Rep. No. BNWL-765. Battelle Northwest, Richland.

Litman, S. and Huggins, W. H. (1963). Proc. IEEE 51, 917–923.

McMaster, R. C. (1959). "Nondestructive Testing Handbook", Vol. II, Section 40. Ronald Press, New York.

McMullen, C. W. (1968). "Communication Theory Principles". Macmillan, New York.

Malinka, A. V. (1966). *Defektoskopiya* **5**, 22–27.

Nilsson, N. J. (1965). "Learning Machines". McGraw-Hill, New York.

Okajima, M., Stark, L., Whipple, G., and Yasui, S. (1963). *Trans. IEEE* **BME-10**, 106–114.

Papoulis, A. (1965). "Probability, Random Variables, and Stochastic Processes". McGraw-Hill, New York.

Philbrick Researches, Inc. (1966). "Applications Manual for Computing Amplifiers". Nimrod Press, Boston.

Renken, C. J. (1968). *In* "Proceedings of the Metallurgical Congress on Nondestructive Testing and Control in the Field of Nuclear Metallurgy and Technology", Paris.

Renken, C. J., Meyrs, R. G. and McGonnagle, W. J. (1958). *AEC Res. and Dev. Rep. No. ANL* 5861. Argonne National Laboratory, Chicago.

Rosie, A. M. (1966). "Information and Communication Theory". Blackie and Son, Glasgow.

Schwartz, M. (1959). "Information Transmission, Modulation, and Noise". McGraw-Hill, New York.

Selin, I. (1965). "Detection Theory". Princeton University Press, Princeton, N.J.

Shannon, C. E. and Weaver, W. (1949). "The Mathematical Theory of Communication Theory". University of Illinois Press, Urbana.

Sklansky, J. (1966). *Trans. IEEE* **AC-11**, 6–19.

Su, K. L. (1965). "Active Network Synthesis". McGraw-Hill, New York.

Van Trees, H. L. (1968). "Detection, Estimation, and Modulation Theory, Part I". John Wiley, New York.

Van Valkenburg, M. E. (1960). "Modern Network Synthesis". John Wiley, New York.

Wainstein, L. A. and Zubakov, V. D. (1962). "Extraction of Signals from Noise" (English Translation). Prentice-Hall, Englewood Cliffs, N.J.

Woodward, P. M. (1953). "Probability and Information Theory, with Applications to Radar". Pergamon Press, Oxford.

CHAPTER 8

Direct-View Radiological Systems

R. Halmshaw

Royal Armament Research and Development Establishment, Fort Halstead, Sevenoaks, Kent, England.

I. INTRODUCTION

Most industrial radiology makes use of radiography-on-film. Direct-viewing systems, i.e. fluoroscopy, have found only a relatively few specialized uses. The reason is that the fluoroscopic image is very dim, so that the observer needs to have some degree of dark-adaptation, and even then the image-brightness is such that the normal visual acuity of the eye is significantly reduced. In the medical field, fluoroscopy is used more widely, partly because it is the only method by which a moving subject can be observed with X-rays without a time-delay, and partly because the problems of thickness penetration are not so severe as in most industrial applications.

The advent of various forms of image intensifier produced a great resurgence of interest in both medical and industrial fluoroscopy. Again, the more immediate applications have been in medical work; this is probably because the commercial market is larger here, but also perhaps because in industrial work comparisons are usually made with radiographs taken with metal intensifying screens, rather than with salt intensifying screens with their relatively poor definition of detail.

Most of the later developments in X-ray image intensifier systems utilize a television chain, and any such system confers several advantages:

(a) The amplification available is virtually unlimited, apart from the signal-to-noise factor (see Section IV).

(b) The image is viewed at high brightness-levels and is available for still and ciné photography and for image storage, e.g. video-tape.

(c) The image-contrast can be varied electronically.

(d) Image-modification techniques, such as background subtraction, may be applied.

II. Types of System

Conventional fluoroscopy, i.e. the direct observation of an X-ray image on a fluoroscopic screen, has developed in two directions. One method uses a high-output X-ray tube, a short distance from the focus to the fluorescent screen, and the specimen close to the screen. In short, the equipment is designed to obtain as bright an image as possible, and image definition is regarded as a secondary factor. The image brightness is usually between 0·2 and 0·01 ft L. ($0.7 - 0.034$ cd/m^2).

The second method uses a fine-focus X-ray tube, the specimen is placed away from the screen and the image is projected several times natural-size onto the screen. This increase in image-size counter-balances the loss in visual acuity, but the use of a fine-focus X-ray tube implies a low X-ray output and therefore a dimmer image, so that dark adaptation of the observer is almost always necessary. The image brightness is likely to be between 0·02 and 0·002 ft L. ($0.07 - 0.007$ cd/m^2).

Both systems have been thoroughly investigated (O'Connor and Polansky 1952; Halmshaw and Hunt, 1957; Criscuolo and Polansky, 1956; Halmshaw, 1966) and Tables I and II give some typical performance data, which will be useful for comparison later with image intensifier systems.

Image intensifier systems can be divided, broadly, into four types, although there are also hybrids.

TABLE I. Image-quality Indicator Sensitivities Obtained in Fluoroscopy, with Wire-type I.Q.I.

Specimen material and thickness (in.)		System A High-brightness fluoroscope	System B Projected image with fine-focus tube
		Sensitivity (%)	
½ in.	Al	5	2·2
1 in.	Al	3	1·5
2 in.	Al	2·3	1·4
0·1 in.	Fe	8	5
0·2 in.	Fe	6·5	4
0·4 in.	Fe	6	3

(1) Image intensifier tubes, in which the X-ray image is converted successively into light and then electrons in a double layer screen; the electrons are then accelerated across a vacuum space on to a smaller screen where the image is reconverted to light. This second image can be viewed directly, or through a television system. Both screens are in a vacuum envelope.

(2) A sensitive television camera (Vidicon, image orthicon, etc.) which views the light image on a fluoroscopic screen and presents the final image on a television monitor.

(3) Special television tubes in which the pick-up surface is directly sensitive to X-rays.

(4) Image intensifier "panels" consisting of photoconductive or electroluminescent layers with an applied electric field.

TABLE II. Contrast/Diameter Sensitivities for Circular Cavities.†
"Chest Fluoroscopy Conditions": 70 kV, 5 mA

Contrast (%)	Diameter (mm)
22	0·7
16	1·0
8	1·8
4	4
1·6	12

† See Sturm and Morgan (1949).

The first image intensifier tubes for use with X-rays were developed concurrently by Teves and Tol (1952) and Coltman (1948). Both types were originally approximately 5 in. diameter tubes, with a circular primary screen and a final viewing screen about $\frac{1}{2}$ in.–1 in. diameter. The brightness-gain, compared with a conventional fluoroscopic screen, is between 300 and 2000 and larger diameter versions of both types are extensively used in medical fluoroscopy. In the earlier models the image was viewed directly with an eyepiece or a mirror system, but more recently the use of a Vidicon television camera tube has become common. This gives the advantages of remote viewing with no radiation hazard to the observer and also provides a further opportunity for image brightness gain and contrast control.

This type of image intensifier has not found extensive use in industrial applications, possibly because it has, to date, been designed for medical use with 60–70 kV X-rays. For this reason, the primary fluorescent layer is kept very thin and is therefore an inefficient convertor of higher-energy X-rays to light, so that the quantum fluctuation limit is soon reached (see Section IV-A). This type of image intensifier has several potential advantages:

(a) The controls are simple; the focusing system requires a simple high-voltage supply and the only controls requiring adjustment are those on the television system, if this is used.

(b) The image brightness on the viewing screen is sufficiently high for film recording, so that a film image can be obtained without any impairment due to a television line-raster.

(c) Fairly high-speed ciné-recording is possible; the filming speed is limited by the decay-time of the screens in the tube and not by the television repetition frequency; the author has filmed from this type of tube at 200 frames/sec, and still greater speeds should be possible. This seems to be a relatively undeveloped application.

The performance of this type of tube, in terms of attainable radiographic sensitivity has been investigated by several workers. Typical data is given in Table III; the advantages of a moderate amount of projective magnification are again evident.

Historically, the type of image intensifier system in which a television camera tube is used to observe the image on a primary fluorescent screen was developed for laboratory use soon after the image intensifier tubes just mentioned (Morgan, 1956). This system appears, at present, to be the type in which there is most scope for development: it will therefore be discussed in detail in subsequent sections (see Sections III and IV, etc.).

Television-type camera tubes, in which the pick-up surface is directly sensitive to X-rays instead of light, have been developed, but only one design

TABLE III. I.Q.I. Sensitivities for a Wire-type I.Q.I., Obtained with a 5 in. Image Intensifier Tube†

Specimen material and thickness (in.)		Sensitivity (%) Natural size	× 2 p.m.
$\frac{1}{2}$ in.	Al	4·2	2·4
1 in.	Al	3·1	1·5
2 in.	Al	2·0	1·2
4 in.	Al	3·0	1·4
0·1 in.	Fe	12·0	5
0·2 in.	Fe	7·0	4·7
0·4 in.	Fe	5·1	3·7
0·6 in.	Fe	4·8	3·1

† See Halmshaw (1966) and Lang (1955).

has become a commercial equipment. This consists of a 1 in. diam. camera tube, with a pick-up surface about $\frac{1}{2}$ in. × $\frac{3}{8}$ in. and it is therefore limited to the examination of very small specimens, or to specimens which can be scanned (McMaster *et al.*, 1962; McMaster *et al.*, 1967; Joinson, 1967). It's characteristics can be summarized:

(a) It requires a very high dose-rate of X-rays on the pick-up surface (10–1000 R/min) and has been used mostly with X-ray energies of 150 keV or less.

(b) The image definition attainable is very good, permitting the $\frac{1}{2}$ in. wide image to be presented on a 17 in. monitor screen; cracks less than 0·001 in. wide have been detected.

Development of a larger-diameter tube of this type would seem to be overdue. Larger-diameter direct X-ray pick-up tubes, up to 9 in. diam. were built some years ago in the U.S.A. but were not marketed.

III. PRINCIPLES OF OPERATION OF IMAGE INTENSIFIER SYSTEMS

All types of image intensifier which produce a brighter image, operate by converting the image into electrons and, at some later stage, back into light. These conversion processes bring in special problems.

It is proposed to illustrate the general principles of operation by detailed consideration of one particular type of equipment—that in which the primary image is formed on a fluorescent screen which converts the X-rays into light and this light image is optically transferrred on to the pick-up surface of a television camera tube; there, the image is converted into a modulated electric current, which, after amplification and possibly other modifying circuitry, modulates another scanning electron beam where the image is re-converted to light on a television display monitor. This type of image intensi-fier system has attractions in that the primary fluorescent screen is not inside a vacuum envelope, so that it can be changed to a different type or size of screen; also, the image is accessible for study at various stages through the system. Although the ensuing discussion will be mainly concerned with this one type of image intensifier system, much of the argument is pertinent to the other types of system mentioned in Section II.

A television-type fluoroscopic system is likely to be limited in its per-formance by one or more of three causes:

(i) Quantum fluctuation limitations (see below).

(ii) Intrinsic performance limitations of some stage of the system, e.g. lack of adequate bandwidth in the electronic circuitry; inadequate resolution of the optical system; television raster lines.

(iii) Radiological factors such as X-ray energy, geometric unsharpness, etc.

According to the operating conditions, any of these factors can become predominant, and in a poorly-designed system neglect of one particular design parameter can produce a serious limitation to performance.

Methods of analysing and studying the successive transfer of the image through the various stages of a system have benefited greatly from techniques developed in other fields of physics. Thus, modulation transfer character-istics can be determined for each stage, and built up to assess the complete system. Methods based on signal-to-noise ratios have also been used.

IV. Determination of Performance

A. Quantum Fluctuation Limitations

The emission of X-ray quanta from the target of an X-ray tube and the absorption processes are random phenomena. Thus, while it is legitimate to give a value to the *average* number of quanta involved at any stage of a pro-cess, this is not an exact number. There is a natural fluctuation in the number of quanta involved, and statistical analysis shows that if we are concerned with N quanta and there are no unequal probabilities, the average fluctuation is $N^{\frac{1}{2}}$, or expressed as a fraction of the total number of quanta involved,

$$\frac{N^{\frac{1}{2}}}{N} = N^{-\frac{1}{2}}.$$

If N quanta are absorbed in unit area of a fluorescent screen, in time t, then when

$N = 100$ quanta, the average fluctuation is $N^{\frac{1}{2}} = 10$, i.e. 10% of N

$N = 10,000$ quanta, ... $= 100$, i.e. 1% of N

$N = 10^6$ quanta, ... $= 10^3$, i.e. $0\cdot1\%$ of N.

If, therefore, one considers a small area of an image on a fluorescent screen, which is slightly different in brightness to the background brightness of the screen, then if this small area is formed as the result of the absorption of, on average, 100 quanta, there will be an average fluctuation in brightness of 10% due to statistical fluctuations in the number of quanta utilized per unit time. If the area is only, say, 5% brighter than the background on average, it will be difficult to detect, because the fluctuations will mask the real brightness difference.

If, however, the same area is formed as the result of the absorption of 10^6 quanta, the statistical fluctuation will be only $0\cdot1\%$ and the 5% brightness difference postulated should be easily observed. The minimum observable contrast depends, therefore, on the number of quanta utilized.

This theory was proposed by Rose (1948) who suggested that the perception of small, low-contrast detail on a uniform background was limited by the random fluctuations in the number of light quanta utilized by the human eye when viewing the image. These fluctuations occur in both space and time.

Sturm and Morgan (1949) were the first to apply this concept to an X-ray image intensifier system. They assumed that because one primary X-ray quantum is converted into a large number of light photons in the fluorescent screen, the random fluctuations in the X-ray quanta would be transmitted through the successive stages even though many more quanta were available at some later stages. The calculations have been detailed in many papers (Oosterkamp, 1954; Fowler, 1960; Vincent and Shemmans, 1966) and lead to a clear conclusion that detail perceptibility is controlled by the stage in the process at which the least number of quanta are utilized. With an image intensifier system having virtually unlimited intensification, the two limiting stages are the number of quanta utilized by the eye of the viewer and the number of quanta absorbed by the primary fluorescent screen; the maximum useful intensification is that which makes these two numbers equal. Additional intensification may increase image brightness but will not improve detail perceptibility.

For this condition, Sturm and Morgan's theory results in a sensitivity equation:

$$d = \frac{200k}{C} \left(\frac{1}{\pi t N} \right)^{\frac{1}{2}}$$ (1)

where d is the diameter of a circular image detail of contrast C (in %),

k is the signal-to-noise ratio required for an image to be discernible: a value between 3 and 5 is usually taken,

t is the integration time of the eye (usually taken as 0·2 sec),

N is the number of quanta originating from each square millimetre of the primary fluorescent screen, per second, at the stage where the number of quanta per unit area is a minimum.

This equation depends on the performance characteristics of the human eye — the quantum efficiency, the integration time, the possibility of a short-time visual memory—but is valuable in indicating the direct dependence of perceptible detail size on N.

It may be argued that under the conditions of viewing a television monitor screen, where the viewing distance and the image brightness can be adjusted to optimum values, the performance limitations of the eye can be eliminated. Knowledge of the eye's quantum efficiency at different brightnesses is, however, still sketchy.

The determination of realistic values of N is not easy: considerable simplifying assumptions are necessary. The primary X-ray beam is heterogeneous, is changed in composition by absorption in the specimen and the different wavelengths have different absorption coefficients in the material of the fluorescent screen; finally, the X-ray-to-light conversion characteristics and the light losses in the screen are imperfectly known. Table IV gives some of the values which have been quoted for different conditions.

Several workers have used quantum fluctuation considerations to show that at some stage an X-ray image intensifier system becomes limited in attainable sensitivity by the number of quanta being utilized, but the uncertainty in magnitude of some of the factors involved makes the determination of the exact point at which this condition is reached, rather doubtful. There is no doubt, however, that for some applications an improvement in performance could only be obtained either by employing a primary fluorescent screen which absorbs more X-ray quanta, or by using some form of image storage.

Hay (1958) has suggested that the theory of quantum fluctuations might be extended to yield a criterion of image quality in the form of an "information index".

TABLE IV. Values of N (Number of Quanta Emitted/mm^2/sec by the Screen)

No. of X-ray quanta absorbed by screen	No. of light quanta emitted by screen	Conditions	Reference
1×10^4	5×10^7	80 kV: " chest "	Banks (1958)
7×10^2	2×10^6	medical	Fowler (1960)
3×10^4	$1 \cdot 7 \times 10^9$	140 kV: 2 in. Al	Halmshaw (1966)
$1 \cdot 3 \times 10^2$	1×10^7	5MV: 7 in. Fe	Noé (1968)
$3 \cdot 5 \times 10^5$	$1 \cdot 7 \times 10^9$	80 kV: " chest "	Sturm and Morgan (1949)
$20\text{–}8 \times 10^3$	3×10^5	5MV: 7 in. Fe	Vincent and Shemmans (1966)
$1 \cdot 5 \times 10^3$	—	90 kV: 10 in. water	Webster and Wipfelder (1964)
$1 \cdot 8 \times 10^4$	—	medical	Morgan (1965)

Contrast is defined as $\Delta B/B$, and the inverse quantity $B/\Delta B = C$ is called contrast sensitivity.

If one now considers the smallest visible square-object detail as having side a, then the resolution r is defined as $a = 1/r$ and the information index as $c.r.$

Then taking the symbols previously used, the number of quanta in the image in the integration time of the eye is

$$a^2 \, t \, N.$$

Then, for the image to be visible,

$$\Delta B \geqslant K(a^2 \, t \, N)^{\frac{1}{2}}$$

$$c.r. = 1/K \, (N(t)^{\frac{1}{2}}. \tag{2}$$

This concept of quantum fluctuation limitations is concerned with perceiving a pattern against a noisy background where the noise is assumed to originate from the quantum nature of light. But in a television-fluoroscopy system it must be assumed that noise can originate elsewhere. In any television system there is likely to be additive noise, independent of the signal strength.

Coltman and Anderson (1960) have analysed noise considerations in a television system and formulated a number of propositions:

(a) "The strength of white noise required to mask an image signal is directly proportional to the linear size of the image." As the viewer can change the gain, brightness and viewing distance, at will, this may be restated as :

"The signal-to-white-noise ratio required for detection of an image is inversely proportional to the linear dimension of the image", which also has a corollary:

"The optimum viewing distance for detection of an image in white noise, is directly proportional to the image size."

(b) "The resolution limit of a system which is limited by white noise can be predicted from a knowledge of noise power per unit bandwidth and the modulation transfer function of the system."

Albrecht and Proper (1967) have further surveyed proposed models of X-ray imaging systems, in terms of signal-to-noise considerations. They show how account can be taken of noise generated in the image-forming system, so that Rose's proposals for the visibility of circular detail can be reformulated:

$$C \geqslant K \left(\frac{N}{S}\right)_{in} A_d^{-\frac{1}{2}} F^{\frac{1}{2}} \gamma \tag{3}$$

and Coltman's equation for sinusoidal patterns:

$$C \geqslant K^1 \left(\frac{N}{S}\right)_{in} R F^{\frac{1}{2}} \gamma \tag{4}$$

where C = threshold object contrast,

$\left(\dfrac{N}{S}\right)_{in}$ = signal-to-noise ratio of input signal referred to 1 cm,

A_d = area of detail,

F = noise factor = $\left(\dfrac{N}{S}\right)_{out}^2 \left(\dfrac{N}{S}\right)_{in}^2 \gamma^{-2}$,

R = spatial frequency,

γ = gradient (gamma).

So far, no account has been taken of image blurring (unsharpness), which in practice is likely to have a considerable influence on the size of the minimum discernible object. The effects of unsharpness would appear to be:

(1) To decrease the contrast of small detail (Halmshaw, 1966).

(2) To produce additional noise (Albrecht, 1965).

Schade (1964) has used the concept of bandwidth and has obtained a sensitivity equation which can be reduced to equation (4) with the addition of a term representing the reduction in contrast of a small detail due to unsharpness. Albrecht and Oosterkamp (1962) have produced a similar relationship for the contrast in a sinusoidal pattern neglecting noise due to unsharpness, and Kuhl (1965) has produced an equation for the perception of circular detail which includes a term for the influence of unsharpness on contrast and which has been shown to agree well with experimental results:

$$ C \geqslant K \left(\frac{N}{S}\right)_{\text{in}} F^{\frac{1}{2}} \gamma \left(\frac{2}{\pi}\right)^{\frac{1}{2}} \frac{1}{\Delta} \ \frac{1}{T_c(\Delta)} , \tag{5} $$

where Δ is the diameter of details, $T_c(\Delta)$ is the relative contrast reduction due to unsharpness in the centre of a detail, (Δ/U), U being the effective total unsharpness.

Thus

$$ C \geqslant K \left(\frac{N}{S}\right)_{\text{in}} F^{\frac{1}{2}} \gamma \left(\frac{2}{\pi}\right)^{\frac{1}{2}} \frac{U}{\Delta^2} . \tag{6} $$

B. Measurement of Image Unsharpness

The concept that a sharp-edged image may be degraded into an unsharp image by various causes and that the amount of blurring can be measured by the width of the band of blurring — the unsharpness (measured as a distance) — was developed many years ago and has proved to be a valuable method of studying radiographic sensitivity (Halmshaw, 1966). It can also be applied to image intensifier systems. Measurements of the unsharpness of fluorescent screens have been made by several workers (Halmshaw, 1966) and are in broad agreement:

High-speed screen (B2. HS); $U_s = 1 \cdot 5$ mm

Fine-grain screen $\qquad U_s = 0 \cdot 6 – 0 \cdot 8$ mm.

The smallest reliable value obtained by the author for any fluorescent screen is $0 \cdot 5$ mm.

It is reasonable to assume that a good optical system on an X-ray image intensifier will not contribute any unsharpness of this magnitude (which corresponds to a bar pattern of limiting resolution of about 8 lines/mm), but the subsequent television system, where the image is split into a raster pattern, may have a considerable effect.

On an X-ray image intensifier used by the author, which utilizes a 1000-line triple-interlace television system, the fluorescent screen was replaced by an optical image of such a brightness that quantum fluctuation effects were negligible and measurements of image unsharpness on the monitor image were:

Edge perpendicular to TV lines $U = 0\cdot8$ mm

Edge parallel to TV lines $U = 0\cdot7$ mm.

Using 60 kV X-rays and a primary fluorescent screen of $U_s = 0\cdot9$ mm, the display image unsharpnesses were measured to be

Edge perpendicular to TV lines $U = 1\cdot2$ mm

Edge parallel to TV lines $U = 1\cdot1$ mm.

These last values are almost the same as the effective sum of the screen unsharpness and the values measured with an optical image, e.g. for the image perpendicular to the TV lines, the total effective unsharpness is $(0\cdot9^2+0\cdot8^2)^{\frac{1}{2}} = 1\cdot2$ mm. If the unsharpness of the fluorescent screen were reduced to $0\cdot6$ mm the total unsharpness would be $(0\cdot6^2+0\cdot8^2)^{\frac{1}{2}} = 1\cdot0$ mm.

These measurements suggest that there is little to be gained by employing a fluorescent screen with a smaller unsharpness, unless at the same time the unsharpness of the rest of the system is reduced; after the fluorescent screen, the limit is almost certainly imposed by the television line raster and the bandwidth of the electronic circuitry. The need for a greater bandwidth if any improvement is to be obtained from a higher line frequency, has been illustrated by Webster et al. (1967) in experiments with different television units attached to the same image intensifier tube.

There is some slight evidence to suggest that the screen unsharpness of any particular fluorescent screen increases slightly with very high energy X-rays. For example, a screen having $U_s = 0\cdot8$ mm with 100 kV and 400 kV X-rays was measured to have $U_s = 1\cdot0$ mm with 5 MV X-rays, possibly due to the additive effective of secondary electrons generated in the screen material: the accuracy of measurement of unsharpness for a fluorescent screen is not however very great, and this change is approximately the same as the estimated maximum experimental error.

If however, with 5 MV X-rays, $U_s = 1\cdot0$ mm, the total unsharpness on the display image should be about $1\cdot3$ mm: measured values have been slightly higher than this, about $1\cdot6$ mm, for reasons which have yet not been determined.

Unsharpness measurements on fluorescent screens, as already stated, are prone to very considerable error, largely because of the shape of the density-distribution curve of the image of a sharp edge, and the influences of this shape on summed unsharpness through a complex system are difficult to determine accurately. A method of image quality assessment which takes account of this, is therefore desirable. The method of contrast transfer, using the line spread function (LSF) appears to have many advantages.

C. Modulation Transfer Function (MTF) Methods

Methods of assessing image quality on the basis of the concepts of contrast transfer are now widely used in optical and photographic work and are likely to be of great value in assessing X-ray image intensifier systems.

If a repeating bar-pattern with a sinusoidally-varying intensity of transmitted X-rays is considered as object, it can be specified by two parameters—frequency and modulation (or contrast). If this pattern is imaged by a system, or transferred through a stage of any imaging system, the pattern frequency is usually unchanged, but the modulation (the contrast between bar and space) is generally reduced. The ratio of modulation before and after is called the Response R and a curve of R for a range of frequencies is the *modulation transfer function* (MTF). In practice it is not necessary to perform the complete experiment with a range of bar patterns of different frequencies, as the MTF can be derived from the distribution of intensity transmitted by a very narrow slit (the line spread function, LSF) by means of a Fourier transform. There is also a relationship between the LSF and the unsharpness curve, in that the former is the plot of the slope of the latter, plotted against distance.

MTF curves can be produced to represent a detail in the object, the performance of a screen, film, optical system or electronic circuit, and are easily combined to produce a resulting value for several stages. Bouwers (1967) and de Winter (1962) have indicated the usefulness of this method of assessment for a Cinelix X-ray intensifier (Fig. 1) and a similar family of curves can be constructed for the display image of an image orthicon/fluorescent system (Fig. 2). The usefulness of these curves is that they give a clear indication of the limiting components of a system. They depend, however, on the accurate determination of the MTF curves for the individual components.

In the case of the Cinelex system (Fig. 1) the limit of performance is controlled by the primary fluorescent screen and the light amplifier tube.

In Fig. 2 the limiting components are the primary fluorescent screen and the image orthicon and it is of interest that if a smaller size of image orthicon is considered, with a coarser television raster, this then becomes the dominant component in limiting performance.

D. The Overall Assessment of an Image Intensifier System

From the foregoing it would seem that MTF measurements can take account of two of the three causes of limitations in performance listed in Section III, provided that the MTF data on individual stages is sufficiently

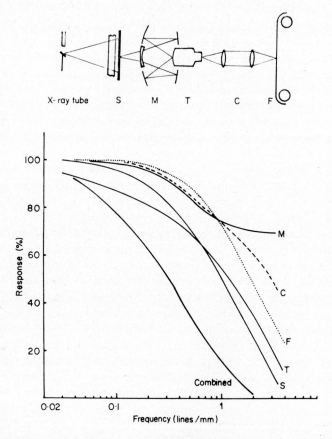

FIG. 1. Modulation transfer function curves for an image intensifier system consisting of a fluorescent screen, mirror optics, light amplifier, second optical system and recording on 70 mm film, based on experimental curves from Bouwers (1967). The thick (lower) curve is the MTF of the complete system.

M—mirror optical system   C—second optical system
S—fluorescent screen   F—film.
T—light amplifier

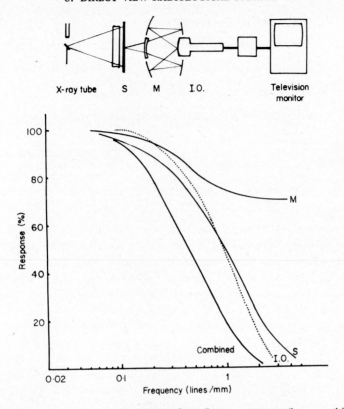

FIG. 2. Modulation transfer function curves for a fluorescent screen/image orthicon/television system, calculated from MTF data for the components.

M—mirror optical system

S—fluorescent screen

I.O.—image orthicon with 1000-line raster.

comprehensive. Such measurements, therefore, need to be combined with quantum fluctuation limitations to provide a complete assessment of the sensitivity of a system.

There have been a few attempts to develop such a comprehensive theory to describe performance.

Morgan (1965) has developed equations based on threshold visual perception, using noise-equivalent passband data based on MTF measurements of the image intensifier system and the human eye. His condition for threshold visual perception is given by

$$C_{r0} = [2K v_{excn}/|A_{cr}(v_\chi)|] [2v_{etcn}/\bar{N}_n]^{\frac{1}{2}} \qquad (7)$$

where $C_{r0} =$ the threshold discernible contrast,

v_{excn} = spatial noise-equivalent passband,

v_{etcn} = composite temporal noise-equivalent passband,

K = threshold signal-to-noise contrast ratio,

$A_{cr}(v_x)$ = composite MTF of the complete visual system,

$\overline{N}_n$ = photons/mm^2/sec.

The "noise-equivalent passband" for a measurement on the x-axis is

$$\int |A(v_x)|^2 \delta v_x = v_{ex}.$$

Using television system data from Coltman and Anderson (1960) and visual system data from DePalma and Lowry (1962), equation (7) is stated to agree well with experimental data. It should be noted that here, K has a small value, about 0·13, instead of a value between 3 and 5 as required by Rose's hypothesis. Compared with Rose's formula, the storage time of the eye is replaced by the temporal noise-equivalent passband, and the dimensions of the object are replaced by two terms, v_{excn} and $A_{cr}(v_x)$. Fig. 3 shows Morgan's predicted values of threshold contrast for an image intensifier system with a gain of $\times 3000$ and two dose-rates on the screen, $3·5 \times 10^{-4}$ and 14×10^{-4} R/sec using medical X-ray energies (c. 70 keV). These dose-rates are lower by a factor of 100–1000 than the dose-rate which is likely to be used for optimum performance in industrial X-ray conditions.

Morgan's theoretical treatment emphasizes several points:

(a) If the resolution capability of an image intensifier system is improved, there may not be much real gain unless higher radiation dose rates can be employed. (It seems probable that for industrial techniques this limitation does not apply.)

(b) With image intensifier systems, large viewing distances should not be employed.

(c) For *stationary* objects, a system with a low temporal frequency response, e.g. a Vidicon, may give superior results.

V. RADIOLOGICAL PERFORMANCE OF IMAGE INTENSIFIER SYSTEMS

There is no standard test object which is widely used in both medical and industrial work to evaluate the performance of an X-ray image intensifier system. The industrial radiologist usually thinks in terms of image-quality

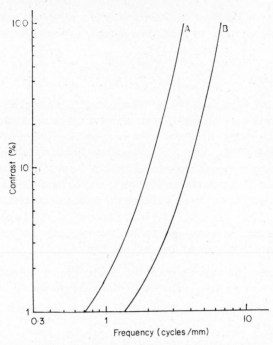

FIG. 3. Predicted values of threshold contrast calculated for conditions where the composite MTF is governed by the image amplifier characteristics; from Morgan (1965).

A—typical image amplifier: exposure rate $3 \cdot 5 \times 10^{-4}$ R/sec
B—image amplifier with doubled passband: exposure rate 14×10^{-4} R/sec.

indicators (penetrameters); these are widely used in both routine and experimental radiography and there is a large quantity of comparison data for film radiography. Such I.Q.I. have virtually no meaning in the medical field where the most widely used test object appears to be the Burger phantom (Burger, 1946) of cylindrical holes of different depths and diameters from which the object characteristics can be expressed as contrast and diameter. The advent of MTF data has revived an interest in bar patterns of different frequencies, although any object shape can be expressed as an MTF curve. Strictly, a bar pattern should be sinusoidal in transmitted intensity and this is difficult to achieve in practice, but for image intensifiers where the resolution is not likely to be much better than 10 lines/mm, direct measurement of MTF with bar test objects of various frequencies seems feasible and is an attractive method of specifying performance. Square-wave bars can be used and the correction to a sinusoidal intensity distribution is not large.

Whatever type of test object is used, the final judgement of what is discernible is by eye. The precise viewing conditions are seldom strictly controlled.

The judgement of the limit of discernibility is a subjective one and likely to vary between observers, so that some spread in sensitivity values must be expected.

A. I.Q.I. Data

Figure 4 shows sensitivity values obtained with a commercial (fluoroscopic screen/large image orthicon) system, using a range of X-ray energies extending into the megavoltage region (Vincent and Shemmans, 1966; Wycherley and Shemmans, 1966). Figure 5 shows data obtained by the author and colleagues, using X-ray energies up to 150 keV from an ultrafine focus X-ray tube with which projective magnification (p.m.) techniques can be employed. By the use of projective image methods the screen unsharpness expressed as a function of object size is reduced by a factor of M, the magnification used, if the tube focus is small enough not to introduce any additional blurring.

Several conclusions can be drawn from Figs. 4 and 5:

(a) Unless p.m. techniques are used, the sensitivities on thin sections are very much poorer than can be obtained with film radiography.

(b) Sensitivities of about 1 % (wire I.Q.I.) can be obtained on steel thicknesses from 3 in. to 12 in.

(c) On the largest thicknesses of steel, using the most powerful X-ray equipments, I.Q.I. sensitivities comparable with those on film are possible.

(d) Fluoroscopy of steel thicknesses up to 14 in. is possible.

(e) With low X-ray energies and thin specimens, the limit to sensitivity is usually the unsharpness of the fluoroscopic screen. Projective magnification techniques can greatly alleviate this loss of definition, and sensitivities which are only about 2–3 times worse than on film are possible, but suitable equipment between 150 keV and 1 MeV does not exist. The maximum steel thickness limit with 150 kV X-rays is about 3 in.

(f) For large thicknesses, the best results are obtained with a high-output X-ray machine: X-ray energy is less important than output.

(g) The limit to performance on thick specimens is usually due to quantum fluctuation effects in the screen and there is a need for a more efficient screen.

Contrast/(minimum perceptible object diameter) curves obtained by various workers are shown in Figs. 6 and 7. These are from medical sources and apply to X-ray energies around 70 keV. Figure 6 shows data determined from quantum fluctuation theory by Sturm and Morgan (1949), for an "ideal" image intensifier operating at a screen dose-rate of 5·3 mR/min. This can be compared with the curves of experimentally-determined sensitivity in Figs. 7 and 8.

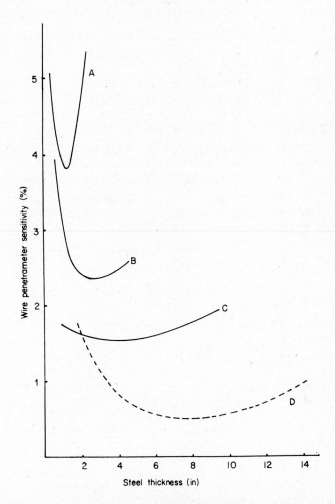

FIG. 4. Wire image-quality-indicator sensitivities for different thicknesses of steel and several X-ray energies, observed on a screen image using Marconi equipment (Vincent and Shemmans, 1966; Wycherley and Shemmans, 1966)

A—400 kV X-rays

B—1000 kV X-rays

C—5 MV X-rays

D—15 MV X-rays (linac, using lead-backed screens).

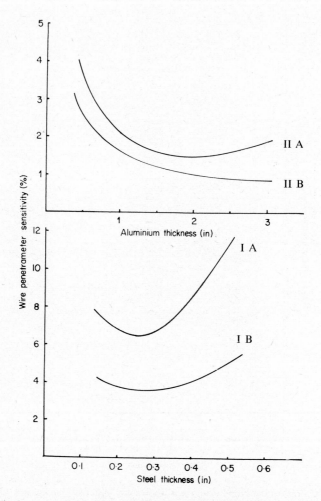

FIG. 5. Wire image-quality-indicator sensitivities, using an ultrafine focus (0·3 mm) 150 kV X-ray set, and projection magnification. The p.m. used is adjusted to an optimum value for each specimen thickness.

I—steel   A—natural size image (1/1)

II—aluminium   B—projective magnification.

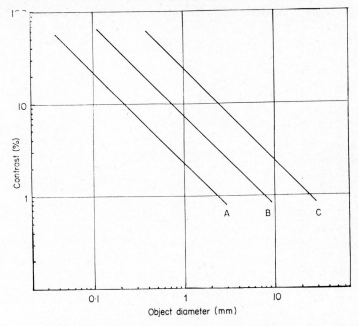

Fig. 6. Contrast/detail size diagram for an "ideal" image intensifier, using 70 kV X-rays. (from Sturm and Morgan, 1949).

A—dose-rate on screen $8 \cdot 8 \times 10^{-3}$ R/sec

B—dose-rate on screen $8 \cdot 8 \times 10^{-4}$ R/sec

C—dose-rate on screen $8 \cdot 8 \times 10^{-5}$ R/sec.

Figure 7 shows data for a 9 in. image intensifier tube coupled to an image orthicon and working at a screen dose-rate of 3·8 mR/min (Webster and Wipfelder, 1964). Essentially the performance here is limited by the dose-rate at the screen and this is well illustrated in Fig. 8, from Garthwaite *et al.* (1963), showing similar data for two higher dose-rates using a screen/image orthicon system. Garthwaite has pointed out that sensitivities seen on a television monitor screen are very dependent on the presence or absence of highlights in the image and that a uniform thickness test block is not representative of typical medical images, where sensitivity may be impaired by glare.

VI. Industrial Applications

Direct viewing systems, in their present state of development, give sensitivities on static subjects which are generally inferior to radiography-on-film, except on very thick specimens. There has therefore been a reluctance to

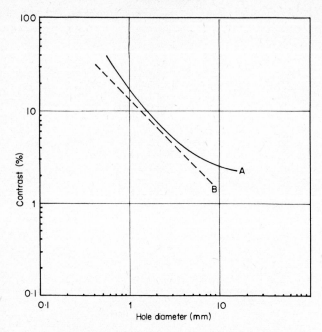

FIG. 7. Contrast-detail curves for a 9 in. image intensifier tube used with an image orthicon. 90 kV X-rays, from Webster and Wipfelder (1964).

A—screen-dose-rate 6.3×10^{-5} R/sec

B—theoretical photon limit, calculated for $K = 2.5$.

introduce such comparatively expensive equipment for what appears to be an inferior result, even if there is a potential saving in consumable material (film). In addition, the poorer sensitivity is usually due to a lack of definition, so that on narrow flaws such as cracks, the discrepancy between film and screen images is still greater. Nevertheless there have been reports of successful routine inspection applications to metal objects not likely to contain cracks and to objects where it is necessary to determine the best orientation of a specimen for interpretation, for example by rotation during viewing. Kobayashi (1968) has reported the use of an image intensifier in a shipyard for the routine inspection of submerged-arc welding in 14 mm thick steel plate. A wire-penetrameter sensitivity of 3–4% was obtained, using × 2 projective magnification with an inspection speed of 1 metre/min.

More important applications, however, have been the use of ciné-fluoroscopy to examine moving systems. Although the image can be viewed on the screen while the event is happening, it is often more convenient to film the image, or video-tape record it, so as to have a record to be studied at leisure.

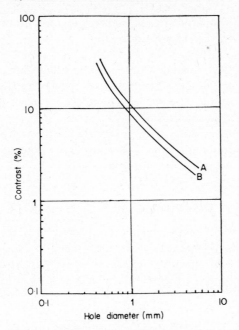

FIG. 8. Contrast/hole size for a fluorescent screen/image orthicon system, as observed on screen (Garthwaite *et al.*, 1963).

A—dose-rate on screen 6×10^{-1} R/sec

B—dose-rate on screen $0 \cdot 8$ R/sec.

One of the first potential uses which was investigated, was to determine the cropping point of hot steel "blooms" during rolling, by locating the ends of internal defects originating from the end of the ingot, as the bloom passed between an image intensifier and a high-energy X-ray source. Several workers (Vincent and Shemmans, 1966; Lutherack *et al.*, 1960) have investigated the attainable sensitivity and shown it to be adequate for this application.

The pouring of liquid metal into a casting mould to study the flow of metal through the ingate, the filling of the mould cavity and the formation of casting defects during solidification, have all been studied in steel, aluminium and some non-metallic cast materials (Halmshaw *et al.*, 1966; Hall, 1967). There is some possibility that under ideal conditions, when maximum radiographic sensitivity is possible, it may be possible to discern the liquid/solid interface (Forstén and Miekk-oja, 1967).

The flow of liquids, the presence of cavitation, etc., can also be studied, if the speed of motion is not too great, and it is of interest that greater filming speeds are possible with simple image intensifier tubes than with a method utilizing a television system, as these latter are normally designed for a 50 or 60 cycles per second repetition rate.

One interesting application has been the study of the electroslag welding process (Lowery, 1968) to study theevents occurring within the slag bath and the effects of various welding parameters on these. The effects of misalignment of the consumable wire and guide and the length of unmelted wire in the slag pool can be clearly shown.

In general, however, fluoroscopy of moving subjects has not yet found widespread industrial application, and many more potential uses are being studied than have actually come into general use.

VII. FUTURE DEVELOPMENTS

The potential lines of development of the fluorescent screen/image orthicon system of X-ray image intensifier has been studied by Noé (1968) and Noé and Halmshaw (1968). From considerations of a complete equipment, stage by stage, the possible lines of development appear to be as follows:

A. X-ray Source

For thin specimens, unless there is a " breakthrough " in the design of grainless fluorescent screens, in order to obtain acceptable sensitivities it will be necessary to use projective magnification, for which an X-ray set with a focus diameter of 0·3 mm or less is necessary. Very fine focus sets operating between 150 keV and 1MeV are not commercially available. With a focus size of 0·3 mm an image magnification of × 6 should be possible before geometric unsharpness effects begin to be noticeable.

For thick specimens, high-energy X-rays are necessary and a high output equipment will give the best results. Linacs with the necessary performance are already marketed, but are expensive.

B. Fluorescent Screen

The requirement here, also, appears to depend on whether the application is to high or low energy X-rays. For low energy X-rays, a screen giving better definition of the image is necessary and single-crystal discs of CsI(Na) seem to offer the best possibilities at present (Noé and Halmshaw, 1968).

For high energy X-rays a screen which will absorb a greater percentage of the X-rays transmitted through the specimen is necessary. Some gain in performance seems possible with thicker fluorescent screens, and single crystal screens should also produce some improvement. A mosaic screen of scintillation fibres, or a liquid-cell scintillator should theoretically give a better screen, but the construction problems are formidable.

C. Optical System

For efficiency of light transfer, the F : 0·68 Schmidt–Bouwers mirror optical system is unlikely to be surpassed, except by a reducing-diameter fibre bundle. The mirror system, however, has to have a central stop and its light transmission is not, therefore, as high as its nominal aperture suggests: furthermore, it is a rigid system designed to work at one value of magnification. There are therefore potential advantages in the use of a wide-aperture lens system, particularly if this is designed to focus different areas of the screen on to the television camera-tube pick-up surface. By this means, the full size of the monitor screen and the full line raster could be used for any size of subject.

D. Television Tube

It seems very probable that the image orthicon is already being replaced by the isocon, with a considerable reduction in noise level, a gain in sensitivity and a greater dynamic range (English Electric Valve Co., 1968). In the image isocon the image section is as in a normal image orthicon tube. When the scanning beam from the electron gun reaches the target, part of the beam is scattered and its magnitude is dependent only upon the charge present at that point on the target: this beam returns to the gun through the steering electrodes into a conventional electron multiplier system. Thus it is the beam of scattered electrons which provides the signal and the magnitude of the signal increases with light level, whereas in the image orthicon it is the specularly-reflected beam which is used and this has its maximum value for zero light input. Defined as

$$\frac{\text{input light for normal operation}}{\text{input light for a just discernible picture}}$$

the dynamic range of an isocon is about 2000 : 1, which is many times better than for an image orthicon.

E. Electronic Circuitry

There appears to have been little assessment of what is worthwhile in the way of circuit improvement, apart from an estimate of the required bandwidth, but several advances are feasible.

The bandwidth must be matched to the line-frequency throughout the circuitry and there seems to be no fundamental reason why the repetition frequency should not be increased to 100 or 150 Hz, to enable faster ciné-recording of moving objects.

Overall contrast-gain is already available, as is automatic brightness control, but local contrast enhancement may have advantages for some applications; the circuitry for this is already available in other applications of television.

F. Image Recording

Ciné-film recording and videotape are already available and can produce images with little or no loss in quality compared with the image on the monitor. Requirements here are the development of simpler, more reliable and less costly recording systems, rather than new methods.

It is important to emphasize that the value of an improvement in performance of any one stage of an image intensifier must be considered in relation to the limitations of other stages. It is for this reason that an analysis of performance stage-by-stage, as described in Section IV, is important.

Little has been said about the comparative merits of the alternative types of direct-viewing systems, other than those of group (b) listed in Section I. It is possible that X-ray-sensitive tubes, or a modification of the group (b) system using an intermediate light amplifier, may prove to have special advantages. So far as is known, these have not yet been assessed for industrial applications.

REFERENCES

Albrecht, C. (1965). *In* "Proceedings of the 2nd Colloquium on Diagnostic Radiological Instrumentation." C. C. Thomas, Springfield, Illinois.

Albrecht, C. and Oosterkamp, W. J. (1962). The evaluation of X-ray image forming systems. *Medica Mundi.* **8,** 106–109.

Albrecht, C. and Proper, J. (1967). The limits of detail rendition in X-ray images; theory and experiment. *Proc. 11th Int. Congr. Radiol.* (Rome) **2,** 1639–1647.

Banks, G. B. (1958). Television pick-up tubes for X-ray screen intensification. *Br. J. Radiol.* **31** (371), 619–625.

Bouwers, A. (1967). *In* "Proceedings of the XIth International Congress of Radiology," Rome.

Burger, G. C. E. (1946). X-ray fluoroscopy with enlarged image. *Philips Tech. Rev.* **8** (11), 321–357.

Coltman, J. W. (1948). Fluoroscopic image brightening by electronic means. *Radiology* **51,** 359–366.

Coltman, J. W. and Anderson, A. E. (1960). Noise limitations of resolving power in electronic imaging. *Proc. Inst. Radio Engrs* **48** (5), 858–865.

Criscuolo, E. L. and Polansky, D. (1956). Improvements in high sensitivity fluoroscopic technique. *Non-destruct. Test.* **14,** 30–32.

DePalma, J. J. and Lowry, E. M. (1962). Sine-wave response of the visual system. *J. opt. Soc. Am.* **52,** 328–335.

de Winter, H. G. (1962). "Zur Optimalisierung der Kontrastübertragung des Schirmbildes." *Proc. 10th Int. Congr. Radiol.* (Montreal), pp. 84–89.

English Electric Valve Co. (1968). Data sheets.

Forstén, J. and Miekk-oja, H. M. (1967). Radiographic observations of the solidification of metals. *J. Inst. Metals* **95**, 143–145.

Fowler, J. F. (1960). Fundamental limits of information content of image intensifying systems. *Br. J. Radiol.* **33** (390), 352–357.

Garthwaite, E., Haley, D. G. and Beurle, R. L. (1963). Television X-ray intensifier sensitivity. *Proc. Instn elect. Engrs* **110** (11), 1975–1978.

Hall, H. T. (1967). Extending the use of X-ray television fluoroscopy. *Metals* **2** (18), 50–51.

Halmshaw, R. (1966). "Physics of Industrial Radiology," Ch. 12. Heywoods, London.

Halmshaw, R. and Hunt, C. A. (1957). Fluoroscopy with an enlarged image. *Br. J. appl. Phys.* **8** (7), 282–288.

Halmshaw, R., Durant, R. L. and Lavender, J. D. (1966). Ciné-radiography of the casting of steel. *Iron Steel, Lond.* **39** (9), 381–387.

Hay, G. A. (1958). Quantitative aspects of television techniques in diagnostic radiology. *Br. J. Radiol.* **31**, 611–618.

Joinson, A. B. (1967). "Television Radioscopy as an Inspection Tool." A.E.R.E. Report M–1859.

Kobayashi, K. (1968). Application of an X-ray fluoroscope to shipbuilding. *Br. J. non-destr. Test.* **10** (4), 70–77.

Kuhl, W. (1965). *In* " Proceedings of the 2nd Colloquium on Diagnostic Radiological Instrumentation." C. C. Thomas, Springfield, Illinois.

Lang, G. (1955). Neue moglichkeiten der röntgendurchleutung von stahl. *Z. Ver. dt. Ing.* **97** (11/12), 347–350.

Lowery, J. F. (1968). "A Radiographic Study of the Electroslag Welding Process". British Welding Research Association Report P/32/68.

Lutherack, W., Flossmann, R. and Fink, K. (1960). Steel bloom piping—continuous X-ray examination during rolling. *Iron Steel, Lond.* **33**, 469–472, 500–504.

McMaster, R. C., Rhoten, M. L. and Mitchell, J. P. (1962). "An X-ray Image Enlargement System." W.A.L. Report 142, 5/1–5.

McMaster, R. C., Rhoten, M. L. and Mitchell, J. P. (1967). The X-ray vidicon television image system. *Mater. Evaluation* **25** (3), 46–52.

Morgan, R. H. (1956). Screen intensification: A review of past and present research with an analysis of future development. *Am. J. Roentg.* **75**, 69–76.

Morgan, R. H. (1965). Threshold visual perception and its relationship to photon fluctuation and sine wave response. *Am. J. Roentg.* **93** (4), 982–997.

Noé, E. (1968). "An Investigation of Methods of Improving the Performance of the 12-in Marconi Image Amplifier with High Energy X-rays (1–5 MeV)," 1st Progress Report. Unpublished RARDE Memo.

Noé, E. and Halmshaw, R. (1968). "An Investigation of Methods of Improving the Performace of an X-ray Television Fluoroscopy Image Intensifier System," 2nd Progress Report. Unpublished RARDE Memo.

O'Connor, D. T. and Polansky, D. (1952). Theoretical and practical sensitivity limits on fluoroscopy. *Non-destruct. Test.* **11**, 10–12.

Oosterkamp, W. J. (1954). Image intensifier tubes. *Acta radiol.* Special Suppl. No. 116.

Rose, A. (1948). Sensitivity performance of the human eye on an absolute scale. *J. opt. Soc. Am.* **38** (2), 196–208.

Schade, O. H. (1964).   An evaluation of photographic image quality and resolving power. *J. Soc. Motion Pict. Telev. Engrs* **73** (2), 81–119.

Sturm, R. E. and Morgan, R. H. (1949).   Screen intensification systems and their limitations. *Am. J. Roentg.* **62** (5), 617–634.

Teves, M. C. and Tol, T. (1952).   Electronic intensification of fluoroscopic images. *Philips Tech. Rev.* **14** (2), 33–68.

Vincent, B. J. and Shemmans, M. J. (1966).   Fluoroscopy with a Marconi 12-in image amplifier at energies up to 5 MeV. *Appl. mater. Res.* **5** (3), 172–180.

Webster, E. W. and Wipfelder, R. (1964).   Contrast and detail perception in television and ciné systems for medical fluoroscopy. *J. Soc. Motion Pict. Telev. Engrs* **73** (8), 617–621.

Webster, E. W., Wipfelder, R. and Prendergrass, H. P. (1967).   High-definition versus standard television in televised fluoroscopy. *Radiology* **88** (2), 355–357.

Wycherley, J. R. and Shemmans, M. J. (1966).   Fluoroscopy with a Marconi 12-in image amplifier. *Appl. mater. Res.* **5** (4), 195–199.

CHAPTER 9

Neutron Radiography †

H. Berger

Argonne National Laboratory,
Argonne, Illinois, U.S.A.

I. INTRODUCTION

Radiography with neutrons, that is the production of photographic or pseudo-photographic images formed as a result of the differential attenuation of neutrons in an object, can be traced essentially to the period of the discovery of the neutron. In the case of neutron radiography, the production of the first pictures was not quite as dramatic as the x-radiographs of hands and other objects taken by Wilhelm Roentgen during the very early days following his discovery of x-rays.   Nevertheless, it is clear that a thorough study of neutron radiography was made in Germany in the 1930's by H. Kallmann and E. Kuhn. This study (Kallmann, 1948) resulted in several patents and one publication, which did not appear until 1948.

Although history does record the fact that neutron radiography studies were made shortly after the discovery of the neutron, it was only in the decade of the 1960's that extensive development and application work

---

† This paper is based on work performed under the auspices of the U.S. Atomic Energy Commission.

in neutron radiography came to pass. Certainly one reason for this is the fact that good sources of neutrons became more available during that period. In any event, radiography with thermal neutrons from reactor sources has now become routine at a great number of nuclear centers. Some work with neutrons in other energy ranges has been reported, including cold, epithermal, and fast neutrons. Each energy range appears to offer some particular advantage for special situations. However, the thermal neutron range seems most generally useful and this energy range will be most thoroughly considered in this report.

The role of neutron radiography in the area of nondestructive testing is similar to that of radiography in general. As a radiographic technique it offers the advantage of an image presentation (for ease of interpretation), and the capability to display relatively small changes in thickness and/or material. It also suffers from the disadvantages normally associated with radiography, such as the inability to reliably reveal cracks which are not aligned with the radiation beam direction, and the inability to yield much information about bonds and bond quality.

There are also important differences between radiography with gamma or x-rays and neutrons, differences which account for the usefulness of neutrography, or radiography with neutrons (Kallmann, 1948; Thewlis, 1956; Berger, 1965). The attenuation of x-rays in materials increases generally with increasing atomic number of the absorbing material. Therefore, it would be very difficult, if not impossible, to image successfully by x-radiography a small light object in an assembly with a thick, high atomic number material. However, the random attenuation of thermal neutrons, when examined as a function of atomic number of absorbing material, may indicate the capability of neutron radiography for producing a successful image. One of the early studies of neutron radiography (Thewlis, 1956), and the first reported investigation to make use of neutron beams from a nuclear reactor for neutron radiography, demonstrated this unusual radiographic capability by showing a successful neutron radiograph of a piece of waxed string on a 5 cm thick lead brick. The high thermal neutron attenuation by the hydrogen in the wax and string, and the low attenuation by the lead makes it a simple matter to produce such a radiograph. Similarly, the high thermal neutron attenuation of several other light elements such as boron and lithium, and the low attenuation of heavy materials such as lead, bismuth and uranium make it possible to obtain useful radiographs of many other potentially difficult material combinations.

Of course, neutron radiography offers a number of additional areas in which it can complement the nondestructive testing capability of x-radiography. One which is somewhat similar to that noted above comes from the

Fig. 1. A positive print of a thermal neutron radiograph of a rectangular piece of cadmium of uniform thickness, approximately 0·6 mm. The high neutron transmission through the lower right area (white area) results from depletion of Cd^{113} from this material, which came from a reactor control element. (Courtesy, Argonne National Laboratory, Argonne, Illinois).

fact that neutron cross sections (Goldberg *et al.*, 1966; Hughes, 1957) vary, sometimes appreciably, between different isotopes of the same element. Therefore, one can radiographically discriminate between different isotopes of the same element. This has proved to be very useful, for example, in showing areas in which Cd^{113} has been depleted from a neutron-bombarded reactor control rod, as shown in Fig. 1. There are many other possible areas in which this unique radiographic capability might be utilized.

A significant advantage presented by neutron radiography, and the one which led to its first continuing application use, is associated with the fact that one can readily detect neutron images by methods which have little or no sensitivity to gamma rays. Therefore, it is possible to obtain good quality neutron radiographs of objects which are highly radioactive. For an x-radiograph of such an object, one encounters many difficulties if one is to avoid excessive fogging of the x-ray film by the radiation emitted from the inspection object. The particular application which has resulted from this discriminatory capability is the neutron radiographic inspection of highly radioactive reactor fuel materials; an example is shown in Fig. 2. Typically, nuclear research organizations must irradiate, in a reactor, new reactor fuel materials and assemblies. After irradiation, these highly radioactive objects must be inspected, preferably prior to disassembly. Neutron radiography has presented a now widely used solution to this problem.

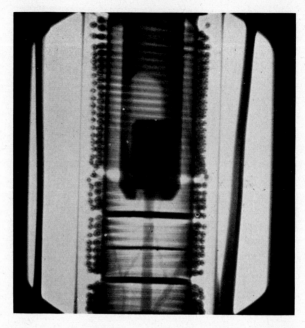

FIG. 2.   A positive print of a thermal neutron radiograph of a highly radioactive (over 10,000 curies) test element. The vertical rod is uranium, heated by the surrounding heating element while the sample was in the reactor. After Barton and Perves (1966). (Courtesy, Centre d'Etudes Nucléaires, Grenoble, France.)

There are many other areas of application of neutron radiography.   A review of many of these will be given following a description of some of the methods used for neutron radiography.

II. Techniques for Neutron Radiography

With relatively few exceptions, neutron radiographic inspection has been performed with multi-energy neutron beams having their peak intensity in the thermal energy range.   These neutron beams have been obtained for the most part from nuclear reactors, and recently, on an increasing basis, from accelerator and radioactive neutron sources.   Specialized methods for neutron radiography have been accomplished in the cold, resonance or epithermal, and fast neutron energy ranges.   Discussion concerning these neutron sources follows in the section below.

Detection methods for neutron radiography are also discussed in a section to follow. However, as an introductory statement, it can be said that the most widely used neutron radiography detection methods make use of photographic film. Typically these methods involve x-ray films used in such a manner that they are exposed directly to the neutron beam, or to a radioactive, image-carrying foil in an autoradiographic technique. The first procedure, in which the film is used in the neutron beam, has been termed the direct exposure method. Convertor materials used with the film enhance the film sensitivity to neutrons. These convertors can be scintillator materials, such as mixtures of a phosphor powder with a lithium or boron-containing material. In this case, the (n, α) reaction in the lithium or boron produces radiation which stimulates the phosphor and the resultant light is recorded by the film. Also used are metal foil convertor materials such as gadolinium, with which advantage is taken of prompt (n, γ) reactions to stimulate the film. The second detection method is known as the transfer technique. The neutron image is detected by a foil of a material which tends to become radioactive easily. Examples are indium, dysprosium, or rhodium, with resultant half-lives of 54 min, 2·3 hours, and 4·4 min respectively. After the neutron irradiation, the image-carrying foil is transferred to a film-loaded cassette, away from the neutron beam, and an autoradiograph reveals the neutron image. These are the primary detection methods which have been used. More complete descriptions of these and other detection methods which have been developed follow in the section on detectors.

A. Sources for Neutron Radiography

Thewlis and Derbyshire (1956) were the first to demonstrate the use of a nuclear reactor beam for neutron radiography. Their investigation produced the first high-quality neutron radiographs; this was possible because the reactor provided a much greater neutron intensity than the accelerator sources used in the earlier studies (Kallmann, 1948; Peter, 1946). Therefore, improved beam collimation could be employed to yield good-quality radiographs. Since that time a great number of nuclear reactors have been used for neutron radiography (Berger et al., 1964–69). The wide variety of reactor types which have been used, such as the graphite-moderated BEPO reactor used by Thewlis (1956) and Thewlis and Derbyshire (1956), the Los Alamos water boiler reactor (Blanks and Morris, 1966), and the swimming pool reactors used at several locations (Rhoten and Cary, 1966; Barton and Perves, 1966; Carbiener, 1966), indicate that the reactor type is not critical. The reactor should be capable of providing a sufficient number of neutrons in the energy range of interest and it should lend itself to the use of a collimator to bring the neutron beam out to a useful radiographic position.

What constitutes a sufficient number of neutrons depends, of course, upon many factors, such as the radiographic quality needed for the particular application and the corresponding degree of collimation required, the intensity and energy of interfering radiation (such as gamma radiation) in the beam, and the amount of time that can be tolerated for the production of the radiograph. Many of these factors are interdependent. Some examples may be useful to illustrate typical values which have proved useful. A thermal neutron intensity in the order of 10^7 neutrons per square centimeter per second (n/cm^2-sec) is sufficient to provide high-quality direct exposure neutron radiographs on fine-grain x-ray film with an exposure time of a few minutes. For example, with a detector consisting of a 12·5 micron thick gadolinium convertor foil and a slow film such as Kodak (U.S.A.) Type R, a film exposure of about 3×10^9 n/cm^2 will provide a good radiograph. This example is given under the assumption that the interfering gamma radiation in the beam is of sufficiently low intensity and/or high energy so that the resultant radiograph will be essentially a neutron radiograph. For

FIG. 3. A negative print of a thermal neutron radiograph of an adhesive bonded, aluminum honeycomb panel. This radiograph was taken by a direct-exposure technique with a gadolinium foil and single emulsion Type R x-Ray film; the total exposure was 3×10^9 n/cm^2. Adhesive fillets at the upright sections and bubble structure in the adhesive can be observed. Several areas showing poor adhesive build-up at the upright sections can be observed in the top, and upper right areas. (Courtesy, Argonne National Laboratory, Argonne, Illinois.)

example, the neutron radiograph shown in Fig. 3 readily shows hydrogenous adhesive material; it was made under conditions indicated above. This hydrogenous material would not be imaged so clearly if the gamma intensity were too high. It has been shown (Berger, 1965; Watts, 1962) that most of the metal convertor foil-film combinations provide about the same film response for a thermal neutron exposure of 10^5 n/cm^2, or for a Co^{60} gamma ray exposure of 1 milliroentgen (mR). Therefore, if the neutron to gamma ratio in the neutron beam is sufficiently better than 10^5 n/cm^2 per mR, the radiograph will be mostly due to neutrons. One can use convertor materials, such as boron or lithium scintillators, which provide a better neutron to gamma ratio response by an order of magnitude. These response ratios are in accord with similar ratio response studies for x-ray film used alone, if one takes into account the increased neutron response which results from the convertor material (Ehrlich, 1960). Although fast neutrons in the beam will degrade the radiograph by contributing secondary and scattered radiation which may be detected, some fast neutrons in the beam can be tolerated for many applications (Berger, 1964).

In general, thermal neutron beams from reactors can provide a sufficiently good neutron to gamma intensity ratio so that direct exposure neutron radiography can be performed. Reactors which have been used for radiography have had peak thermal neutron intensities in the reactor in the 10^{10} to 10^{14} n/cm^2-sec range. Collimation of the beam, in order to bring out a useful radiographic beam, results in an intensity reduction at the radiographic location. This reduction varies with the degree of collimation (Barton, 1968a; Barton et al., 1969). A relatively coarse collimation system, providing a collimator length to width ratio of 10 : 1, results in an intensity loss of about 10^3. Another order of magnitude is lost if the collimation is improved to provide a length to width ratio of about 30 : 1. This type of collimation provides useful results for a wide range of radiographic applications. For applications requiring better radiographic sharpness quality, a collimator length to width ratio of 100 : 1 can provide quality very similar to that yielded by good x-radiography; the loss in neutron intensity from collimator input to the output is a factor of 10^5 in this case. Collimator length to width ratios as high as 400 have been used.

Radiography collimation systems (Berger, 1965) to bring thermal neutron beams from a moderator have varied from simple straight collimators (Berger, 1964) to Soller slit techniques in order to provide beams with smaller angular divergence (Blanks et al., 1967; Morris, 1963), and more recently to a divergent collimator (Barton and Perves, 1966; Barton, 1967a). The last approach has been shown to provide resolution capabilities equivalent to the other collimation techniques for the same collimator length to input width ratio, assuming that the collimator is constructed with a good neutron absorbing material. The divergent approach offers the significant advantage that one can radiograph a much larger area in one exposure. Therefore, this type of collimator is coming into much greater use. This is particularly true (Berger *et al.*, 1964–69) in moderators surrounding accelerator or radioactive neutron sources, and in pool reactors.

The collimation systems discussed above also apply to other types of moderated neutron sources. Indeed, interest in non-reactor sources for thermal neutron radiography has very much increased in recent years (Berger *et al.*, 1964–69). Sources of fast neutrons, of the accelerator and radioactive type, have been shown to be capable of providing useful thermal neutron beams for radiography (Barton, 1968a). The radioactive sources Sb^{124}–Be, a (γ, n) type, and the (α, n) sources, Am^{241}–Be and Am^{241}–Cm^{242}–Be, have been successfully used thus far (Berger *et al.*, 1964–69; Barton *et al.*, 1969; Warman, 1965; Cutforth, 1968; Barton and Boutaine, 1968). An example of a neutron radiograph taken with a radioactive source is shown in Fig. 4. Other radioactive sources also appear to be useful for neutron radiography (Tyufyakov, 1968). There is somewhat less interest in sources of the (γ, n) type because the intense gamma radiation background forces

FIG. 4. A positive print of a transfer thermal neutron radiograph of a stepped uranium test object taken with an Sb–Be neutron source. The steps observed on the print have thicknesses of 2 mm (top), 1·5 mm, 1 mm and 0·75 mm. The two holes in the two upper steps are twice the step thickness, and the step thickness in diameter. The third hole in the thinner steps is four times the step thickness in diameter. The object is on natural uranium, 9·35 mm thick. The original film showed 2% contrast. The radiograph was taken with a 270 curie Sb–Be source. The $5 \times 5 \times 52·5$ cm collimator provided a thermal neutron intensity at the dysprosium detector of $2·1 \times 10^4$ n/cm²-sec. After Cutforth (1968). (Courtesy, Argonne National Laboratory, Idaho Falls, Idaho, reprinted with the permission of the American Society for Nondestructive Testing.)

one to use detection methods which are not sensitive to gamma radiation. On the other hand, several (γ, n) sources, notably Sb–Be, yield neutrons of relatively low energy, thereby presenting an advantage in moderation. For Sb–Be, for example, the 24 kev neutrons produced can be moderated to yield a peak thermal neutron intensity in the moderator of about $100 \, n/cm^2$-sec per $10^4 \, n/sec$ yielded from the source.

Sources of the (α, n) type are attractive for neutron radiography because of the much lower background of gamma radiation. Spontaneous fission sources are also good future possibilities (Stetson, 1966); Cf^{252}, if it can be made available at reasonable prices, could prove to be a very useful, high yield and high specific activity source (Barton et al., 1969; Reinig, 1968a and b; Barkalov et al., 1970). The high specific activity will be an advantage for fast neutron radiography where the source would be used directly. Also, a point source will give a higher peak thermal neutron flux in a moderator than a large volume source of the same neutron yield, because of the favorable geometry in the moderation system. For Cf^{252}, one can obtain about 130 thermal n/cm^2-sec in the moderator per 10^4 n/sec yielded from the source.

The important characteristics of several of the more promising radioactive sources are tabulated in Table 1. Generally, the peak thermal neutron flux in the moderator surrounding such a fast neutron source is about 10^{-3} times the fast neutron yield from the source, for the case with a flux-depressing collimator in the moderator . Therefore, for a 10 : 1 collimation system, the total factor between fast neutron yield and thermal neutron intensity at the radiographic location approaches 10^{-6}. For a thermal neutron beam intensity of 10^4 n/cm^2-sec, one must start with a source having a total yield in the order of 10^{10} n/sec. This means that the source will be a kilocurie source. Of course, one can do radiography with smaller sources if transfer detection methods are not necessary, but the exposure times become long (Barton, 1968a). Under some conditions, several hour or overnight exposures can be quite feasible.

The same losses between fast neutron yield and thermal neutron beam intensity apply to fast neutron sources of the accelerator type. The relatively low cost, high yield accelerators employing the $T(d, n) \, He^4$ reaction to yield 14 Mev neutrons have been most discussed for possible use as neutron radiography sources (Berger et al., 1964–69; Barton, 1967b, 1968a; Barton et al., 1969; Spowart, 1968a; Kawasaki, 1968). Such accelerators have demonstrated continuous operation at total 4π yields of 10^{11} neutrons per second (n/sec) or more for many hours by the use of rotating targets, or of tubes employing gas mixtures to replenish the tritium used in the target. An

TABLE I

Some Radioactive Sources Potentially Useful For Neutron Radiography

Source	Reaction	Half-Life	Cost [a] in Thousands of Dollars	Average Neutron Energy [b] Mev	Neutron Yield (n/sec-gram)	Gamma Dose (rads/hr at one metre [c])	Gamma Ray Energy Mev	Comments
Sb^{124}–Be	(γ, n)	60 days	25	0.024	2.7×10^9	4.5×10^4	1.7	short half-life and high gamma background, available in high intensity sources, low neutron energy is an advantage for thermalization
Po^{210}–Be	(α, n)	138 days	20	4.3	1.28×10^{10}	2	0.8	short half-life; low gamma background
Pu^{238}–Be	(α, n)	89 years	310	~4	4.7×10^7	0.4	0.1	high cost, long half-life
Am^{241}–Be	(α, n)	458 years	1500	~4	1×10^7	2.5	0.06	easily shielded gamma output, long half-life, high cost

Am^{241}–Cm^{242}–Be	(α, n)	163 days	—	~4	$1 \cdot 2 \times 10^{9}$ (80% Am, 20% Cm)	low	0·04, 0·06	increased yield over Am^{241}–Be for relatively little more cost but with a short half-life
Cm^{242}–Be	(α, n)	163 days	—	~4	$1 \cdot 46 \times 10^{10}$	0·3	0·04	high yield source, but half-life is short
Cm^{244}–Be	(α, n)	19 years	200[d]	~4	$2 \cdot 4 \times 10^{8}$	0·2	0·04	long half-life, low gamma background area attractive. Source is expensive and has a lower neutron yield than many other transplutonic sources. It can also be used as a spontaneous fission source, with about half the neutron yield
Cf^{252}	Spontaneous Fission	2·65 years	20[e]	2·3	3×10^{12}	2·9	0·04, 0·1	very high yield source, present cost is high but projected future cost may make it attractive

(a) Cost of the radionuclide only is given ; see Reinig (1968a). The cost is normalized to a source total yield of 5×10^{10} n/sec.
(b) See Reinig (1968a) and Rubbino et al. (1966) for spectral information. The low average energy of neutrons emitted from Sb–Be does represent an advantage for moderation. One can gain a peak thermal neutron flux in the moderator of about an order of magnitude for these lower energy neutrons as compared to fast neutron sources yielding neutrons in the Mev range.
(c) The gamma ray dose is normalized to a neutron yield of 5×10^{10} n/sec.
(d) The cost is based on a proposed cost of $1000 per gram.
(e) The cost is based on a conservative future cost estimate of $1000 per milligram.

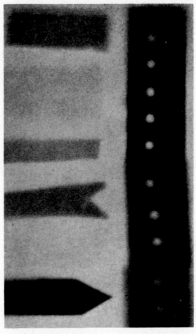

FIG. 5. A positive print of a thermal neutron radiograph of a test object taken with an accelerator neutron source, having a total fast neutron yield of 4×10^{10} n/sec. The radiograph shows at top left, 3·7 mm plastic, a slight image of 3·7 mm of lead directly below, and, continuing in the downward direction, plastic, indium, lead (which does not show) and cadmium at the bottom left. On the right side is a strip of cadmium containing holes; the spacing of the cadmium to the detector varies. After Barton (1967b).

example of a thermal neutron radiograph taken with a 14 Mev neutron generator is shown in Fig. 5.

A similar low-cost accelerator involves the reaction $D(d, n)\,He^3$; this has a reasonable yield of neutrons having an energy of about 3 Mev. Although the yield from this reaction is typically less than that available from the $T(d, n)\,He^4$ reaction (often by a factor of 100 times), it offers a possible advantage in long continuous operation. Also, the 3 Mev neutrons are easier to moderate than the 14 Mev (d, T) neutrons. Other accelerator neutron reactions (Burrill and MacGregor, 1960) have smaller neutron yields and/or require voltages much higher than the 100 to 400 kev accelerating voltages typically used for the above two reactions. The accelerators, therefore, become more expensive. However, these may offer advantages in certain circumstances. The Van de Graaff accelerator (Burrill, 1963) and the cyclotron (Fleischer, 1966) are examples of accelerators which can provide high neutron yields (10^{12} n/sec or more), but at higher investment cost. These accelerators offer the advantage of versatility in that, in many machines, the accelerated particle can be a proton, deuteron, or other charged particle. Therefore, a great variety of neutron production reactions becomes possible. Some of these, such as the widely used $Li^7(p, n) Be^7$ reaction, offer the possibility of choosing a useful neutron energy by proper adjustment of the accelerating voltage (Marion and Fowler, 1960).

It is, of course, also possible to use an electron accelerator of sufficiently high voltage to produce neutrons by the (γ, n) reaction. Beryllium targets have been most used for this reaction; the threshold gamma ray energy is 1·66 Mev. Examples of the neutron production capability of this reaction are given by the reports of Moses and Saldick (1956) and Wagner et al. (1960). In the first report, a thermal neutron flux of 10^7 n/cm²-sec was obtained in a 22·5 cm cube of beryllium bombarded by x-rays of 2 Mev, at a beam current of 250 microamperes. In the latter report, the higher thermal neutron flux of 8×10^7 n/cm²-sec was obtained in a 15 cm cube of beryllium, bombarded by 3 Mev x-rays at 100 microamperes. Although these neutron production reactions have a significant disadvantage due to the high x-ray background, they offer the advantage that neutron radiography can be performed with high energy accelerators already in use for x-radiography. Therefore, they offer an x-radiographer the choice of using neutron radiography when the occasion demands it. Successful work of this type, with a 5·5 Mev electron linear accelerator for inspection of metal-jacketed explosive devices, has recently become known (Halmshaw, 1968).

Beryllium containing moderators are very useful to employ with these (γ, n) reactions because neutrons continue to be produced in the moderator as the gamma rays are absorbed. Data indicating the thermal neutron flux in beryllium moderators surrounding an Sb–Be source are shown in Fig. 6. However, for the charged particle neutron production reactions, light moderators such as light or (preferably because of the smaller attenuation by the deuterium) heavy water, paraffin or plastic have proved very useful. There appears to be no advantage for using the more expensive beryllium moderators in these cases. Mourvaille (1968) has reported good thermal neutron yields in a heavy water moderator, and results almost as good (only about 20% less) for a combined heavy and light water moderator. In this case, the target of a 14 Mev neutron generator was surrounded by a 55 mm thickness of heavy water; the remainder of the moderator was light water. For a total fast neutron yield of 10^{10} n/sec, a peak thermal neutron flux of $2·4 \times 10^7$ n/cm²-sec was measured in the heavy water–light water moderator. Further improvements may also result from the use of a medium thickness of uranium around the neutron accelerator target. Such a material will add neutrons to the moderator by fission and $(n, 2n)$ reactions. Another advantage

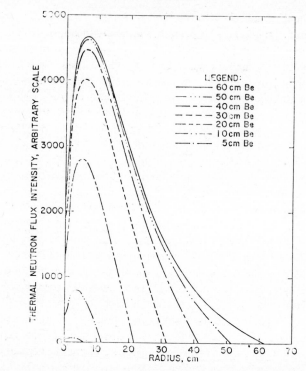

FIG. 6. Thermal neutron flux distributions for the Sb-Be system with 1 cm radius Sb and various thicknesses of beryllium. The peak thermal neutron flux occurs within the 10 cm radius; collimators to bring out the most useful thermal neutron beam should originate at the peak flux location. A total beryllium radius of 40 cm gives about 90% of the possible thermal neutron flux. After Cutforth (1968). (Courtesy, Argonne National Laboratory, Idaho Falls, Idaho, reprinted with the permission of the American Society for Nondestructive Testing.)

of this type of assembly is the reduction of much gamma and x-radiation leakage from the accelerator into the neutron radiographic beam.

Thus far we have considered only sources for thermal neutron radiography. Radiography with neutrons of other energies is possible and sources should be considered. The characteristics of neutron radiography in these various energy ranges are summarized in Table II.

Although radiography with fast neutrons has received relatively little research attention, it may be appropriate to start with this technique because the work that has been done has utilized the accelerator and radioactive sources of the preceding discussion (Criscuolo and Polansky, 1961; Anderson *et al.*, 1964; Tochilin, 1965; Gorbunov and Pekarskii, 1966; Wood, 1967). The ready availability of essentially point sources of fast neutrons is one of

TABLE II

Energies of Neutrons Used For Radiography

Type	Energy Range	Comment
COLD	Below 0·01 ev	This energy range is characterized by high cross sections thereby decreasing the transparency of most materials, but also increasing the efficiency of detection. A particular advantage may be reduced scatter, for example, in iron at energies below 0·005 ev.
THERMAL	0·01 to 0·3 ev	Good discriminatory capability between different materials, and ready availability are advantages that have made radiography with thermal neutrons most used.
EPITHERMAL	0·3 to 10,000 ev	One can achieve excellent discrimination for particular materials by working at an energy of resonance. More generally, the advantages of this method are greater transmission and less scatter in samples containing materials such as hydrogen, and enriched reactor fuel materials.
FAST	10 kev to 20 Mev	This range is attractive because good "point" sources of fast neutrons are available. As yet it has been little studied. At the lower energy end of the spectrum, fast neutron radiography may be able to perform many of the inspections now done with thermal neutrons, but with an arrangement (such as a panoramic technique) which may be attractive to industrial users.

the major advantages associated with this technique. By using the fast neutrons directly, one does not have to moderate and collimate, as is the case when the neutrons from these sources are used for thermal neutron radiography. An example of a radiograph taken with a fast neutron generator of the $T(d, n)$ He^4 type is shown in Fig. 7. For fast neutron radiography, physically small, high-yield neutron sources, such as Cf^{252}, would be particularly useful. Also the accelerator sources, such as $Li^7(p, n)$ Be^7, in which one could vary neutron energy over a wide range,

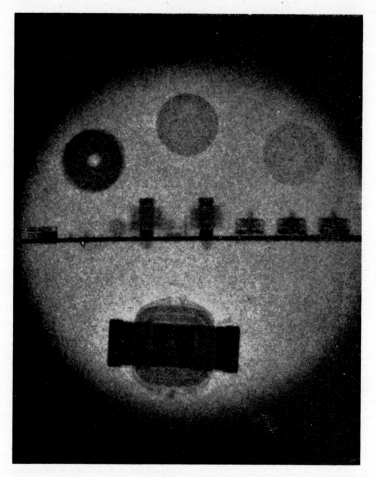

FIG. 7. A positive print of a fast neutron radiograph of a transformer (lower area), a circuit board (middle) and three polyethylene cylinders (upper area). The radiograph was taken at a distance of 5·5 metres from a 4×10^{10} n/sec yield neutron generator (14 Mev); the detector was Kodak Royal Blue x-Ray film and Radelin T I-2 fluorescent screens. The radiograph represents an unknown mixture of fast and lower energy neutrons, and gamma rays. The exposure was 12 min. After Wood (1967). (Courtesy, Kaman Nuclear Division, Kaman Sciences Corp.)

become especially attractive. In that case, one could choose an optimum energy to detect the material of interest in the particular application. For example, lithium could be detected readily if the neutron beam energy corresponded with the lithium resonance at 250 kev.

Beams in the epithermal neutron range have also been used for neutron radiography (Barton, 1965a, b and c; Atkins, 1965; Spowart, 1968b).

These beams have been obtained from nuclear reactors and have involved both relatively wide spectrum beams obtained by filtration (usually with cadmium) and monochromatic beams obtained by diffracting the reactor beam from a crystal. In either case the resultant radiograph was normally a resonance energy neutron radiograph because advantage was taken of a resonance response of a detector such as indium (resonance at 1·46 ev). The reduced scatter one obtains for such radiographs and the enhanced capability for the detection of particular materials (at resonances) makes this technique very useful. In Fig. 8, an example of an epithermal neutron radiograph of an enriched reactor fuel demonstrates the advantage one gains by the decreased attenuation of the fuel material for the higher energy, epithermal neutrons.

Cold neutron beams have also been obtained from a nuclear reactor source, in this case by using bismuth filters (Barton, 1965b, d and e). The increased efficiency of detection of these lower energy neutrons because of the increased cross section of the detector materials (often lithium or boron-containing scintillators) tends to make up for the decreased neutron intensity in the beam. A cold neutron intensity of 3×10^5 n/cm^2-sec (of energy less than 0·005 ev) from the Herald reactor at Aldermaston in England, provided reasonable exposure times. An advantage that has been shown for cold neutron radiography is that one could inspect large thicknesses of iron or steel more usefully, because of the decreased scatter in iron for neutrons of energy less than 0·005 ev.

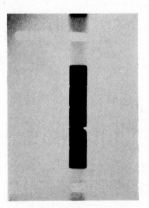 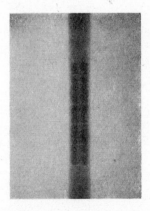

FIG. 8. Positive prints of neutron radiographs of plutonium fuel pellets, 1 cm diameter. The radiograph at the left is a thermal neutron radiograph, taken by a transfer technique with dysprosium. At the right is an epithermal neutron radiograph taken with a cadmium filtered reactor beam, and an indium transfer method. The improved transmission of epithermal neutrons allows one to visualize the annulus in the fuel. After Spowart (1968b). (Courtesy, UKAEA, Dounreay Experimental Reactor Establishment, Scotland.)

B. Detectors for Neutron Radiography

As indicated briefly in the introductory material, most neutron radiography detectors make use of photographic film either directly in the beam or in an autoradiographic technique (Berger, 1962a). For the direct exposure method

TABLE III

Properties of Some Useful Thermal Neutron Radiography Detector Convertors

Material	Isotope	Useful Reaction	Cross-Section for Thermal Neutrons BARNS	Half-Life
Lithium	Li^6	$Li^6(n, \alpha)H^3$	910	—
Boron	B^{10}	$B^{10}(n, \alpha)Li^7$	3,830	—
Rhodium	Rh^{103}	$Rh^{103}(n)Rh^{104m}$	11	4·5 min
		$Rh^{103}(n)Rh^{104}$	139	42 sec
Silver	Ag^{107}	$Ag^{107}(n)Ag^{108}$	35	2·3 min
	Ag^{109}	$Ag^{109}(n)Ag^{110}$	91	24 sec
Cadmium	Cd^{113}	$Cd^{113}(n, \gamma)Cd^{114}$	20,000	—
Indium	In^{115}	$In^{115}(n)In^{116m}$	157	54 min
		$In^{115}(n)In^{116}$	42	14 sec
Samarium	Sm^{149}	$Sm^{149}(n, \gamma)Sm^{150}$	41,000	—
	Sm^{152}	$Sm^{152}(n)Sm^{153}$	210	47 hours
Gadolinium	Gd^{155}	$Gd^{155}(n, \gamma)Gd^{156}$	61,000	—
	Gd^{157}	$Gd^{157}(n, \gamma)Gd^{158}$	254,000	—
Dysprosium	Dy^{164}	$Dy^{164}(n)Dy^{165m}$	2,200	1·25 min
		$Dy^{164}(n)Dy^{165}$	800	140 min
Gold	Au^{197}	$Au^{197}(n)Au^{198}$	98·8	2·7 days

there are many prompt (n, γ) reaction materials which can be used as convertor foils to enhance the neutron response of the film itself. Materials which yield short half-life radioactivities can also be used directly with the film, since the activities decay and stimulate the film during and shortly after the neutron exposure. Materials which undergo prompt (n, α) reactions, such as Li^6 and B^{10}, have also been shown to be useful (Wang *et al.*, 1962; Sun *et al.*, 1956). In Table III, many of the properties of useful convertor materials for either direct exposure or transfer method thermal neutron radiography are summarized.

The (n, α) materials can be used directly with film (Berger, 1963), or in the faster technique in which the alpha emitter is intermixed with a phosphor powder; the resulting neutron scintillator is then used with a film sensitive to light. Since the range of the alpha particle is very short, the effective thickness of an alpha convertor used directly with film, or the effective absorption of a loaded emulsion, is relatively low. However, if the alpha emitter is mixed with a phosphor powder, such as $ZnS(Ag)$, the light stimulated from phosphor grains adjacent to the alpha emitting materials provides a very fast neutron detection method. These scintillators, when employed with a fast, light-sensitive film, such as Kodak (U.S.A.) x-Ray Type F, are

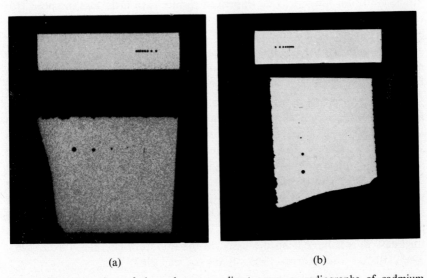

(a) (b)

FIG. 9. Negative prints of thermal neutron direct exposure radiographs of cadmium (upper) and gadolinium (lower) test objects. Fig. 9a was taken with a scintillator, Fig. 9b was taken with a gadolinium metal foil convertor. The holes in the cadmium object are 0·5 mm in diameter; the spacing of the holes decreases to a value of about 30 microns. (Courtesy, Argonne National Laboratory, Argonne, Illinois.)

in the order of 30 times faster in thermal neutron response as compared to a metal conversion foil combination used with a fast x-ray film (Berger, 1962a, 1965, 1968a). A detector exposure of only 10^5 n/cm² can yield a useful radiograph with the fast scintillator techniques. The speed advantage is sometimes offset by the fact that such images have a grainy appearance, as shown in Fig. 9. This is due to the phosphor grains, and in some cases to statistical variations because of the short exposure. The use of a slower film can improve the latter limitation.

The best-quality neutron radiographs in terms of resolution and contrast sensitivity have been obtained with slower response metal conversion foils. Gadolinium foils have been most used for direct exposure thermal neutron radiography. Convertor foil thickness of 12·5 to 25 microns provide good-quality, high-resolution results. Gadolinium is useful because of its high thermal neutron cross section and because the emission of a relatively low energy internal conversion electron (70 kev or less) provides high-resolution radiographs (because of the short range of the electron in the emulsion). Best results have been obtained by using a single gadolinium metal foil as a back screen (on the side of the film opposite from the entering neutron beam). Total detector exposures in the order of 3×10^9 n/cm² provide high-quality radiographs on fine-grain x-ray film (such as Kodak-U.S.A. Type R). Single-emulsion films provide improved results with little loss in speed (for the identical emulsion) because most of the film response occurs in the emulsion facing the gadolinium foil. Very few electrons emitted from the gadolinium penetrate beyond the adjacent emulsion. This technique also offers an advantage in relative neutron-gamma response for this detector, because of the decreased gamma response of the single-emulsion film.

Increased speed to neutrons can be gained by employing two convertor foils, one on each side of a double emulsion film, as is the normal case in x-radiography. Double gadolinium screens (25 microns on the neutron source side and 50 microns on the backside of the film) increase the speed about 50% over that of a single gadolinium back screen technique. A useful double screen method, which provides about twice the speed of a single gadolinium back screen, makes use of a 250 micron thick rhodium front screen and a 50 micron gadolinium back screen. Speed, resolution and contrast characteristics of these and other convertor materials in various thickness ranges have been discussed in the literature (Berger, 1962a, 1963, 1965; Blanks and Morris, 1966; Blanks et al., 1967).

For transfer neutron radiography, a method which can provide true neutron radiographs even in the presence of intense gamma radiation, the most common convertor materials have been indium (in the thickness range 50 to 250 microns), dysprosium (125 to 250 microns), and to a lesser extent gold (50 to 125 microns) and rhodium (125 to 250 microns in thickness). Indium is relatively inexpensive, has a convenient half-life, and provides reasonable speed. In a typical reactor beam intensity of 10^7 thermal n/cm²-sec, a 5 to 10 minute exposure of indium can provide good results when transferred to a medium-speed x-ray film for several half-lives. With fast films, indium transfer methods can be used with thermal neutron intensities in the order of 5×10^4 n/cm²-sec. Dysprosium, with its greater cross section and longer half-life, can provide useful neutron radiographs at even lower neutron intensities; radiographs made with intensities as low as 6×10^3 n/cm²-sec have been demonstrated (Berger, 1968a). Gold, with its very long half-life, is not as convenient to use as the above transfer materials. However, it does appear to offer a small improvement in radiographic resolution as compared to other transfer materials (Berger, 1963). Rhodium has been a very convenient transfer material to use with reactor beams (a beam intensity in the order of 5×10^6 n/cm²-sec seems necessary for good rhodium transfer results). The short half-life can provide radiographic results very quickly on fast films. This has proved useful for alignment radiographs with Polaroid films (Berger, 1965, 1968b). Figures 10 and 11 show some properties of the rhodium-Polaroid film technique, and a sample radiograph.

In terms of resolution and contrast, these photographic detector methods for thermal neutron radiography provide results comparable to those

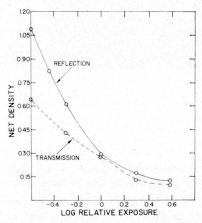

FIG. 10. Densities of Polaroid Type TLX Film after transfer exposure to a thermal neutron activated rhodium foil. The thermal neutron intensity was 2×10^7 n/cm²-sec. A time of 40 sec was permitted between the end of the neutron exposure and the beginning of the autoradiographic decay. The decay period in each case was one minute. A fluorescent screen, Radelin Type UD, was used with the autoradiographic exposure to speed the result. A one-minute neutron exposure, typically the exposure used to produce neutron radiographs by this method, was used as the 0 log relative exposure base. (Courtesy, Argonne National Laboratory, Argonne, Illinois.)

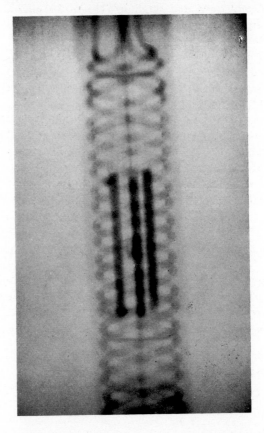

Fig. 11. A print of a Polaroid film, rhodium transfer thermal neutron radiograph taken by the method described in Fig. 10. The radioactive reactor fuel specimens (three dark vertical lines) can be easily observed within the surrounding heating coil. (Courtesy, Argonne National Laboratory, Argonne, Illinois.)

obtained with gamma and x-radiography. High-contrast resolution results with a thin gadolinium convertor foil are in the order of 10 microns or better (Berger, 1963). Resolution properties of the other photographic detectors are somewhat poorer because of the greater range in the emulsion of the radiation emitted from the convertor. Nevertheless, high-contrast resolution capabilities are in the order of 25–50 microns for many of the detection methods (Berger, 1963; Inouye *et al.*, 1965). Contrast results have been obtained, for the most part (Berger and Kraska, 1964; Schultz and Leavitt, 1965), with metal samples and known changes in thickness, as is usually done in x-radiography. On this basis, contrast sensitivities in the order of

1-3% have been demonstrated for materials such as steel, lead, and natural uranium. This type of inspection, however, is only one of the areas in which neutron radiography may be useful. Therefore, studies are in progress to prepare radiographic test objects which may be more appropriate for contrast studies in neutron radiography (Barton, 1965f, 1967c, 1968b) since contrast between different materials is often of special concern in neutron radiography applications.

For the detection of neutrons in other energy ranges, essentially the same techniques as outlined above have been used. Scintillators containing lithium and boron have provided good results for cold neutron radiography. In the resonance energy region, indium has been the primary detector because of its excellent response at the neutron energy of 1·46 ev. Other detectors for resonance neutron radiography are listed in Table IV. In the limited work

TABLE IV

Potentially Useful Detectors For Epithermal Neutrons

Material	Isotope	Half-Life	Neutron Energy of Resonance
Indium	In^{115}	54 min	1·46 ev
Gold	Au^{197}	2·7 days	4·9 ev
Tungsten	W^{186}	24 hours	18·8 ev
Lanthanum	La^{139}	40 hours	73·5 ev
Manganese	Mn^{55}	2·56 hours	337 ev

that has been done with fast neutron radiography, the detection methods have involved scintillators used either with counters or film (Criscuolo and Polansky, 1961; Anderson et al., 1964; Tochilin, 1965; Gorbunov and Pekarskii, 1966; Wood, 1967; Polansky and Criscuolo, 1964). Although the direct scintillator-film method does show response to other radiations in the fast neutron beam, such as gamma rays and lower energy neutrons, this technique does provide a fast response technique which will be useful in some circumstances. The most widely used detection method has been the fluorescent intensifying screens normally used for x-radiography and a fast, light-sensitive x-ray film; an example is shown in Fig. 7. The (n, p) reaction of the hydrogen in the emulsion and in the plastic binders for the phosphor in the fluorescent screen provides the response. Total exposures in the order of 10^7 n/cm^2 have provided good results with 14 Mev neutrons (Criscuolo and Polansky, 1961; Wood, 1967; Polansky and Criscuolo, 1964). Film used without the fluorescent intensifying screens requires a longer exposure by one or more orders of magnitude (Ehrlich, 1960).

Another detection possibility for fast neutrons is a transfer technique. Threshold response detectors employing a reaction such as $Cu^{63}(n, 2n)\,Cu^{62}$ which responds to neutrons of energy above about 11 Mev, for example, can provide useful transfer fast neutron radiographs if the neutron intensity is sufficiently high (in the order of 10^7 n/cm^2 sec or more). In this example, it is appropriate to point out that the detector would show little response to scattered neutrons when used with a 14 Mev neutron generator, because most of the scattered neutrons would have an energy less than the threshold response of the detector. Other threshold detector possibilities and their characteristics are given in Table V.

A different type of detection possibility for fast neutrons, and perhaps for neutrons of other energies, is the track-etch technique (Berger, 1968a;

TABLE V

Potentially Useful Threshold Detectors For Fast Neutron Radiography

Material	Reaction	Half-Life	Threshold Neutron Energy For Reaction	Cross Section For Reaction In Millibarns at 3 Mev	14 Mev
Sulphur	$S^{32}(n, p)P^{32}$	14 days	1·7 Mev	200	220
Phosphorus	$P^{31}(n, p)Si^{32}$	2·65 hours	1·8 Mev	80	140
Aluminum	$Al^{27}(n, p)Mg^{27}$	9·5 min	2·6 Mev	low	65
	$Al^{27}(n, \alpha)Na^{24}$	15 hours	6·1 Mev	—	116
Silicon	$Si^{28}(n, p)Al^{28}$	2·3 min	4·5 Mev	—	329
Iron	$Fe^{56}(n, p)Mn^{56}$	2·58 hours	5·0 Mev	—	120
Copper	$Cu^{63}(n, 2n)Cu^{62}$	10 min	11 Mev	—	460

Fleischer *et al.*, 1965; Becker, 1966; Furman *et al.*, 1966; Berger and Kraska, 1967). Dielectric materials such as plastics, glass and mica have been used to display tracks of charged particles by selective etching of the areas damaged by the particle. Neutron images, such as the thermal neutron radiograph shown in Fig. 12, have been prepared by this technique. In this case, the detector was a sheet of cellulose nitrate plastic exposed to the neutron beam next to a U^{235} foil. Thermal neutrons produce fission in the foil and the fission fragments damage the plastic so that the etch reveals the image. The method is attractive for fast neutron radiography because threshold

FIG. 12. A print of a track-etch thermal neutron radiograph of several cadmium objects. The object at the right is the same cadmium test object described in Fig. 9. At the left are shadows of 25, 50 and 75 micron thick cadmium, top to bottom, respectively. The total exposure for this radiograph on cellulose nitrate film (with U^{235}) was 10^9 n/cm². (Courtesy, Argonne National Laboratory, Argonne, Illinois.)

fission detectors such as Th^{232} or U^{238}, or some of the (n, p) or (n, α) materials listed in Table V, could provide useful fast neutron images without response to gamma rays or to lower energy neutrons. For visual images, as opposed to the counting of tracks, as in the case for track-etch applications in dosimetry, geological ageing, etc., one needs several million tracks per cm^2. For fission fragment detectors, one can calculate the necessary exposure from the relationship given by Becker (1966), that one obtains $1 \cdot 3 \times 10^{-5}$ tracks per neutron per barn. A good-quality, track-etch radiograph such as that shown in Fig. 12 will have a track density of about 6×10^6 tracks/cm^2; a thermal neutron exposure of about 10^9 n/cm^2 is required for a plastic exposed with an enriched U^{235} foil. The fission cross section for U^{235} for thermal neutrons is about 580 barns. For fast neutron detection by this method, however, the low cross sections indicate that total exposures of 10^{11} n/cm^2 or more will be necessary.

A dynamic response detection method for thermal neutron radiography has been demonstrated recently at several laboratories. This opens up a wide variety of application possibilities for neutron radiography, since the technique can now be considered for motion study problems and for inspections on a continuous, moving basis. The detection methods all employ neutron scintillators and television techniques. One can observe a neutron scintillator directly with a sensitive television camera and display useful neutron images (Farny et al., 1968). A second approach is to employ a neutron scintillator optically coupled to a light image intensifier tube and a television system (Spowart, 1968a; Kawasaki, 1968). In this case, the television camera can be a relatively inexpensive Vidicon type, because of the high gain yielded by the light image intensifier. This is also the case if the neutron scintillator is incorporated directly into an image intensifier tube (Niklas et al., 1966; Anon, 1966; Berger, 1966).

The neutron television system can provide contrast sensitivities in the order of 6%, can follow high-contrast motion as fast as 15 meters/min with little image blur, and can display high-contrast details in the order of 0·125 to 0·50 mm. These values are dependent upon reasonable statistical properties for the image presentation (Berger, 1966, 1968a). This implies an incident thermal neutron intensity of 10^6 to 10^7 n/cm^2-sec for picture presentation at normal television frame rates (25 to 30 frames/sec), or longer integration time for an image if lower neutron intensities are used. An example of the effect of the statistical variation on the quality of a neutron television image is shown in Fig. 13. Motion detection, however, is the most important property of the neutron television approach. An example of a motion sequence photographed with a motion picture camera from the television monitor is shown in Fig. 14.

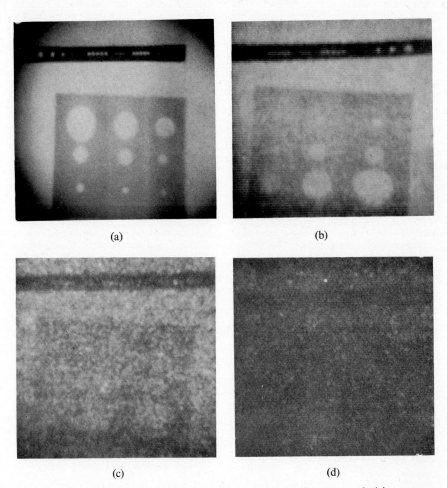

(a)   (b)

(c)   (d)

FIG. 13. Photographs of the television screen of a thermal neutron television system. Fig. 13a shows a cadmium test strip containing holes (upper). The series of holes near the middle are 1 mm dia., spaced at 0·5 mm. Below that is a Masonite stepped object containing holes in each step. This photograph was taken with a 30 frame per second TV system; the neutron intensity was 2×10^7 n/cm²-sec. Figures 13b, c, and d show the same objects (in a slightly different orientation) but at neutron intensities of 2×10^6, 10^5 and 10^4 n/cm²-sec respectively. The time for each picture was about a TV frame time. (Courtesy, Argonne National Laboratory, Argonne, Illinois.)

Other detection methods (Berger, 1965) for neutron images have been discussed in the literature. Many of these require further development work before they can be considered practical detectors for neutron radiography. The spark counter (Watts, 1962) has been investigated as a neutron image

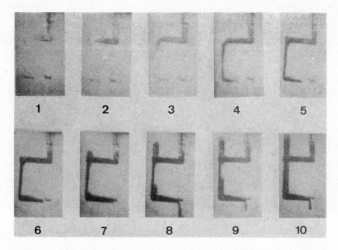

Fig. 14. Prints of a neutron television sequence showing the pouring of Wood's metal into an aluminum mould (mould dia. was about 0·63 cm). The time for this sequence of pictures was 0·66 seconds. (Courtesy, Argonne National Laboratory, Argonne, Illinois.)

detector. The counter is an assembly containing an enriched B^{10} layer located near an array of wires maintained at high potential. Alpha particles emitted from the boron upon neutron bombardment ionize the gas in the counter and a spark results in the region of the ionization. The sparks can be photographed to yield an image. The counter investigated by Watts (1962) had a resolution of about 1 mm (as set by the spacing of the wires). A neutron exposure of a few seconds in a thermal neutron intensity of 5×10^5 n/cm^2-sec could yield an extended grey scale image. The method is attractive because it is sensitive and because it offers excellent discrimination against gamma rays.

A thermoluminescent method for detecting thermal neutron images has also been described (Berger, 1968a; Kastner et al., 1967). The light released by an extended area thermoluminescent neutron detector (such as LiF), upon heating after exposure, can be photographed to display the image. At the present state of development, this technique requires a total thermal neutron exposure of about 2×10^9 n/cm^2. The images have a resolution capability of about 125 microns and display a contrast sensitivity of 10%. The method is of interest because one can accumulate image information for long periods, as is the case for direct exposure photographic methods. However, the low density of many of the thermoluminescent detection materials may offer this technique an advantage in terms of relative neutron-gamma response. Another advantage of the thermoluminescent approach, as compared to direct exposure scintillator methods, is that one can accumulate image information long enough to obtain a large light release, thereby eliminating reciprocity-law failure problems which may be encountered with scintillators at low intensities (Berger, 1962b).

Other detection methods (Berger, 1965), such as mechanical scanning with a masked or small neutron detector, appear feasible, but little neutron radiography work with such methods has been reported. Detectors such as scintillation counters or semiconductors could be used for this purpose. Although the image would take some time to scan (the time would be determined by the area and resolution desired), the technique would offer advantages in discrimination capability against gamma rays and against scattered neutrons. Interest in this technique appears to be increasing, particularly for the difficult problem of detecting neutrons of higher energy in biological radiographic applications, where neutron scatter is a significant problem (Brown and Parks, 1968; Flynn, 1968).

III. Applications of Neutron Radiography

A number of applications of neutron radiography have already been illustrated in many of the figures in earlier parts of this report and, more important, a brief account of the potential areas of application was given in the introductory material. At the present time, the radiographic examination of irradiated reactor fuel material remains the most widely used area of application for neutron radiography (Barton, 1967b; Barton and Perves, 1966; Barton et al., 1969; Beck and Berger, 1964; Berger and Beck, 1963; Carbiener, 1966; Cutforth and Aquino, 1967; Ogawa et al., 1966; Spowart, 1968b). The primary advantage for using neutron radiography for this inspection is that the radioactivity of the sample causes no problem except for personnel shielding during handling. Shielding problems are easily solved (Beck and Berger, 1964), however, and the increasing use of swimming pool reactors for this application reflects the convenient shielding problem solution offered by that approach (Barton and Perves, 1966; Carbiener, 1966). Hot cell examinations of fuel with radioactive or accelerator neutron sources also promises to be a very useful technique (Barton, 1967b; Barton et al., 1969; Cutforth and Aquino, 1967). At least one such facility, employing an Sb–Be source, is in operation (Cutforth and Aquino, 1967).

Another advantage for using thermal neutron radiography for this inspection is that one obtains good neutron transmission through many heavy reactor fuels. The half-value-layer for thermal neutrons for natural uranium (Thewlis, 1956; Berger and Kraska, 1964) is in the order of 1·3 cm. This advantage is no longer true for the inspection of enriched uranium or plutonium fuels, however. In this case, the use of epithermal neutrons offers a much superior technique (Barton *et al.*, 1969; Spowart, 1968b), not only because of the improved transmission but also because of the reduced scatter for the higher energy neutrons. The improvement in image quality was demonstrated in Fig. 8.

The possible use of neutron radiographic inspection of other radioactive devices has been suggested, but no real applications of this type are known to this author. Certainly as the technology of non-reactor neutron sources improves, this technique may find applications in hot cell inspections of welds on sealed radioactive sources, and for similar inspection problems.

The next most widely used area of application involves the neutron radiographic capability for imaging light materials made from hydrogen, lithium or boron within heavy assemblies. This is especially attractive since such inspections often present insurmountable problems for x-radiography techniques because of the high attenuation of the surrounding heavy metal. Many of the small explosive devices used in the aerospace field have provided a stimulus for such neutron radiography work (Rhoten and Cary, 1966;

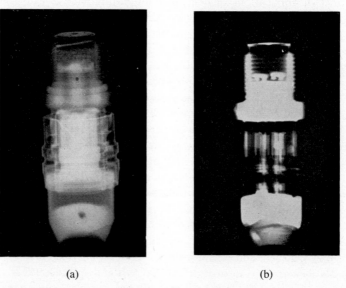

(a)   (b)

FIG. 15. Negative prints of radiographs of an explosive device. Fig. 15a is a thermal neutron radiograph; it shows the explosive (dotted material) through the threaded region in the upper part of the sample. The tilted white line inside the upper stainless steel cap is the neutron image of filter paper. The x-radiograph, Fig. 15b, does not show the explosive or paper. However, it does image the metal components in the central area, components which are masked by the neutron attenuation of surrounding plastic on the neutron radiograph. (Courtesy, Argonne National Laboratory, Argonne, Illinois.)

Burkdoll, 1968; Berger and Drexelius, 1967). These explosive devices typically contain hydrogen, and in many cases boron (Burkdoll, 1968). The explosives are often encased in stainless steel or lead, a fact which makes x-ray inspection difficult. An example which illustrates the capability of neutron radiography for such inspections, as compared to x-radiography,

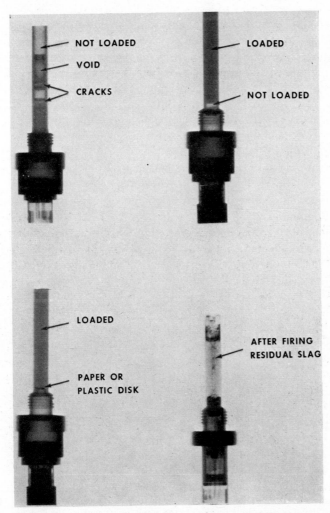

FIG. 16. Positive print of neutron radiographs of several pyrotechnic time delays. In each case the explosive material is enclosed in a metal case. It is interesting to observe that one can locate voids, cracks and foreign material in the explosive, and that even the residual explosive remaining after firing can be observed (as in the lower right). (Courtesy, General Electric Co., Pleasanton, California.)

is shown in Fig. 15. Examples of the types of information one can gain from neutron radiographic examinations of explosive devices are shown in Fig. 16.

There are many other similar application problems which involve hydrogenous material (Berger, 1965) in the form of rubber (Garrett and Morris, 1967), plastics, wood, or adhesives (Perry, 1967). The problem of inspecting adhesive bonded devices, such as are used in aerospace components, is an important one. Figure 3 illustrated the sensitivity of neutron radiography for visualizing thin layers of adhesives on metals, and for the important visualization of the build-up of adhesive along upright sections of a honeycomb structure. Although it is true that the presence of such a fillet of adhesive does not guarantee that a good bond has been made, the absence of such a fillet almost always indicates the lack of a good bond. In addition to the detection of these adhesive fillets, the bubble-like structure of the neutron image of the adhesive gives much information about the wetting and cure of the adhesive resin (Perry, 1967).

Recently, work involving the thermal neutron detection of hydrogenous material has been accomplished in two difficult and useful areas. One of these is the radiographic detection of potentially dangerous hydride formations in titanium assemblies (Basl et al., 1968). Observation of the hydride was accomplished with a high degree of confidence in titanium test samples as thick as 0·63 cm, and containing 500 to 1000 ppm hydride contamination. In actual titanium hardware, hydride deposits were radiographically detected in flat, threaded, and weld areas. Some examples of neutron radiographs

FIG. 17. Positive prints of neutron radiographs of titanium weld specimens. The titanium in each case was 0·63 cm thick, and approximately 7·5 cm in length. Hydride can be detected by the dark areas in the welds (at the small, upper ends of the samples) in samples CP-4 (left) and CP-3 (second from right). The very dark deposits observed on sample CP-4 were later analyzed to be due to hydride concentrations of 500 to 1000 ppm. Deposits having concentrations less than 100 ppm were observed. (Courtesy, General Electric Co., Pleasanton, California.)

of titanium weld specimens showing the detection of hydride, are given in Fig. 17. Similar results have been reported for hydride detection in zirconium (Berger, 1965; Chountas and Rauch, 1968). The second interesting application concerns the study of hydrogen diffusion into deuterated materials (Chountas and Rauch, 1968). This promises to be useful for the study of chemical, osmotic, and biological processes.

Of course the high sensitivity of thermal neutrons to hydrogen has offered the possibility that neutron radiography might be useful in some medical and biological problems (Berger, 1965; Anderson *et al.*, 1964; Atkins, 1965; Barton, 1965b; Brown and Parks, 1968; Flynn, 1968). There appear to be three types of such problems for which neutron radiography may offer some advantage (Barton, 1967d). One centers on the fact that neutrons are easily transmitted through bone, while x-rays in the medical energy region are highly attenuated by bone. Therefore, neutron techniques offer the possibility for radiographic viewing of tissue which might be masked by bone in an x-ray examination. Also, neutrons may offer some advantage in examining bone itself, especially in cases in which there is relatively little tissue to mask the neutron image.

The use of neutron techniques also opens the possibility for many different types of contrast agents for the study of biological systems (Atkins,

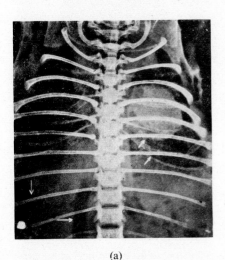 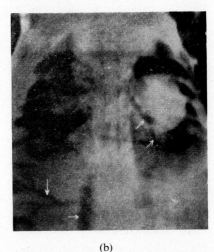

(a)   (b)

Fig. 18. Negative prints of radiographs of the chest of a dead rat. The x-radiograph at the left shows much detail of the backbone and ribs. The thermal neutron radiograph, right, shows little of the bone, but it does show the tissue hidden by the bone and also shows air in several arteries and veins (small white arrows). After Brown and Parks (1969). (Courtesy, Savannah River Laboratory, Aiken, S. Carolina, reprinted with the permission of the Am. J. of Roentgenology, Radium Therapy and Nuclear Medicine.)

1965).   Absorbing materials which could be introduced into the system include compounds made from materials such as boron, cadmium, samarium, europium and gadolinium.   These would serve the same purpose as do barium salt contrast agents in medical x-radiography.   However, a recent study (Brown and Parks, 1969) indicated that, in many cases, a negative contrast agent, one which displaces material and absorbs less rather than

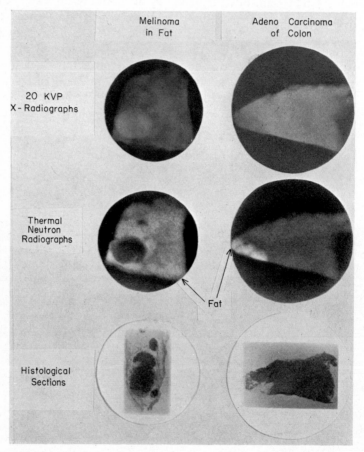

FIG. 19. Negative prints of radiographs, and photographs of deuterated tumor specimens. The improved radiographic differentiation of the tumor area from the surrounding fatty tissue area results from the take-up of D_2O by the tumor. On the x-radiographs (upper row) it makes little difference, while on the thermal neutron radiographs (middle row) the increased neutron transparency of the D_2O as compared to H_2O outlines the tumor very well, in agreement with the histological sections shown in the lower row of the illustration. After Parks and Brown, private communication, 1969. (Courtesy, Savannah River Laboratory, Aiken, South Carolina.)

more, may be more useful in many biological problems. The displacement of hydrogenous material by air (or other gases) or by deuterated material appears to offer a real advantage. Figure 18 illustrates the neutron radiographic visualization of air in a biological system, even through bone areas, as compared to x-radiography (Brown and Parks, 1969).

A third possible biological advantage of neutron techniques concerns the high sensitivity to hydrogen. It is possible that one could differentiate radiographically between normal and cancerous tissue if the hydrogen content difference was sufficiently great (Barton, 1967d).

Radiographic differences, apparently due to hydrogen content variations, have been observed between neutron and x-radiographs for tumor specimens (Brown and Parks, 1969). A more reliable technique for tumor differentiation (in some cases), calls for the deuteration of the suspect tissue sample, followed by radiography. Tumor tissues readily deuterate while fatty tissues do not; therefore, after deuteration the tumor appears more transparent to thermal neutrons than the surrounding fatty tissue. An example is shown in Fig. 19 (Brown and Parks, 1968).

Since much of the thermal neutron attenuation of hydrogen is caused by scatter, it is difficult to obtain good radiographic images of thick hydrogenous samples (Berger and Kraska, 1964). The use of anti-scatter grids, as are often used in medical x-radiography, offers some help for this problem by eliminating many of the scattered rays which would otherwise reach the detector (Berger, 1965). Boron and cadmium have been used for the absorbing grid lines while aluminum, which is essentially transparent to thermal neutrons, has made a good material for the transmitting regions (Atkins, 1965; Brown and Parks, 1969; Parks and Brown, 1969). Some advantage can also be gained by using epithermal neutrons since the scatter cross section for hydrogen is somewhat lower. In that case, too, a grid with a material such as indium for the absorbing lines offers an improvement in image quality (Parks and Brown, 1969).

Boron, lithium, cadmium, indium and several rare earth materials are some other elements that have good attenuation characteristics for thermal neutrons, a fact which makes these materials particularly susceptible to neutron radiographic inspection. The observation of the distribution of boron compounds in a mixture to be used for a reactor control element is one example (Berger, 1965) of the type of problem which readily lends itself to neutron radiographic inspection. Similarly, one could hope to observe the distribution of boron in a boron–steel sample. Boron itself, also, can be usefully inspected under some circumstances. In Fig. 20, for example, neutron images of several boron filament composite material samples are presented. Boron filaments, 0·1 mm in diameter, can be readily observed

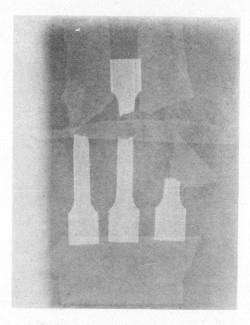

FIG. 20. A negative print of a thermal neutron radiograph of boron-filament composite material samples. The 0·1 mm dia. boron filaments are contained in an aluminum matrix. The breaks in the filaments in the sample at lower right are easily detected, as are the filament spacing variations on the sample at the left. The odd shapes are neutron images of masking tape used to hold the samples in place. (Courtesy, Argonne National Laboratory, Argonne, Illinois.)

in a metal matrix of aluminum or nickel. The filaments can be inspected for straightness, uniformity, or breakage, for composite material thicknesses up to perhaps three or four layers of boron fibers. It has also been possible to observe the interaction of boron with the matrix material in annealed samples (Holloway *et al.*, 1967).

Useful observations of lithium in heavy metal corrosion samples have also been made by neutron radiography, as shown in Fig. 21. Samples of lithium in containers of materials such as molybdenum or tungsten can be easily visualized by thermal neutron radiography. It is expected that comparisons of radiographs made periodically during the environmental tests to observe the corrosive effects of the lithium on the capsule material will be useful for helping to determine good stopping points for the corrosion tests.

One could readily inspect cadmium-plated objects to determine the cadmium thickness by neutron techniques (see Fig. 12). The high cross

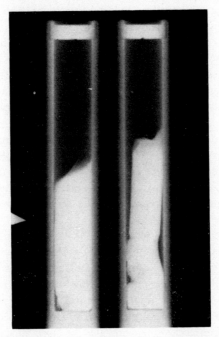

FIG. 21. A negative print of a thermal neutron radiograph of a molybdenum jacketed lithium corrosion sample. The lithium inside each of the two capsules can be readily observed. (Courtesy, Argonne National Laboratory, Argonne, Illinois.)

section of cadmium has also made it a useful material for alloy studies with other lower cross section metals. Cadmium alloys have been studied by neutron microradiographic techniques in two investigations (Epplesheimer and Arment, 1965; Fahmy, 1967). Contact radiographic techniques with thin samples inside the exposure cassette have yielded neutron images having sufficient sharpness and contrast so that viewing magnifications of about 100 times can be used.

In Fig. 1 it was shown that one can radiographically discriminate between Cd^{113} and other, lower cross section cadmium isotopes. Similarly one can study and observe other high cross section isotopes in the presence of lower cross section isotopes of the same material. Studies of this type have involved cadmium, gadolinium and the radiographic discrimination between U^{238} and U^{235}. This latter capability has led to the suggestion that neutron radiography might be useful in nuclear safeguard applications (Boland, 1968).

The use of contrast agents in neutron radiography has been discussed above in relation to medical problems. Of course, this technique can also be applied to industrial problems. An example concerns the inspection of turbine blades (Berger, 1965; Watts, 1962). These are often made from dense materials, thereby making it difficult to observe the integrity of the open cooling channel in the blade by x-ray techniques. It is so difficult that, in some cases, the channel is filled with mercury, as a contrast agent. For neutrons, the relatively easy transmission through the heavy metal, and the possible use of water as a contrast agent, very much simplifies the problem.

The fact that neutrons are easily transmitted through heavy metals such as lead, bismuth and natural uranium, has led to the suggestion that neutrons could be used to inspect large thicknesses of such materials. Good contrast (1 to 3% in many cases) can be observed, and exposure times can be short even for fairly large thicknesses (Berger and Kraska, 1964; Schultz and Leavitt, 1965). For lead, for example, the half-value-layer for a thermal neutron beam is about 2·7 cm. Thicknesses up to 30 cm have been examined (Kraska and Berger, 1968). Scatter problems do occur when the radiographic sample thicknesses become great. For that reason, it may be advisable to employ a neutron energy where scatter cross sections are low, as in the case of cold neutrons and steel (Barton, 1965d and e).

Thus far, the discussion of applications has centered on static problems for which a radiographic exposure time of several minutes or longer causes no difficulty. Neutron techniques can also be used in dynamic inspection problems. The observation of the flow of liquids in metal containers is one problem which readily lends itself to a dynamic neutron radiographic technique. Such a neutron study has been made in order to observe the flow of water in a steel heat-pipe (Moss, 1967). Similarly the motion capability of the neutron television technique may make it useful for the study of the process of casting heavy metals, such as uranium reactor fuel pins. The high attenuation of the uranium would make it difficult to employ established x-ray methods for such studies. With neutrons, transmission problems are less critical. An example of a casting sequence recorded by neutron television was shown in Fig. 14.

The television approach also makes it feasible to consider inspections on a highly repetitive basis by passing objects continuously before the neutron camera. The explosive device shown in Fig. 15, for example, could be examined by such a technique to reveal the presence or absence of the explosive. There are perhaps many applications such as that for which the contrast and resolution capabilities of the television technique would not impose a limitation.

The television technique is one method for stopping action at some point in order that a dynamic problem may be studied. Another possible technique is a short radiation pulse. One can obtain very short neutron pulses in the fast neutron energy range. In the thermal energy range, pulse widths less than several milliseconds are difficult to obtain. Nevertheless, this may prove to be a useful technique for some applications.

This brief review of applications is meant to emphasize the types of problems for which neutron radiography can be used. Certainly many other application problems involving such diverse items as castings (Radwan and Sikorska, 1965), brazed objects (Kanno, 1965), art objects (Barton, 1965g) and electronic components (Wilkinson, 1968) have been studied. However, the basic concepts of these problems are similar to those application problems discussed in this review. For new and current application studies the reader is referred to the literature and to the Neutron Radiography Newsletter (Berger *et al.*, 1964–69).

IV. FURTHER RESEARCH ON NEUTRON RADIOGRAPHY

Neutron radiography is such a relatively young field of endeavour that advances in many areas are certain to come about. Although radiography with thermal neutrons is now a well-established technique, it does, for the most part, depend upon nuclear reactors as a source of neutrons. Future work will have to emphasize the development of non-reactor neutron sources if industry is to really make use of this technique. One possibility is the development of fast neutron radiography in order to make use of the sources which are now available to industry. However, work on detectors, techniques and application studies remains to be done. The methods for moderating and collimating neutrons from non-reactor sources in particular, seem to hold the important key to the possible industrial use of thermal neutron radiography. Techniques must be optimized by proper choice of moderator material and geometry, collimator design and position, and hopefully by the use of some intensification technique such as an $(n, 2n)$ or fission reaction, in order to obtain in the thermal neutron radiographic beam all the neutrons which are possible.

In this connection, with source intensities which may be low, improved detection methods also merit attention. The use of separated isotopic conversion layers of Gd^{157} for direct exposure techniques, and Dy^{164} for transfer exposures, offers good potential for improvement in speed of detection. In the case of gadolinium, the improved speed for thermal neutron detection should also reduce effects of secondary radiation such as gamma rays, thereby leading to improved contrast. The use of enriched Dy^{164} transfer foils should make it possible to detect thermal neutron intensities approaching 10^3 n/cm²-sec. Scintillators which have better detection efficiency or which make better statistical use of the available neutrons will play an important role in the utilization of small sources for neutron radiography. Such detectors are especially attractive because of the good relative neutron-gamma response, and because the detector is capable of long integration time. Other improvements in scintillator techniques, such as the reduction of the effects of multiple light scattering in glass scintillators and the corresponding image quality improvement, are already the subjects of current studies (Spowart, 1968c).

The reactor is certain to continue to be a major source of neutrons for radiography. The reactor's high neutron intensity, which means that good collimation can be conveniently used, will probably continue to present an advantage over other neutron sources. Also, the reactor can be conveniently used to improve techniques for radiography with cold and epithermal neutrons, both of which are techniques which have been relatively little explored. As the demand for neutron radiography increases, one would think that special reactors, designed specifically for neutron radiography, would become available. Such reactors could provide many beams for multi-inspection problems and also could be optimized for cold or epithermal neutrons for those special applications.

The dynamic technique also can be improved, particularly by the development of systems offering improved relative neutron-gamma response and resolution. The latter property would be especially attractive for industrial applications. Such systems would probably have to include storage capability in order to minimize statistical variation problems. Storage times of a few seconds or even minutes would not seriously degrade the usefulness of such an inspection technique for many industrial problems.

It is clear that neutron radiography has earned a place in the group of nondestructive testing methods. At the present time, the use of neutron radiography is limited to special application problems and to special organizations which have neutron sources (mostly reactors) available. However, it is also clear that techniques for the use of non-reactor neutron sources for radiography are now receiving much research attention and that useful non-reactor neutron source techniques have already been demonstrated. Only the future holds the key to the possibility of expansion of the use of neutron radiography to more general industrial and/or medical use.

V. ACKNOWLEDGEMENTS

The author is grateful to Argonne National Laboratory and the U.S. Atomic Energy Commission for granting permission to publish this paper, and to use many illustrations. Acknowledgment is also given to the many other organizations who granted permission for illustrations.

REFERENCES

Anderson, J., Osborn, S. B. and Tomlinson, R. W. S. (1964). Fast neutron radiography in man. *Br. J. Radiol.* **37**, 957.

Anon. (1966). "Convertisseur d'Image Pour Neutrons Thermique." Report DDP 3645, Compagnie Générale de Télégraphie Sans Fils, Domaine de Courbeville (Essonne), France.

Atkins, H. L. (1965). Biological application of neutron radiography. *Mater. Evaluation* **23**, 453.

Barkalov, S. N., Moshkin, G. N., Tyufyakov, N. D., Shtan, A. S. and Yaskevich, V. S. (1970). Cf^{252} neutron source. *Atomn. Energ.*, to be published.

Barton, J. P. (1965a). Radiography with resonance energy neutrons. *Physics Med. Biol.* **10**, 209.

Barton, J. P. (1965b). "A Comparison Between Thermal, Epithermal and Subthermal Neutrons For Radiography." Proc. Symp. on Physics and Nondestructive Testing, Sept. 1965, published by IIT Research Inst., Chicago, Illinois, pp. 110–143.

Barton, J. P. (1965c). Neutron radiography using a crystal monochromator. *J. scient. Instrum.* **42**, 540.

Barton, J. P. (1965d). Radiographic examination using cold neutrons. *Br. J. appl. Phys.* **16**, 1051.

Barton, J. P. (1965e). Radiographic examination through steel using cold neutrons. *Br. J. appl. Phys.* **16**, 1833.

Barton, J. P. (1965f). Contrast sensitivity in neutron radiography. *Appl. Mater. Res.* **4**, 90.

Barton, J. P. (1965g). Radiology using neutrons. *Stud. Conserv.* **10** (4), 135.

Barton, J. P. (1967a). Divergent beam collimator for neutron radiography. *Mater. Evaluation* **25**, 45A.

Barton, J. P. (1967b). Toward neutron radiography of radioactive objects in hot cells. *Trans. Am. nucl. Soc.* **10**, 443.

Barton, J. P. (1967c). "Neutron Radiography Newsletter." Am. Soc. for Nondestructive Testing, Evanston, Illinois, No. 7.

Barton, J. P. (1967d). "Neutron Radiography—Biological Aspects." Neutron Radiography Newsletter, Am. Soc. for Nondestructive Testing, No. 7.

Barton, J. P. (1968a). "Neutron Radiography Using Non-Reactor Sources." Neutron Radiography Newsletter, Am. Soc. for Nondestructive Testing, Evanston, Illinois, No. 8, pp. 7–13.

Barton, J. P. (1968b). Argonne National Laboratory, Argonne, Illinois, private communication.

Barton, J. P. and Boutaine, J. L. (1968). Initial development of neutron radiography in France. *Isotopes Radiation Technol.* **5** (3), 214. English translation of French paper originally published in Supplément au Bulletin d'Information de A.T.E.N., No. 66, p. 4 (July–Aug. 1967).

Barton, J. P. and Perves, J. P. (1966). Underwater neutron radiography with conical collimator. *Br. J. non-dest. Test.* **8**, 79.

Barton, J. P., Berger, H. and Cutforth, D. C. (1969). "Future Potentialities of Neutron Radiography." Proc. 16th Conference on Remote Systems Technology, American Nuclear Society, Hinsdale, Illinois, 1969, pp. 276–288.

Basl, G., Halchak, J. and Hagemaier, D. (1968). "Detection of Titanium Hydride By Neutron Radiography." Materials and Processes, Rocketdyne Division of North American Rockwell Corp., Canoga Park, Calif.

Beck, W. N. and Berger, H. (1964). "A Shielded Enclosure For Neutron Radiographic Inspection of Encapsulated Irradiated Specimens." Argonne National Laboratory, Argonne, Illinois, Report ANL–6799.

Becker, K. (1966). Nuclear track registration in dosimeter glasses for neutron dosimetry in mixed radiation fields. *Hlth. Phys.* **12**, 769.

Berger, H. (1962a). Comparison of several methods for the photographic detection of thermal neutron images. *J. appl. Phys.* **33**, 48.

Berger, H. (1962b). Photographic detectors for neutron difffraction. *Rev. scient. Instrum.* **33**, 844.

Berger, H. (1963). Resolution study of photographic thermal neutron image detectors. *J. appl. Phys.* **34**, 914.

Berger, H. (1964). Characteristics of a thermal neutron beam for neutron radiography. *Int. J. appl. Radiat. Isotopes* **15**, 407.

Berger, H. (1965). "Neutron Radiography—Methods, Capabilities and Applications." Elsevier, Amsterdam.

Berger, H. (1966). Characteristics of a thermal neutron television imaging system. *Mater. Evaluation* **24**, 475.

Berger, H. (1968a). Recent progress in neutron imaging. *Br. J. non-dest. Test.* **10**, 26.

Berger, H. (1968b). "A Fast Imaging Alignment Method For Neutron Radiographic Inspection of Radioactive Samples." Third Czechoslovak Conf. on Nondestructive Testing, Gottwaldov, Sept. 1968, Dom Techniky CSVTS, Bratislava, 1968, pp. 3–7.

Berger, H. and Beck, W. N. (1963). Neutron radiographic inspection of radioactive irradiated reactor fuel specimens. *Nucl. Sci. Engng* **15**, 411.

Berger, H. and Drexelius, V. W. (1967). "Neutron Radiographic Inspection of Ordnance Components." Fifth Symp. on Electro-explosive Devices, Franklin Institute, Philadelphia, Pennsylvania, 1967.

Berger, H. and Kraska, I. R. (1964). Neutron radiographic inspection of heavy metals and hydrogenous materials. *Mater. Evaluation* **22**, 305.

Berger, H. and Kraska, I. R. (1967). A track-etch, plastic-film technique for neutron imaging. *Trans. Am. nucl. Soc.* **10**, 72.

Berger, H., Barton, J. P. and Ogawa, K. (1964–69). "Neutron Radiography Newsletter." Am. Soc. for Nondestructive Testing, Evanston, Illinois (2 issues per year).

Blanks, B. L. and Morris, R. A. (1966). Experiments with foil-film combinations and collimators for neutron radiography. *Mater. Evaluation* **24**, 76.

Blanks, B. L., Garrett, D. A. and Morris, R. A. (1967). "Improved Resolution Neutron Radiography." Proc. Fifth Int. Conf. on Nondestructive Testing, Montreal, May 1967. The Queens Printer, Ottawa, pp. 242–247.

Boland, J. F. (1968). Radiography—a tool for nuclear materials safeguards. *Nucl. News* **11** (8), 48.

Brown, M. (1968). Medical College of Georgia, Augusta, Georgia, and Parks, P. B. (1968), Savanah River Laboratory, Aiken, South Carolina, private communication.

Brown, M. and Parks, P. B. (1969). Neutron radiography in biological media–techniques, observations, and implications. *Am. J. Roentg.* **106**, 472–485.

Burkdoll, F. B. (1968). Nondestructive inspection by neutron radiography. *Space/Aeronaut.* **49** (5), 117.

Burrill, E. A. (1963). "Neutron Production and Protection." High Voltage Eng. Corp., Burlington, Mass., 36 pages.

Burrill, E. A. and MacGregor, M. H. (1960). Using accelerator neutrons. *Nucleonics* **18**, 64.

Carbiener, W. A. (1966). Nondestructive examination of radioactive material using neutron radiography. *Nucl. Applic.* **2**, 468.

Chountas, K. and Rauch, H. (1968). Neutron radiographic inspections of metal adhesions, alloys, active fuel elements, diffusion of H into Zr and diffusion of H_2O–D_2O. *Atomkernenergie* **13**, 444.

Criscuolo, E. L. and Polansky, D. (1961). "Fast Neutron Radiography." Proc. Missiles and Rockets Symp., U.S. Naval Ammunition Depot, Concord, Calif., pp. 112–115.

Cutforth, D. C. (1968). On optimizing an Sb–Be source for neutron radiographic applications. *Mater. Evaluation* **26**, 49.

Cutforth, D. C. and Aquino, V. G. (1967). Neutron radiography in the EBR-II fuel cycle facility using an isotopic neutron source. *Trans. Am. Nucl. Soc.* **10**, 442.

Ehrlich, M. (1960). Sensitivity of photographic film to 3 MeV neutrons and to thermal neutrons. *Hlth Phys.* **4**, 113.

Epplesheimer, D. S. and Arment, M. (1965). Neutron microradiography of a cadmium–tin alloy. *Nature, Lond.* **207**, 69.

Fahmy, A. A. (1967). "Neutron Radiography and Microradiography Studies." Proc. Cairo Solid State Conference, Sept. 1966, Plenum Press, New York, pp. 537–551.

Farny, G., Barbalot, R., Huvey, M. and Couton, P. (1968). "Neutroscopie." Report DPE/SPE/68–452, Centre D'Etudes Nucléaires, Saclay, France.

Fleischer, A. A. (1966). "A Practical Source of Charged Particles and Fast Neutrons for Activation Analysis and Production of Neutron-Deficient Isotopes." Proc. Seventh Conf. on Radioisotopes, Japan Industrial Forum, Inc., Tokyo, 1966, pp. 198–205.

Fleischer, R. L., Price, P. B. and Walker, R. M. (1965). Neutron flux measurements by fission tracks in solids. *Nucl. Sci. Engng* **22**, 153.

Flynn, M. J. (1968). "Medical Applications of Neutron Research Program." University of Michigan, Ann Arbor, Mich., Dept. of Nuclear Engineering, Report MAN–I.

Furman, S. C., Darmitzel, R. W., Porter, C. R. and Wilson, D. W. (1966). Track-etching, some novel applications and uses. *Trans. Am. nucl. Soc.* **9**, 598.

Garrett, D. A. and Morris, R. A. (1967). The application of high resolution neutron radiography to industrial inspection problems. *Trans. Am. Nucl. Soc.* **10**, 444.

Goldberg, M. D., Mughabghab, S. F., Purohit, S. N. and Magurno, B. A. (1966). "Neutron Cross Sections." Report BNL-325, Second Edition, Brookhaven National Laboratory, Upton, Long Island, New York.

Gorbunov, V. I. and Pekarskii, G. Sh. (1966). Fast neutron radiography of composite materials. *Defectoscopy*, pp. 188–189.

Halmshaw, R. (1968). Royal Armament Research and Development Establishment, Fort Halstead, Kent, private communication.

Holloway, J. A., Sturhke, W. F. and Berger, H. (1967). Low-voltage and neutron-radiographic techniques for evaluating boron-filament metal-matrix composites. *Trans. Am. Nucl. Soc.* **10**, 445.

Hughes, D. J. (1957). "Neutron Cross Sections." Pergamon Press, London and Paris.

Inouye, T., Ogawa, K. and Iwanaga, M. (1965). Image analysis of thermal neutron radiographs. *J. appl. Phys. Japan* **34**, 801.

Kallmann, H. (1948). Neutron radiography. *Research, Lond.* **1**, 254. This publication lists many of the early Kallmann–Kuhn patents.

Kanno, A. (1965). The nondestructive testing of brazed joints. *J. Nondest. Insp.* **14**, 103.

Kastner, J., Berger, H. and Kraska, I. R. (1967). "A Thermoluminescent Image Detection Method For Neutron Radiography." Proc. Fifth Int. Conf. on Nondestructive Testing, Montreal, May 1967. The Queens Printer, Ottawa, pp. 132–134. See also *Nucl. Applic.* **2**, 252 (1966).

Kawasaki, S. (1968). Thermal neutron television system using a high yield neutron generator. *Nucl. Instrum. Meth.* **62**, 311.

Kraska, I. R. and Berger, H. (1968). Experimental neutron radiographic scatter factors. *Mater. Evaluation* **26**, 187.

Marion, J. B. and Fowler, J. L. (eds.) (1960). "Fast Neutron Physics, Part I: Techniques." Interscience, New York and London, pp. 1–176.

Morris, R. A. (1963). "An Investigation of Neutron Collimators and Their Application in Neutron Radiography." Unpublished thesis, M.S., Physics, University of New Mexico.

Moses, A. J. and Saldick, J. (1956). Electron accelerator used for producing neutrons. *Nucleonics* **14**, 118.

Moss, R. A. (1967). The application of neutron radiography to heat-pipe diagnostics. *Trans. Am. Nucl. Soc.* **10**, 445.

Mourvaille, H. (1968). "Thermalisation De Neutrons de 14·7 MeV Produits Par Un Accelerateur De Faible Energie." Centre D'Etudes Nucléaires, Grenoble, France, Section Des Accelerateurs, Report NT/ACC/68–03.

Niklas, W. F., Dolon, P. J. and Berger, H. (1966). "A Thermal Neutron Image Intensifier." Photo-Electronic Image Devices (edited by J. D. McGee, D. McMullen and E. Kahan), *Adv. Electronics Electron Phys.* **22B**, 781–800.

Ogawa, K., Wakabayashi, N. and Kawasaki, S. (1966). "Neutron Radiographic Observation of Cracks and Structural Changes in UO_2 Pellets". Third Japan–U.S. Information Exchange Meeting on Ceramic Nuclear Fuels, Atomic Fuel Corp., Tokai-Mura, Ibaraki-Ken, Japan, 1966, Paper No. 58.

Parks, P. B. and Brown, M. (1969). Antiscatter grids for low energy neutron radiography. *Radiology* **92**, 178–179.

Perry, E. (1967). Boeing Co., Wichita, Kansas, private communication. See also "Neutron Radiography Newsletter," Am. Soc. for Nondestructive Testing, Evanston, Illinois, No. 8, April 1968, pp. 2–3.

Peter, O. (1946). Neutron radiography. *Z. Naturf.* **1**, 557.

Polansky, D. and Criscuolo, E. L. (1964). "Radiographic Aspects of Fast Neutron Detection." Proc. Neutron Radiography Session, 24th Nat. Conf. Am. Soc. for Nondestructive Testing, Am. Soc. for Nondestructive Testing, Evanston, Illinois.

Radwan, M. and Sikorska, D. (1965). The possibility of applying neutron radiography for testing inclusions in magnesium castings. Zesz. probl. Nauki pol. 26, 149.

Reinig, W. C. (1968a). Advantages and applications of Cf^{252} as a neutron source. Nucl. Applic. 5, 24.

Reinig, W. C. (1968b). Californium-252: A new isotopic source for neutron radiography. Mater. Evaluation 26, 33A, Abstract.

Rhoten, M. L. and Cary, W. E. (1966). Neutron radiography of pyrotechnic cartridges. Mater. Evaluation 24, 422.

Rubbino, A., Zubke, O. and Meixner, C. (1966). Neutrons from (α, n) sources. Nuovo Cim. 44B, 178.

Schultz, A. W. and Leavitt, W. Z. (1965). The neutron radiography of uranium and lead. Mater. Evaluation 23, 324.

Spowart, A. R. (1968a). Mobile unit for neutron radiography. Nucl. Engng, Lond. 13, 429.

Spowart, A. R. (1968b). The advantages of epicadmium neutron beams in neutron radiography. Nondestructive Testing 1, 151.

Spowart, A. R. (1968c). "Improvements In The Image Quality Of Radiographs with Glass Neutron Scintillators and Plastic Beta Particle Detectors." The Reactor Group, H.Q. Risley, Warrington, Lancs., Dounreay Report TRG 1690 D.

Stetson, R. L. (1966). Transplutonics–promising neutron sources for research. Nucleonics 24, 44.

Sun, K. H., Malmberg, P. R. and Pecjak, F. A. (1956). High efficiency slow neutron scintillation counters. Nucleonics 14, 46.

Thewlis, J. (1956). Neutron radiography. Br. J. appl. Phys. 7, 345.

Thewlis, J. and Derbyshire, R. T. P. (1956). "Neutron Radiography." UKAEA Report, AERE M/TN 37, Harwell, England.

Tochilin, E. (1965). Photographic detection of fast neutrons: Application to neutron radiography. Physics Med. Biol. 10, 477.

Tyufyakov, N. D. (1968). "Neutron Sources and the Possibility of Their Application in Neutron Defectoscopy." Proc. Second Scientific and Technical University Conference on Radiation Methods of the Nondestructive Quality Control of Materials and Finished Products, Tomsk, Oct. 1968, to be published.

Wagner, C. W., Camanile, V. A. and Guinn, V. P. (1960). Techniques of chemical research with the electron van de Graaff. Nucl. Instrum. Meth. 6, 238.

Wang, S. P., Shull, C. G. and Phillips, W. C. (1962). Photography of neutron diffraction patterns. Rev. scient. Instrum. 33, 126.

Warman, E. A. (1965). Neutron radiography in field use. Mater. Evaluation 23, 543.

Watts, H. V. (1962). "Research on Neutron Interactions in Matter as Related to Image Formation." Report ARF-1164-27, Armour Research Foundation, Chicago, Illinois.

Wilkinson, C. D. (1968). General Electric Co., Pleasanton, Calif., private communication.

Wood, D. E. (1967). Fast neutron radiography with a neutron generator. *Trans. Am. nucl. Soc.* **10**, 443.

CHAPTER 10

A Communication Network Approach to Sub-millimetric Wave Techniques in Nondestructive Testing

W. H. B. COOPER

Electronics and Applied Physics Division, Atomic Energy Research Establishment, Harwell, Berkshire, England

Principal Symbols*

General		Electrotechnics	
ω	angular frequency	E	electric field
		H	magnetic field
p	$\dfrac{d}{dt}$	μ	permeability
		ε	permittivity
$\Gamma\ \alpha+j\beta$	propagation coefficient	σ	conductivity
α	attenuation coefficient	P	power flow

---

* All units are S.I.

General (contd.)

β	phase-change coefficient
Z_0	characteristic impedance
$Z(p)$	impedance parameter
$Y(p)$	admittance parameter
v	velocity of phase propagation
k	modal number

Electrotechnics (contd.)

J	current density
I	current
V	potential difference
ϕ	magnetic flux
B	magnetic flux density
n	electron density
τ	mean collision time
m	mass

Heat Flow

θ^0	temperature
f	heat flow-rate
k	thermal conductivity
ρ	density
S	specific heat

Acoustics

$\tilde{u}$	particle velocity; fluctuating
$\tilde{p}$	pressure; fluctuating
p_0	pressure; steady
γ_0	ratio of specific heat capacity at constant volume to that at constant pressure.

I. Introduction

Wave-like phenomena are utilized in N.D.T. in a variety of ways to locate discontinuities including flaws, cracks, and departures from homogeneity. Ultrasonic waves and electromagnetic waves in loss-free media are classical wave processes, whereas heat and electromagnetic waves in conducting media are so heavily damped that they propagate as a diffusion process. All these transport mechanisms have much in common so that it is convenient to refer to them and to others which fall into this broad category as " wave-like " in nature.

This chapter is concerned with the use of wave-like phenomena to gather information about the properties of media rather than to change the nature of the media as happens with, say, eddy-current heating, ultrasonic cleaning and radio therapy. The objective is to acquire this information without significantly disturbing the medium.

So the basic commodity is " information ", or more precisely the signals bearing the information. But the information content of a signal is, more often than not, partially obscured because of signal distortion and extraneous interference including " unwanted signals " and noise. The development of techniques and equipments which sort order out of this confusion and the invention of appropriate theories and concepts which seek to explain how such devices work have for a long time been the concern of Electrical Communication and Radar Engineers.

Many of the techniques and concepts developed in these fields are directly applicable to the wave-like techniques of N.D.T. and this chapter sets out to explain how two of these concepts, the Field Impedance and the Transfer Function, can be applied to many of the wave-like processes of N.D.T.

All physical wave-like processes are characterized by two dependent variables and the independent variables of space and time. Pressure "$\tilde{p}$" and velocity "$\tilde{u}$" are the commonly-used dependent variables in longitudinal waves in fluids, concentration "c" and flow rate "f" in chemical diffusion "waves", electric "E" magnetic "H" fields in electric waves and temperature θ^0 and flow rate "f" in heat diffusion waves. Those who work in these areas know that it is the boundary conditions which complicate both the theory and practice of wave work. Fundamentally the boundary conditions determine *the ratio* of the two dependent variables at all points throughout the system and so it is convenient to give this ratio a name. It is called the "Field Impedance" in the case of electromagnetic waves and the inverse of this ratio is called the Field Admittance. It will avoid confusion in applying this concept generally if the equivalents of the "E" variable and the "H" variable are identified as follows:

TABLE I

Type of "Wave"	Dependent Variables in MKS Units				Field Impedance
Electromagnetic Waves	Electric Field	E volts/m	Magnetic Field	H amps/m	E/H
Heat Waves	Temperature	θ° Degrees Kelvin	Flow Rate	f watts/m^2	θ°/f
Longitudinal Acoustic Waves	Pressure	$\tilde{p}$ newtons/m^2	Particle Velocity	$\tilde{u}$ m/sec	$\tilde{p}/\tilde{u}$

Strictly, some of the quantities are scalars and others are vectors but this is unimportant in a one-dimensional analysis. The field impedance is a complex quantity.

It turns out that, in all wave-like propagation in one-dimensional semi-infinite media, the field impedance is independent of position. We shall call it the "Characteristic Impedance" of the medium. This "Characteristic Impedance" together with the "Propagation Constant" completely specify the properties of the medium devoid of boundary conditions.

II. ONE-DIMENSIONAL ELECTROMAGNETIC WAVES IN LOSS-FREE MEDIA

Perhaps the simplest way of introducing the impedance concept is in terms of the propagation of electromagnetic waves along coaxial cables of the kind used for interconnecting generators, transducers and amplifiers which are so widely used throughout the N.D.T. field.

In the first instance the conductors are assumed to be perfect containers of electromagnetic fields so that conductivity must be infinite. The potential difference between the conductors gives rise to an electric field E_r and the current flowing along them results in a magnetic field H_θ (Fig. (1a)).

Power-flow into and along the cable can be " explained " by accepting the statement that:

$$P = E_r \times H_\theta \quad \text{watts/m}^2. \tag{1}$$

This is an application of one of the more fundamental " laws " of electromagnetism. Poyntings law of energy flow is often claimed to be the most comprehensive and exhaustive expression of all the experimental results gathered in the sphere of electromagnetism.

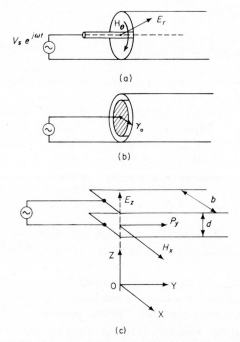

FIG. 1. The coaxial cable (a) reduced to a parallel plate transmission line (c).

It is important to note that the transport of electrical energy always takes place in the medium outside the conductors. This is so even with direct current since the conductors only guide the power-flow, they do not convey it.

The geometry of Fig. 1(a) is unnecessarily complicated for a quantitative explanation of wave propagation because the fields vary with radius. If the radius of the inner conductor is increased until it is very nearly the same as that of the outer cylinder (Fig. 1(b)), then the simple model of Fig. 1(c) is approached in which $b = 2\pi r_0$ and d is the difference in radii. The cylinders have been slit longitudinally and flattened out but the field distributions remain unchanged. Furthermore the field distributions are now uniform over the XZ plane so that they vary only with y, the direction of propagation, and of course with time

$$P = E_z \times H_x \quad \text{watts/m}^2. \tag{1a}$$

All that can be said now is that the two fields E_z and H_x are interrelated in accordance with the two remaining "laws" which underlie the theory of electromagnetism:

$$\frac{\partial E}{\partial y} = -\mu \frac{\partial H}{\partial t} = -\mu p H, \tag{2}$$

which is Faraday's law of flux linkages, $d\phi/dt$ applied to Fig. 1(c), and

$$\frac{\partial H}{\partial y} = -\varepsilon \frac{\partial E}{\partial t} = -\varepsilon p E, \tag{3}$$

which is Ampere's law relating current to magnetic field but including Maxwell's remarkable intuition which identified displacement current with conduction current so far as the associated magnetic field is concerned. Suffixes have been omitted because they are no longer needed and "p" replaces $\partial/\partial t$. The time variable is commonly suppressed by considering all impressed fields to be of time form $\exp(j\omega t)$ but the Transfer Function concept removes that restriction.

Accepting for a moment the assertion that the Characteristic Impedance E/H does not vary with space then

$$E/H = Z_0(p). \tag{4}$$

It may well vary with time and hence with p. Substitution of this relationship into eqns. (2) and (3) reveals that:

$$Z_0(p) = (\mu/\varepsilon)^{\frac{1}{2}}, \tag{5}$$

which happens to be time-independent, but also that:

$$\frac{\partial E}{\partial y} + p(\varepsilon\mu)^{\frac{1}{2}} E = 0,$$ (6)

so that

$$E = A \exp\left[-p(\varepsilon\mu)^{\frac{1}{2}} y\right].$$ (7)

A is obviously the value of E where $y = 0$. This "infinite line" result is too well established to require further comment but it is invariably and, of course, correctly derived via the second-order-space equation which results from combining eqns. (2) and (3) without the information which eqn. (5) provides. The significance of eqns. (5) and (6) is that they demonstrate that the wave transport process is fundamentally represented by a first-order ordinary differential equation in one (space) dimension; it is the essence of simplicity. Indeed, all the transport processes considered in this chapter fall into a similar category so that it becomes profitable in terms of mental effort, ink and paper to generalize eqns. (2) to (7).

Before going on to the generalized approach it is perhaps worth noting that the Characteristic Impedance of the *medium*, eqn. (5), can be converted into the Characteristic Impedance of the Cable Model (Fig. 1(c)) by integrating E over length d to produce the voltage between the plates and integrating H along b to give the current along the plates. In this way the Characteristic Impedance of the cable model (Fig. 1(c)) becomes:

$$Z_0(p)\frac{d}{b} \quad \text{ohms}$$

$$= \frac{120\pi}{\sqrt{2}}\frac{d}{b} \quad \text{ohms for a Polythene dielectric.}$$

III. A GENERALIZED FORM FOR THE BASIC EQUATIONS OF THE TRANSPORT PROCESS

Equations (2) and (3) may be written:

$$\frac{\partial E}{\partial y} = -Z(p)\,H,$$ (8)

$$\frac{\partial H}{\partial y} = -Y(p)\,E.$$ (9)

Repeating the first exercise:
Characteristic Impedance of the medium

$$Z_0(p) = \left(\frac{Z(p)}{Y(p)}\right)^{\frac{1}{2}}, \tag{10}$$

$$E = E_s \exp\{-[Z(p)\,Y(p)]^{\frac{1}{2}}\,y\}, \tag{11}$$

and of course
$$H = \frac{E_s}{Z_0(p)} \exp\{-[Z(p)\,Y(p)]^{\frac{1}{2}}\,y\}. \tag{12}$$

It is convenient and conventional to write:
Propagation Coefficient

$$\Gamma_0(p) = [Z(p)\,Y(p)]^{\frac{1}{2}}. \tag{13}$$

If names are needed for the symbols $Z(p)$ and $Y(p)$ they may be called "Impedance Parameter" $Z(p)$ and "Admittance Parameter" $Y(p)$. Re $\Gamma_0(j\omega)$ is called the attenuation coefficient "α" in népers per metre and $I_m \Gamma_0(j\omega)$ is called the phase coefficient "β" in radians per metre.

The Wavelength $\qquad \lambda = 2.\pi/\beta$ metres.

The Velocity of Phase Propagation

$$v = \omega/\beta \quad \text{metres/sec.}$$

The Group Velocity $\qquad v_g = d\omega/d\beta$ metres/sec.

We have now arrived at a stage where the mechanism of the transport process can be understood simply by inspecting the basic equations, invariably two in number, of the System; equations like (8) and (9). Two relationships are needed because there are two (unknown) dependent variables.

If one is concerned with stimuli of the form $\exp(j\omega t)$ then p is simply replaced by $j\omega$ but for more sophisticated time functions it is usually convenient to proceed beyond eqn. (11) by referring to a table of Fourier or Laplace Transforms (Campbell and Foster, 1951; Pipes, 1946). This is a powerful way of extending the scope of ones applied mathematics without becoming too involved in the rigour, the complexities and niceties of the pure mathematics.

IV. ELECTROMAGNETIC WAVES IN CONDUCTING MEDIA

If the plates of Fig. 1(c) are of finite conductivity, σ, then the current flowing along them parallel to OY will give rise to an electric field E_y. This field together with H_x will, according to Poyntings law, see eqn. (1), cause power to flow along the OZ direction, i.e. into the plates.

According to Faraday's law:

$$\frac{\partial E_y}{\partial z} = -\mu \frac{\partial H_x}{\partial t} = -\mu p H_x \tag{14}$$

so that

$$Z(p) = \mu p \quad \text{see eqn. (8),}$$

and from Ampere's law:

$$\frac{\partial H_x}{\partial z} = -\sigma E_y \tag{15}$$

so that

$$Y(p) = \sigma \quad \text{see eqn. (9).}$$

Hence

$$Z_0(p) = \left(\frac{\mu p}{\sigma}\right)^{\frac{1}{2}} \quad \text{see eqn. (10),} \tag{16}$$

and

$$E_y = A \exp\left[-\Gamma_0(p)z\right] \quad \text{see eqn. (11)} \tag{17}$$

where

$$\Gamma_0(p) = (\mu\sigma p)^{\frac{1}{2}}. \tag{18}$$

This transport process is in fact one of classical diffusion.

The technique of combining eqns. (17) and (7) to give the overall transfer function of the cable including, of course, the loss due to finite conductivity of the containing walls, is treated simply and extensively in the electrical communication literature (Guillemin, 1947; Ramo and Whinnery, 1944). It is irrelevant to the present line of thought which seeks only to explain the fundamentals of the transport mechanism.

A conductor can be energized by placing it in the path of an electromagnetic wave, or more simply by surrounding it with a coil of wire carrying a current, as in Fig. 2. In all cases the transport mechanism is identical with eqns. (14) to (18) although the conducting strip might be a part of the core of a trans-former or inductor. The strip might equally be the sample in an eddy-current tester. If the tester works with sinusoidal waveforms—$\exp(j\omega t)$—then p is replaced by $j\omega$, but if so-called "pulsed waveforms" are employed eqn. (17) must be regarded as a transfer function.

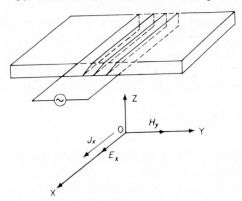

FIG. 2. A conductor energized by a plane wave.

In terms of sinusoidal excitation the real part of $\Gamma_0(j\omega)$—eqns. (18) and (17)—represents attenuation of the field as it propagates into the metal. This "skin effect" is of course common to all diffusion processes so that "eddy-current testing" tends to be a surface phenomenon unless the testing frequency is very low, in which case the wavelength in the metal arising from the imaginary part of $\Gamma_0(j\omega)$ tends to be too long for small defects to be observable. For copper $\sigma = 6 \times 10^7$ and $\mu = 4\pi \, 10^{-7}$ and from eqn. (17) the imaginary part of $\Gamma_0(j\omega)$ shows that the wavelength at 10 kHz is 4 mm. The attenuation for all diffusion processes is $2\pi 8.68$ dB $\doteqdot 55$ dB for a distance of one wavelength; 4 mm at 10 kHz in copper.

It is interesting and may be, in the long run, technologically very important to view this transport process from a microscopic standpoint.

V. ELECTROMAGNETIC WAVES IN CONDUCTING MEDIA FROM A MICROSCOPIC POINT OF VIEW

Returning to Fig. 2 Faraday's law remains unchanged

$$\frac{\partial E_x}{\partial z} = -\mu p H_y,$$

so

$$Z(p) = \mu p. \tag{19}$$

But the microscopic picture demands that Ampere's law be expressed rather differently:

$$\frac{\partial H_y}{\partial z} = -J_x - \varepsilon_0 \frac{\partial E_x}{\partial t}$$

$$= -J_x - \varepsilon_0 \, p E_x. \tag{20}$$

J_x is the current density which, in eqn. (15), was written as σE but this ignored the microscopic mechanism of conduction. The displacement current term $\varepsilon_0 pE$ was omitted from eqn. (15) because it would have been an unnecessary complication but now it is necessary to take care of the free space between the atoms by allowing a displacement current density to flow equal to $\varepsilon_0 \partial E_x/\partial t$.

Microscopically:

$$J_x = nqv_x, \tag{20a}$$

where n is the number of free electrons per cubic metre, q is the electronic charge, here assumed positive to avoid unnecessary confusion, and v_x is the average or drift velocity of the electrons due to the electric field E_x. The equation of motion of each electron is:

$$mpv_x + \frac{mv_x}{\tau} = qE_x. \tag{21}$$

The first term is the force needed to accelerate the electron; τ is the mean collision time of the drift conduction process, so the second term is the force needed to keep the electron moving; the mass m of the electron is in fact the effective or wave mechanical mass and may be somewhat different from the isolated electron mass.

Therefore:

$$J_x = \frac{\tau nq^2 E_x}{m(1+p\tau)} \tag{22}$$

$$= \frac{\sigma E_x}{(1+p\tau)}. \tag{23}$$

For copper at room temperature τ is of the order 10^{-14} sec so that the "lag" term of eqn. (23) is significant only at exceedingly high frequencies—around 10^{13} Hz. The lag comes about, of course, due to the inertia of the electrons. From eqns. (20) and (23)

$$Y(p) = \frac{\sigma}{(1+p\tau)} + \varepsilon_0 p \tag{24}$$

$$= \frac{\sigma + \varepsilon_0 p + \varepsilon_0 \tau p^2}{(1+p\tau)}, \tag{25}$$

so from eqns. (19) and (25)

$$\Gamma_0(p) = \left(\frac{\mu p(\sigma + \varepsilon_0 p + \varepsilon_0 \tau p^2)}{(1+p\tau)} \right)^{\frac{1}{2}}. \tag{26}$$

At very low cisoidal frequencies this reduces to eqn. (18), as one might expect, but a resonance is present at very high frequencies. This is simply demonstrated by considering $j\omega\tau > 1$. From eqn. (25):

$$Y(j\omega) = \frac{\dfrac{nq^2}{m} + \dfrac{\varepsilon_0 \, j\omega}{\tau} - \varepsilon_0 \, \omega^2}{j\omega}$$

and at the particular frequency ω_0 such that:

$$\frac{nq^2}{m} - \varepsilon_0 \, {\omega_0}^2 = 0,$$

$$ {\omega_0}^2 = \frac{nq^2}{m\varepsilon_0}, $$

$$Y(j\omega) \rightarrow \varepsilon_0/\tau,$$

and

$$\Gamma_0(j\omega) \rightarrow \left(\frac{j\omega\mu\varepsilon_0}{\tau}\right)^{\frac{1}{2}}.$$

This frequency ω_0 is called the plasma frequency of the electron "gas". ω_0 is typically in the region 10^{16}, i.e. 10^{15} Hz, resulting in $\Gamma_0(j\omega_0) \doteqdot (j/\tau)^{\frac{1}{2}}$ which of course falls with increasing conductivity so that the metal shows a tendency to become transparent at the plasma frequency. There is some experimental support for this microscopic model but there would be little justification for discussing it here were it not for the fact that it provides a simple introduction to the next topic.

VI. ELECTROMAGNETIC WAVES IN A CONDUCTING MEDIUM IN THE PRESENCE OF A STRONG STEADY MAGNETIC FIELD

Consider now Fig. 3 in which the specimen is energized by two orthogonal windings and placed in a steady magnetic field B_{oz}. Concentrating firstly on $E_x \times H_y$, the situation is identical with eqns. (19) and (20) so that

$$Z(p) = \mu p \tag{27}$$

$$\frac{\partial H_y}{\partial z} = -J_x - \varepsilon_0 \, p E_x. \tag{28}$$

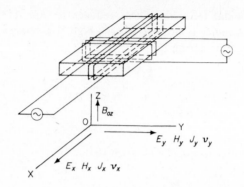

FIG. 3. A conductor energized by a circularly-polarized field and steady magnetic field B_{oz}.

It so happens that the analysis is considerably simplified and at the same time the effect dramatically demonstrated by assuming that $\tau \to \infty$. The attenuation under these conditions tends to infinity without a magnetic field at all engineering frequencies. This assumption enables the drift term to be omitted so that:

$$J_x = nqv_x \quad \text{as before,} \tag{29}$$

but

$$mpv_x = q(E_x + v_y B_0). \tag{30}$$

The last term arises due to the force exerted by the magnetic field; Fleming's left-hand rule,

$$\text{and} \quad mpv_y = q(E_y - v_{x} B_0). \tag{31}$$

Therefore

$$v_x = \frac{p/(\omega_c) E_x + E_y}{B_0[(p/(\omega_c))^2 + 1]} \tag{32}$$

$$\omega_c{}^2 = \left(\frac{B_0 q}{m}\right)^2.$$

From inspection, the transform of eqn. (32) shows that the electron will oscillate continuously at ω_c which is called the cyclotron frequency.

If now the two generators in Fig. 3 are adjusted to be equal in amplitude but not in phase quadrature then:

$$E_x = E_0 \exp(j\omega t); \quad Ey = E_0 \exp[j(\omega t + \tfrac{1}{2}\pi)].$$

Therefore

$$E_y = jE_x.$$

From eqn. (32)

$$v_x \fallingdotseq \frac{jE_x}{B_0} \quad \text{provided} \quad \omega \ll \omega_c, \tag{33}$$

in which case:

$$J_x = \frac{jnqE_x}{B_0}, \tag{34}$$

and from eqn. (28)

$$Y(j\omega) = j\left[\omega\varepsilon_0 + \frac{nq}{B_0}\right]. \tag{35}$$

Hence

$$\Gamma_0(j\omega) = j\left(\left(\omega\varepsilon_0 + \frac{nq}{B_0}\right)\omega\mu\right)^{\frac{1}{2}}. \tag{36}$$

This represents a wave propagation with no attenuation. The steady magnetic field has changed the transport process from one of infinite to zero attenuation.

A simple physical explanation of this phenomenon follows directly from eqns. (32) and (33) and their counterparts in the orthogonal directions. Due to the presence of the steady magnetic field the electron motion is always *normal* to the direction of the electric field, whereas in the drift process, which underlies the theory of conductivity, the electron motion is along the direction of the electric field.

It is interesting to consider the kinematics of the electron motion in a little more detail. Presumably the cyclotron rotational energy is of the same order as the Fermi level, say $10\,\text{eV}$ or 10^{-18} newton-metres.

If

$$B_0 = 1 \cdot 0 \quad \text{and} \quad \omega_c \fallingdotseq 10^{11}$$

$$\frac{m}{2}(\omega_c r)^2 = 10^{-18},$$

where r is the radius of the cyclotron motion.

Therefore

$$r \fallingdotseq 10^{-5}\,\text{m}.$$

This is the motion due to thermal energy.

The low-frequency motion due to the impressed fields E_x and E_y may be estimated so:

$$|v_x| = \frac{E_x}{B_0} \quad \text{see eqn. (33).}$$

Therefore

$$|x| = \frac{E_x}{\omega B_0},$$

say

$$E_x = H_y Z_0(j\omega),$$

$$Z_0(j\omega) = \left(\frac{Z(j\omega)}{Y(j\omega)}\right)^{\frac{1}{2}}$$

$$= \left(\frac{\omega \mu B_0}{nq}\right)^{\frac{1}{2}} \quad \text{see eqns. (27) and (35)}$$

$$\doteqdot 10^{-5} \quad \text{at} \quad \omega = 10^5, \qquad B_0 = 1\cdot0,$$

$$|x| = \frac{H_y Z_0}{\omega B_0} = \frac{H_y 10^{-5}}{10^5}$$

$$= H\, 10^{-10} \quad \text{metres.}$$

Even if H_y is as large as 1000 amp/m, $|x|$ is $\frac{1}{100}$ of r. So the radius of the impressed motion is trivial compared with that of the thermal motion.

VII. The Wave-like Properties of Heat Flow in Solids

If the heat flow across a surface is $f(x)$ watts/m^2, the temperature distribution $\theta^0(x)$ and conductivity k then

$$\frac{\partial \theta^0}{\partial x} = -f/k. \tag{37}$$

This static relationship expresses the fact that heat flow is proportional to temperature gradient.

If heat flow and temperature vary with time it is necessary to find a dynamic relationship between these two dependent variables. By equating the rate at which an elemental volume of the solid is gaining heat due to the net flow across its walls to the rate at which the total quantity of heat energy within the volume is changing with time, we find:

$$\frac{\partial f}{\partial x} = -S\rho \frac{\partial \theta^0}{\partial t}.$$ (38)

S is specific heat in joules per kg per degree and ρ is density in kg/m^3.

In terms of the impedance concept (see eqns. (8) and (9) and Table I),

$$Z(p) = \frac{1}{k},$$ (39)

$$Y(p) = pS\rho,$$ (40)

$$\Gamma_0(p) = \left(\frac{pS\rho}{k}\right)^{\frac{1}{2}},$$ (41)

$$Z_0(p) = \left(\frac{1}{pS\rho k}\right)^{\frac{1}{2}}.$$ (42)

The transport process is, of course, one of classical diffusion.

As the following example indicates, micro wavelengths correspond to surprisingly low frequencies in metals.

Copper:

$$S = 420 \qquad \rho = 9000$$

$$k = 380 \quad \text{all in MKS units,}$$

$$\beta = I_m \, \Gamma_0(j\omega)$$

$$= \left(\frac{\omega S\rho}{2k}\right)^{\frac{1}{2}} \quad \text{radians/metre,}$$

$$\lambda = \frac{2\pi}{\beta}$$

$$\doteqdot 3\cdot5\,\text{mm} \quad \text{at} \quad 100\,\text{Hz}.$$

VIII. LONGITUDINAL ACOUSTIC WAVES IN FLUIDS AND SOLIDS

A. Fluids

The small-signal linearized approximations are derived firstly from the equation of motion:

$$\frac{\partial \tilde{p}}{\partial x} \doteqdot -\rho_0 \frac{\partial \tilde{u}}{\partial t}$$ (43)

$$\doteqdot -p\rho_0 \tilde{u},$$ (44)

so that

$$Z(p) = \rho_0 p \quad \text{see eqn. (8),} \tag{45}$$

and secondly from a combination of the equation of continuity and the gas laws:

$$\frac{\partial \tilde{u}}{\partial x} = -\frac{1}{\gamma_0 p_0} \frac{\partial \tilde{p}}{\partial t} \tag{46}$$

$$= -\frac{p}{\gamma_0 p_0} \tilde{p},$$

$$Y(p) = \frac{p}{\gamma_0 p_0} \quad \text{see eqn. (9),} \tag{47}$$

where

ρ_0 = mean density, p_0 = mean pressure,

$\tilde{p}$ = fluctuating pressure, $p = \partial/\partial t$.

γ_0 = ratio of specific heats.

Hence

$$Z_0(p) = (\rho_0 \gamma_0 p_0)^{\frac{1}{2}}, \tag{48}$$

$$\Gamma_0(p) = \left(p^2 \frac{\rho_0}{\gamma_0 p_0} \right)^{\frac{1}{2}},$$

$$\Gamma_0(j\omega) = j\omega \left(\frac{\rho_0}{\gamma_0 p_0} \right)^{\frac{1}{2}}. \tag{49}$$

Therefore

$$\beta = \omega \left(\frac{\rho_0}{\gamma_0 p_0} \right)^{\frac{1}{2}},$$

and

$$v = \frac{\omega}{\beta}$$

$$= \left(\frac{\gamma_0 p_0}{\rho_0} \right)^{\frac{1}{2}}.$$

The term $\gamma_0 p_0$ is often called the " Bulk Modulus ".

B. Solids

The expressions are identical with eqns. (43) to (49) except that $\gamma_0 \, p_0$ is replaced by Youngs Modulus E. Some approximate but typical values for phase velocity v and characteristic impedance Z_0 in MKS units are:

	v	Z_0
Steel	6×10^3 m/sec	50×10^6
Copper	5×10^3 m/sec	45×10^6
Aluminium	5×10^3 m/sec	15×10^6
Mercury	$1 \cdot 5 \times 10^3$ m/sec	20×10^6
Water	$1 \cdot 5 \times 10^3$ m/sec	$1 \cdot 5 \times 10^6$
Air	350 m/sec	350

IX. REFLECTIONS DUE TO DISCONTINUITIES

Transport processes considered in the earlier sections were confined to semi-infinite media so that the wave-like properties could be explained, devoid of the complications which arise from reflections and multiple-reflections.

In Section II the one-dimensional Electromagnetic Field picture was developed starting with the co-axial cable, but concentrating on the fields E and H. In this Section the network concepts will demand initially a concentration on V the voltage between the plates, i.e. $E_z . d$, and I the current flowing along the plates; $I = H_x \, b$ in Fig. 1(c).

This approach to the "understanding" of boundary value problems is likely to appeal most strongly to those versed in both the theory and practice of electric networks but many such people are likely to be working in NDT fields.

Returning to Fig. 1(c) we note that

$$V . I = E \times H \, db \quad \text{watts} \quad \text{see eqn. 1(a).} \tag{50}$$

It is not surprising therefore that circuit engineers persist with the notion that the power flows "down the wires" rather than in the space between them. If the parallel plate line of Fig. 1(c) is left open circuit at its far end then at very low frequencies the line looks like a capacitance of

$$C = \frac{\varepsilon_0 \, b}{d} \quad \text{farads/m.} \tag{51}$$

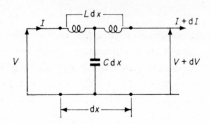

FIG. 4. An element of coaxial line or parallel plate line.

If the line is short-circuited at its far end it looks like an inductance

$$L = \frac{\mu_0 d}{b} \quad \text{henries/m.}$$

By applying Kirchoff's laws to the element network of Fig. 4 (Guillemin, 1947) we find that:

$$\frac{\partial V}{\partial x} = -j\omega L I, \tag{52}$$

and

$$\frac{\partial I}{\partial x} = -j\omega C V. \tag{53}$$

These equations are identical in form to eqns. (2) and (3) or (8) and (9).

Having introduced the transmission network concept there is in fact no need to retain either " p " or " $j\omega$ " in this picture which is, after all, concerned immediately only with the space variable.

This leads naturally to the time-independent or " Leaky-line " model in which the series element is a resistance of R ohm/m and the shunt element is a conductance of G mhos/m (Fig. 5). It is of course necessary to return to

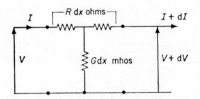

FIG. 5. An element of leaky line. $\Gamma_0 = (RG)^{\frac{1}{2}}$, $R_0 = (R/G)^{\frac{1}{2}}$.

$Z(p)$, $Y(p)$ or $Z(j\omega)$ and $Y(j\omega)$, eqns. (8) and (9), in place of R and G in order ultimately to bring time or frequency, "p" or "$j\omega$", back into the picture. This $R.G.$ model is nevertheless adequate for dealing with all one-dimensional boundary value–impedance mismatch problems. The model can be extended quite simply to two-dimensions—Section X. From Fig. 5:

$$\frac{dV}{dx} = -RI, \tag{54}$$

$$\frac{dI}{dx} = -GV. \tag{55}$$

Following on the lines of Section III for the infinite line

$$R_0 = \left(\frac{R}{G}\right)^{\frac{1}{2}}, \tag{56}$$

$$V = V_s \exp\left[-(RG)^{\frac{1}{2}} x\right], \tag{57}$$

$$\Gamma_0 = (RG)^{\frac{1}{2}}. \tag{58}$$

Considering now the general case where reflections are important, eqns. (54) and (55) become:

$$\frac{d^2 V}{dx^2} = RGV,$$

or

$$\frac{d^2 V}{dx^2} - \Gamma^2 V = 0. \tag{59}$$

Therefore

$$V = A \cosh \Gamma_0 x + B \sinh \Gamma_0 x \tag{60}$$

$$I = -\frac{G}{\Gamma_0}\left[A \sinh \Gamma_0 x + B \cosh \Gamma_0 x\right]$$

$$= -\frac{1}{R_0}\left[A \sinh \Gamma_0 x + B \cosh \Gamma_0 x\right]. \tag{61}$$

The constants A and B are determined by the terminating—conditions—impedances.

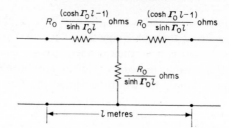

FIG. 6(a). Equivalent " Γ " network for leaky line of length l.

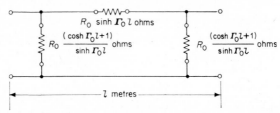

FIG. 6(b). Equivalent " π " network for leaky line of length l.

If, for example, V is V_s at $x = 0$ and $V = 0$ at $x = l$, i.e. in a short circuit at $x = l$, then by substitution in eqn. (60)

$$A = V_s,$$
$$B = -V_s \coth \Gamma_0 l.$$

There is, however, another, often very rewarding, way of looking at the boundary condition problem. It is shown in the many texts on electrical communication (Cooper, 1962; Guillemin, 1947) that a line like Fig. 5 of length l metres can be represented by either an equivalent " Γ " or " π " network as in Fig. 6.

These networks therefore provide a complete mathematical/pictorial representation of the one-dimensional medium devoid of boundary conditions. It is important to note that the shunt elements are resistances (impedances) and not conductances (admittances) as in the elemental networks.

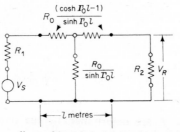

FIG. 7. Line or medium of length l with " mismatch " at each end.

Consider now the case of a line (medium) of length l driven from a generator R_1 and terminated in a load or some other medium of resistance (field impedance) R_2 (Fig. 7). The current distribution in this network is readily determined—it is:

$$V_R = \frac{V_s R_2}{\left(\dfrac{R_1 R_2}{R_0} + R_0\right) \sinh \Gamma_0 l + (R_1 + R_2) \cosh \Gamma_0 l} \tag{62}$$

$$= \frac{V_s}{2} \exp(-\Gamma_0 l) \quad \text{when} \quad R_1 = R_2 = R_0. \tag{63}$$

This equation is consistent with eqn. (11), the half appears due to the finite resistance $R_1 = R_0$ of the generator. Equation (62) can be manipulated into another form which turns out to be most revealing:

$$V_R = \frac{2V_s \exp(-\Gamma_0 l) \dfrac{R_0 R_2}{(R_0 + R_1)(R_0 + R_2)}}{1 - \exp(-2\Gamma_0 l) \dfrac{(R_0 - R_1)(R_0 - R_2)}{(R_0 + R_1)(R_0 + R_2)}}. \tag{64}$$

Each term in this expression has some physical significance associated with the reflections and multiple reflections arising from the boundary conditions, $\exp(-\Gamma_0 l)$ being the main wave propagation of eqn. (11). Commencing with the numerator:

$\dfrac{R_0}{(R_0 + R_1)}$ represents the voltage mismatch loss at the sending end.

$\dfrac{R_2}{(R_0 + R_2)}$ represents the voltage mismatch loss at the receiving end.

Now considering the denominator:

$\dfrac{(R_0 - R_1)}{(R_0 + R_1)}$ represents the reflected voltage at the sending end.

$\dfrac{(R_0 - R_2)}{(R_0 + R_2)}$ represents the reflected voltage at the receiving end.

Hence

$$\exp\left(-2\Gamma_0 l\right)\frac{(R_0-R_1)(R_0-R_2)}{(R_0+R_1)(R_0+R_2)} \tag{65}$$

represents the voltage loss ratio due to one trip around the system. Therefore the denominator in eqn. (64) represents the sum of an infinite number of trips around the system corresponding to $1/1-\mu\beta$ in closed-loop system language. Clearly the multiple reflections can be regarded as waves travelling backwards and forwards or as waves travelling around the " loop ".

If Re $\Gamma_0 l$ is unity or greater, then eqn. (65) cannot exceed $\exp(-2) = 0\cdot14$, in which case eqn. (64) is accurate to $\pm14\%$ neglecting the eqn. (65) term. Considerations of this kind can enormously simplify rough calculations. The mismatch losses and reflection factors can in fact be deduced from first principles as follows:

Figure 8 represents a generator R_1 feeding a load R_2; the voltage " wave " transmitted into R_2 is:

$$V_T = \frac{V_s R_2}{R_1 + R_2}.$$

This can be regarded as two waves, an incident wave $V_I = V_s/2$ which is the voltage transmitted into a matched load, and a reflected wave V_R such that:

$$V_I + V_R = V_T.$$

This expression of continuity of V at a discontinuity between two media (impedances) corresponds to the continuity of the tangential components of E and H at electromagnetic boundaries.

Hence

$$V_R = V_T - V_I$$

$$= \frac{V_s(R_2-R_1)}{2(R_2+R_1)}.$$

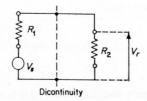

Dicontinuity

FIG. 8. Reflections at a mismatch.

Therefore

$$\frac{\text{The Reflected Voltage Wave, } V_R}{\text{Incident Voltage Wave, } V_I} = \frac{(R_2 - R_1)}{(R_2 + R_1)},$$

$$\frac{\text{The Transmitted Voltage Wave, } V_T}{\text{Incident Voltage Wave, } V_I} = \frac{2R_2}{R_1 + R_2}.$$

The dependent variables throughout this section have been V voltage and I current but the identity of form between eqns. (2), (52) and (54) on the one hand and between eqns. (3), (53) and (55) on the other suggests that E, say, can replace V and H can replace I simply as symbols and the electrical circuit networks immediately become field networks directly applicable to all transport processes.

X. THE IMPEDANCE CONCEPT APPLIED TO TWO-DIMENSIONAL MEDIA

In Section II, Power flow $P_y = E_z \times H_x$ in the dielectric along the direction OY was considered and in Section IV attention was confined to Power flow $P_z = E_y \times H_x$ into the metallic plates. There is no reason why E_y should not exist in the dielectric even with perfectly conducting plates. This will give rise to:

$$P_y = E_z \times H_x$$

and

$$P_z = E_y \times H_x$$

simultaneously. Faraday's and Ampere's laws for these three dependent variables are:

$$\frac{\partial E_z}{\partial y} - \frac{\partial E_y}{\partial z} = -\mu \frac{\partial H_x}{\partial t}, \tag{66}$$

$$\frac{\partial H_x}{\partial y} = -\varepsilon \frac{\partial E_z}{\partial t}, \tag{67}$$

$$\frac{\partial H_x}{\partial z} = \varepsilon \frac{\partial E_y}{\partial t}. \tag{68}$$

These equations reduce to:

$$\frac{\partial^2 H_x}{\partial y^2} + \frac{\partial^2 H_x}{\partial z^2} = \Gamma_0^2 H_x, \tag{69}$$

where
$$\Gamma_0{}^2 = j\omega\mu . j\omega\varepsilon,$$

or
$$\Gamma_0{}^2 = p\mu . p\varepsilon.$$

Equation (69) is solved in many text books (Ramo and Whinnery, 1944) so there is no need to repeat the detail here. The forward, OY, travelling wave solution with symmetrical sending end energization is

$$H_x = A \exp\left[-(\Gamma_0{}^2 + k^2)^{\frac{1}{2}} y\right] \cos kz, \tag{70}$$

so from eqn. (67)

$$-\varepsilon p E_z = -(\Gamma_0{}^2 + k^2)^{\frac{1}{2}} H_x.$$

Therefore
$$\frac{E_z}{H_x} = \frac{(\Gamma_0{}^2 + k^2)^{\frac{1}{2}}}{\varepsilon p}. \tag{71}$$

The mode factor "k" can have any value depending upon the lateral boundary conditions (Ramo and Whinnery, 1944). Equation (71) is the modal characteristic impedance corresponding to eqns. (5) and (10). It is in fact the same as eqn. (5) for $k = 0$ which is an acceptable solution to Fig. 9 if the plates are energized by wires connected to a generator as in Fig. 1.

Returning to eqn. (70), for all finite values of k, the two-dimensional wave system attenuates frequencies such that

$$\omega^2 \mu\varepsilon < k^2,$$

i.e.
$$\omega < \frac{k}{(\mu\varepsilon)^{\frac{1}{2}}}.$$

This is the familiar cut-off frequency of guided wave systems. The modal propagation factor can be written

$$(\Gamma_0{}^2 + k^2)^{\frac{1}{2}} = [(p\mu + k^2/p\varepsilon)\,p\varepsilon]^{\frac{1}{2}}, \tag{72}$$

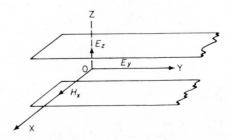

FIG. 9. Two-dimensional waves in a parallel plate system.

and the modal characteristic impedance:

$$\frac{(\Gamma_0{}^2 + k^2)^{\frac{1}{2}}}{\varepsilon p} = \left(\frac{(p\mu + k^2/p\varepsilon)}{p\varepsilon} \right)^{\frac{1}{2}}. \tag{73}$$

These equations are the same as eqns. (13) and (10) if

$$Z_k(p) = p\mu + k^2/p\varepsilon,$$

and $$Y(p) = p\varepsilon. \tag{74}$$

Hence $$Z_k(p) = p\mu + k^2/Y(p). \tag{75}$$

Therefore, to convert a one-dimensional solution into a two-dimensional solution eqns. (8) to (13) can be used, $Z_k(p)$ is obtained simply by adding $k^2/Y(p)$ to $Z(p)$.

In this example we started with two E (Electric fields) and one H field. Had we chosen one E field and two H fields we should find that:

$$Z(p) = p\mu \quad \text{as in one-dimension}, \tag{76}$$

and

$$Y_k(p) = p\varepsilon + k^2/p\mu$$

$$= p\varepsilon + k^2/Z(p). \tag{77}$$

Once the modal propagation coefficient and modal characteristic impedance have been determined they can be used directly in the networks of Section IX.

Referring to Table I and Section VIII the two-dimensional acoustic system will contain one pressure and two component velocities. Its two-dimensional modal characteristics will therefore follow the pattern of one E field and two H fields since E corresponds to pressure $\tilde{p}$ and H corresponds to velocity $\tilde{u}$.

XI. Some Electrical Analogues of the Transport Processes of Table I

For those who find it profitable to think in terms of electric circuit analogues, the following diagrams, Figs. 10 to 21, portray the differential elements of the transport processes described in this chapter.

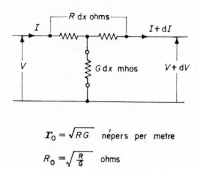

$$\Gamma_0 = \sqrt{RG} \quad \text{nepers per metre}$$

$$R_0 = \sqrt{\frac{R}{G}} \quad \text{ohms}$$

Fig. 10. Element of a leaky line.

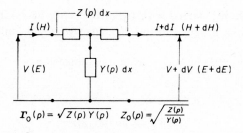

$$\Gamma_0(p) = \sqrt{Z(p)\,Y(p)} \qquad Z_0(p) = \sqrt{\frac{Z(p)}{Y(p)}}$$

Fig. 11. Generalized element of line or medium.

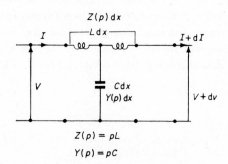

$$Z(p) = pL$$

$$Y(p) = pC$$

Fig. 12. The coaxial or parallel plate line.

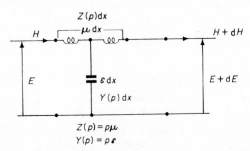

FIG. 13. Plane electromagnetic waves in loss-free medium.

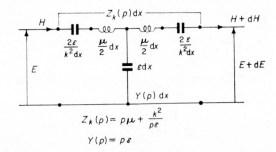

FIG. 14. Two-dimensional electromagnetic waves in loss-free medium (two E and one H).

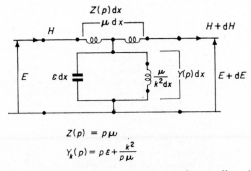

FIG. 15. Two-dimensional electromagnetic waves in loss-free medium (one E and two H).

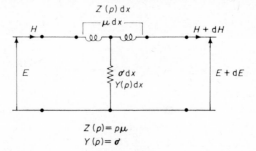

$$Z(p) = p\mu$$
$$Y(p) = \sigma$$

Fig. 16. One-dimensional electromagnetic waves in conducting medium (macroscopic model).

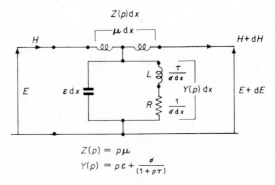

$$Z(p) = p\mu$$
$$Y(p) = p\varepsilon + \frac{\sigma}{(1 + p\tau)}$$

Fig. 17. One-dimensional electromagnetic waves in conducting medium (microscopic model).

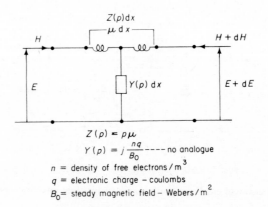

$$Z(p) = p\mu$$
$$Y(p) = j\frac{nq}{B_0} \text{---- no analogue}$$

n = density of free electrons / m^3
q = electronic charge – coulombs
B_0 = steady magnetic field – Webers / m^2

Fig. 18. Electromagnetic waves in conducting medium approaching infinite conductivity with steady magnetic field B_0.

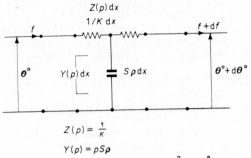

$$Z(p) = \frac{1}{K}$$

$$Y(p) = pS\rho$$

$K =$ Thermal conductivity — Watts / m^2 per θ°/m
$S =$ Specific heat – Joules per kg per degree
$\rho =$ Density – kg/m^3

FIG. 19. One-dimensional heat diffusion.

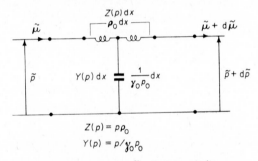

$$Z(p) = p\rho_0$$

$$Y(p) = p/\gamma_0 P_0$$

$\tilde{p} =$ Fluctuating pressure $\tilde{\mu} =$ Particle velocity
$\rho_0 =$ Mean density $\gamma_0 =$ Ratio of specific heats
$P_0 =$ Mean – steady pressure

FIG. 20. One-dimensional longitudinal acoustic waves.

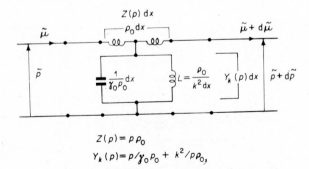

$$Z(p) = p\rho_0$$

$$Y_k(p) = p/\gamma_0 P_0 + k^2/p\rho_0,$$

FIG. 21. Two-dimensional longitudinal acoustic waves.

REFERENCES

Campbell, G. A. and Foster, R. M. (1951). "Fourier Integrals for Practical Applications." Van Nostrand, New York and Macmillan, London.

Cooper, W. H. B. (1962). Diffusion of electricity. *Electron. Technol.* **39**, 279–285 and 297–301.

Guillemin, E. A. (1947). "The Classical Theory of Long Lines, Filters and Related Networks", Communication Networks, Vol. II. Wiley, New York.

Pipes, L. A. (1946). "Applied Mathematics for Engineers and Physicists." McGraw-Hill, New York.

Ramo, S. and Whinnery, J. R. (1944). "Fields and Waves in Modern Radio." Wiley, New York.

CHAPTER 11

Multiparameter Eddy Current Concepts

H. L. LIBBY

Battelle Memorial Institute, Pacific Northwest Laboratory,
Richland, Washington, U.S.A.

I. INTRODUCTION

This paper describes multiparameter or multivariable eddy current test concepts and techniques. Application of these techniques results in the extraction of more information from eddy current nondestructive tests than can be obtained by conventional methods.

Serviceability requirements of materials and parts become more strict and specialized problems arise as metals technology advances. To meet new nondestructive testing needs, it is becoming more important to treat the tests in ways which lead to better use of available information. The development of the multiparameter eddy current test (Libby, 1966) has been one result of a study of broadband electromagnetic testing methods (Libby, 1958; Libby and Cox, 1961; Libby and Atwood, 1963). It has been shown (Libby and Atwood, 1963; Libby and Wandling, 1968) that the application of multiparameter principles yields test performances not possible by other techniques.

The purpose of this paper is to explain the principles of the multiparameter eddy current test method from several points of view, to show its relation to single frequency tests, to report recent developments in the theory of the test and to show some test results. Explanations and examples given are based mainly upon the use of nonferrous test specimens. Although the principles of the multiparameter tests are generally applicable to the examination of ferrous materials, a treatment of the effects of magnetic permeability and magnetic saturation are beyond the scope of this paper.

Three concepts of the multiparameter test are given:

1. A generalization of the single frequency phase discrimination technique.

2. An algebraic approach comprising the solution of a set of simultaneous equations.

3. A geometrical interpretation involving vector space.

In the new approach multiple frequency or pulsed currents drive the eddy current test coil. The test coil output signal, having been modulated by the test specimen parameters (or test variables), is then amplified, applied to spectrum filters, and demodulated. The demodulated signal components are next recombined in selected linear combinations giving the read-out of individual test parameters on separate output channels.

This method promises to help solve the problem of interpreting eddy current test signals which arises due to the nature of the eddy current test.

A. The Eddy Current Test

The eddy current nondestructive test is used in the metals industry to perform a variety of measuring, sorting, and flaw detection functions. It is based upon electromagnetic induction principles, its techniques being founded on the basic discoveries of Faraday and Maxwell. The method has been highly developed by specialists in the metallurgical and electronic fields to meet the needs of industry.

Metals technology advances place new and more exciting demands on test techniques. Thus, new approaches are needed to meet special problems.

1. *Indirect Nature of the Eddy Current Test*

Some test problems arise due to the indirect nature of the eddy current test. Eddy currents are caused to flow in the metal test specimen by placing it in the a.c. magnetic field of an induction coil. The flow of currents is affected by the electrical properties and shape of the test specimen or by the presence of discontinuities or defects. The current flow, in turn, makes its effect observable by affecting the electrical impedance of the exciting coil (or by the related effect of altering the induced voltage of the coil). Thus, the test specimen effects may be observed by monitoring the test coil impedance. The eddy current test is indirect in that it does not measure directly any specific characteristic of the specimen. Rather it responds to some weighted function of the current flow which is indirectly related to test specimen conditions. This weighted function depends on the test coil design, the operating frequency, and the test specimen.

The result is that if several test conditions vary it becomes difficult or perhaps impossible to identify the individual test conditions by observing the test signal in the conventional test. The effectiveness of the conventional single frequency eddy current test is limited to the identification of only one or two test conditions.

2. *Skin Effect*

A major factor determining the selection of the test frequency is the skin effect. This is caused by the mutual interaction of currents and results in the currents being concentrated near the surface into which energy flows from the excitation source. The effect varies as the square root of the frequency, and currents are increasingly concentrated near the surface as frequency increases. Thus, the effective depth of eddy current test sensitivity can be controlled by test frequency selection. In addition, the electrical phase angle of the eddy currents lags increasingly with increases in test depth below the surface, test frequency, and test specimen conductivity. The varying phase angles of currents are reflected in defect and thickness signals and play an especially important role in the multiparameter test.

II. THREE CONCEPTS OF THE MULTIPARAMETER PRINCIPLE

The essence of the multiparameter eddy current test is that multiple frequency or pulsed currents drive the test coil assembly whose output signals are analyzed making use of their multivariable aspects. These test coil output signals, having been modulated by the test specimen parameters (or test variables), are amplified, applied to spectrum filters and demodulated.

The demodulated signal components are then recombined in selected linear combinations giving the read-out of individual test parameters on separate output channels.

The principles of the multiparameter test may be explained in many different ways. As an aid to better understanding of the basic principles of the multiparameter test, three explanations follow. The first explanation relates the new technique to a generalization of the phase discrimination method (Foerster, 1954; McMaster, 1959) used in single frequency eddy current tests. This explanation is given because of the appeal it has by virtue of the wide use of the phase discrimination technique. The second explanation is based on elementary algebra and at first appears simpler and less abstract than the others. Matrix and vector space concepts characterize the third explanation. The most general and abstract of the three, these vector concepts result in compact notation. As might be expected, each of the explanations has a common underlying basis—that of separating or decoupling variables in a multivariable system.

The test was developed based on concepts of the third explanation—that of matrix algebra and vector space theory. The vector space approach is very abstract and thus more difficult to visualize than other, less general explanations. However, it is the most powerful.

The explanations are made with the assumption of small signal conditions but are valid in principle for large signal conditions. Large signal conditions bring special problems and are discussed later.

A. Generalization of the Phase Discrimination Technique

The multiparameter test technique may be presented as a generalization of the phase discrimination technique, extending it to operate simultaneously at two or more frequencies. The explanation will be made by first showing the principle of the phase discrimination technique as used with one test frequency. It will be shown how two variables or parameters can be separated at a single frequency.

1. Extension to More Parameters

By adding a second test frequency and showing the resulting three-dimensional signal, the method will then be extended to three parameters. Extension to four or more parameters cannot easily be shown graphically, but it can readily be inferred as a logical result of increasing the number of variables and dimensions over that already shown for 2 and 3 dimensions.

The electrical impedance or voltage output of an eddy current probe-type test coil assembly for several test conditions at a test frequency ω_1 is shown in the phasor (phase vector) diagram in Fig. 1.

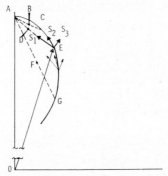

FIG. 1. Phasor (phase vector) diagram showing lift-off and sub-surface defect signals—probe-type test coil.

2. *Phasor diagrams*

It is helpful in the interpretation of the phasor diagram to recall that it is a compact way of showing the relative phase angles and amplitudes of a.c. sinusoidal quantities. Each phasor diagram constructed usually depicts conditions for a specific frequency. The length and direction of a straight line drawn between any two points on the diagram are proportional to the amplitude and describe the relative phase angle, respectively, of a sinusoidal quantity at the specific frequency. Thus, the phasor diagram, itself a rather static looking thing, gives a condensed display of dynamic quantities.

3. *Effect of Parameter Variations*

In Fig. 1 the values are assumed for purposes of explanation and are in approximate agreement with some measured quantities. The phasor OA represents the coil output signal for the unloaded condition (no test specimen adjacent to coil). Consider a probe coil with its axis normal to a test specimen, moving toward that specimen and finally coming to rest on the surface. During the motion of the probe, the test coil output follows some locus such as ABC, ADE, or AFG depending upon the electrical conductivity of the test specimen. These are often called "lift-off loci". The locus $ACEG$ is the specimen conductivity locus for the particular test coil in its position adjacent to the nonferrous test specimen. Consider the point E associated with the locus ADE. The vector OE represents the coil output signal when the coil is adjacent to a particular specimen which we assume for the moment to be without defects within the coil's zone of sensitivity. Now consider the effect on the coil signal of the variation of the three following test parameters:

(1) p_1, a lifting of the probe a small distance away from the surface;

(2) p_2, the presence of a defect within the metal a short distance below the surface;

(3) p_3, a defect within the metal at a greater distance below the surface.

Lifting the probe from the surface will result in a signal S_1 lying along the lift-off locus ADE. Defects beneath the surface intefere with eddy currents flowing in the specimen. Due to the skin effect, these currents flow at increasingly greater lagging phase angles with increasing depth below the surface. Thus, a defect at a fairly shallow depth might give a signal S_2 while a defect at greater depth might give a signal S_3. Note that the phase angle of S_2 is lagging with respect to the lift-off signal S_1. Similarly, the signal S_3 lags with respect to signal S_2. This is the situation for the test frequency ω_1. A little later we will discuss the effect on the relative phase angle of these test signals at some different, higher test frequency. The main effect will be seen to be an increase in lagging phase angle of the defect signals. This occurs due to the increased phase angle lag of the current at a given depth as the frequency increases. However, before we consider increasing the test frequency we will first free ourselves of unneeded portions of Fig. 1 and discuss the phase discrimination technique, so often used in the single frequency technique.

The phasor signal OE can be compensated by adding a signal $-OE$ to it, effectively removing the reference zero from point O to point E. Figure 2 depicts the resulting phasor arrangement. Here the origin is labeled O_E to remind us that it is still directly related to point E in Fig. 1.

4. *Phase Discrimination in the Two Parameter Case*

For our immediate purposes of discussing the phase discrimination technique we show signals S_{11} and S_{21} only, as only two parameters can be separated using this method. The amplitude-phase detector used in a

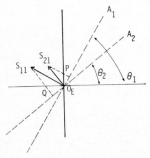

Fig. 2. Eddy current test signals—single frequency phase discrimination technique—one or two parameter separation.

single frequency eddy current test is so connected that its output gives a signal proportional to the component of the test signal in phase with a reference or switching signal. For example, let θ_1 in Fig. 2 be the reference signal phase angle and let it be adjusted so that the reference signal is in quadrature with respect to the lift-off signal S_{11}. Axis A_1 is called the axis of Phase Detector 1.

The Phase Detector 1 output is proportional to the projection of S_{11} upon Axis A_1. With Axis A_1 adjusted to be in quadrature with respect to the lift-off signal, S_{11}, its projection on A_1 is very small or zero. Thus, this adjustment discriminates against the lift-off signal. In contrast, signal S_{21}, having projection component $O_E P$ along A_1, will give indications in the Phase Detector 1 output. The foregoing describes the phase discrimination technique applied to discriminate against a signal (S_{11}) having a particular phase angle and reading out a component of a signal (S_{21}) having a different phase angle. This component is proportional to both the signal amplitude and the cosine of the phase angle between the signal and the reference signal.

Similarly, the signal S_{21} may be discriminated against and a component of signal S_{11} read out by using a second phase detector with reference signal phase angle θ_2 and phase detector axis A_2. In this case the read-out component of S_{11} is proportional to $O_E Q$.

Thus, operating with one test frequency, we have read one signal with one phase detector, and a second signal with a second phase detector. This is done even when the two signals may occur simultaneously.

5. Effect of a Third Parameter

If we add a third signal we cannot read it out separately with this system. This is because we are operating in two dimensions and are asking for separation of three variables. This theory will be discussed in detail later.

6. Extension of Principle to Higher Dimensions

The multiparameter test is based on extending the phase discrimination technique, so far described, into the third, fourth, and higher dimensions. This is done by increasing the number of applied test frequencies so that the number of dimensions required to describe the signal increases thus permitting the handling of a greater number of parameters. For example, four parameter separation requires the use of two frequencies. However, for simplicity consider the separation of three parameters using two test frequencies. By using only three signal components and three parameters, the graphical construction can be made in the more familiar three space.

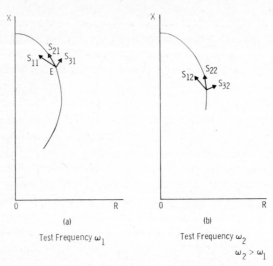

FIG. 3. Phasor diagrams showing lift-off and sub-surface defect signals for two frequencies.

Figure 3 shows the additional signal information needed for the analysis. Figure 3(a) reproduces a portion of the phasor diagram in Fig. 1 for frequency ω_1 and permits ready comparison with the new phasor diagram, Fig. 3(b), for a higher test frequency, ω_2.

In Fig. 3(b) the defect signal phasors S_{22} and S_{32} have been rotated clockwise (lagging phase) to show the effect of the lesser skin depth at the higher frequency.

The test signal information of Fig. 3 has been transferred to Fig. 4. The signal patterns have been rotated counterclockwise so that the phasors S_{11} and S_{12} are aligned with the axis of abscissas. Exact scale values for these phasors (signals) have been established to facilitate later comparisons and explanatory analyses. The scale values are in approximate agreement with some experimental values.

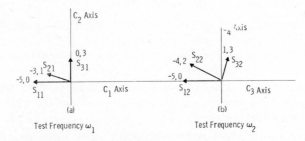

FIG. 4. Eddy current test signals—two frequency system $\omega_2 > \omega_1$.

Note that the projections of the various signals on the C_1 and C_2 axes could be obtained from two phase detectors operating at frequency ω_1 having quadrature switching signals adjusted to the required reference phase angle. Likewise, the signal projections on the C_3 and C_4 axes could be obtained from phase detectors operating at frequency ω_2.

7. Three-dimensional Case

The signal components along the C_1, C_2 and C_3 axes of Fig. 4 are now used to make the diagram of Fig. 5 which shows the signal phasors in three dimensions. S_1, S_2, and S_3 identify the three signals of interest using parameter rather than frequency designating suffixes. The components along the C_4 axis of Fig. 4(b) are not used in this three-dimensional construction, although they would be needed for separation of four parameters. The algebraic sign of the components along the C_1 and C_3 axes of Fig. 4 have been reversed for constructing Fig. 5. This makes the figure easier to visualize with no loss in generality. Axis A_{21} is constructed normal to the plane containing S_1 and S_2. Next, axes A_{32} and A_{31} are constructed normal to the planes containing the signal pairs S_3, S_2, and S_3, S_1, respectively.

8. Phase Discrimination in Three Dimensions

A generalized explanation of the phase discrimination principle to more than two dimensions can now be made. Suppose we wish to read an output proportional to the signal S_3 independently of the signals S_1 and S_2. The signal S_3 has a component Oa along axis A_{21}. However, signals S_1 and S_2 have no component in this direction because axis A_{21} was constructed to be

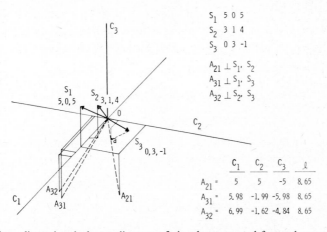

FIG. 5. Three-dimensional phasor diagram of signal constructed from phasors in Fig. 4.

normal to both S_1 and S_2. Thus, by reading the signal along axis A_{21} an output proportional only to signal S_3 is obtained.

Similarly, it can be shown that signals S_1 and S_2 have components along only axes A_{32} and A_{31}, respectively. Thus, outputs proportional to S_1, S_2 and S_3 can be obtained separately by reading signal projections (components) along the specially selected axes A_{32}, A_{31} and A_{21}, respectively. The principle can be extended to higher dimensions by analogy.

We will defer the discussion of actual instrumental methods of reading the signal projections along the chosen axes. Means to accomplish this should become clear as we proceed with the descriptions of the algebraic and vector space representations.

B. Algebraic Representation

The principles underlying the multiparameter eddy current tests can also be explained by showing that the method involves the solution of a set of simultaneous algebraic equations. Referring to Fig. 4, and recalling that the signal projections on the axes C_1, C_2, C_3, and C_4 are the outputs of the amplitude-phase detectors of an eddy current instrument, equations for these outputs can be written. As linear (small signal) conditions have been assumed, the phase detector output for each axis (or channel) is equal to the sum of the signal projections upon that axis. Assigning a coefficient a_{ij} to each signal relates the value of the signal projection to the particular parameter value p_i. For example, the projection of the signal S_{12} on the C_2 axis, disregarding the other signal contributions, becomes

$$a_{12}p_2 = c_2, \tag{1}$$

where
$$a_{12} = 1.$$

Taking the other signals into consideration, a set of simultaneous equations may be written. Note that the phase detector outputs, C_1, C_2, C_3, and C_4 are each given by the sum of the signal projection upon the respective axes, i.e.

$$\left.\begin{aligned} a_{11}p_1 + a_{12}p_2 + a_{13}p_3 &= c_1, \\ a_{21}p_1 + a_{22}p_2 + a_{23}p_3 &= c_2, \\ a_{31}p_1 + a_{32}p_2 + a_{33}p_3 &= c_3. \end{aligned}\right\} \tag{2}$$

The values of the coefficients a_{ij} for the three-parameter example given in Fig. 4 may be read from the scale values

$$\left.\begin{aligned} c_1\,a_{11} &= -5 & a_{12} &= -3 & a_{13} &= 0 \\ c_2\,a_{21} &= 0 & a_{22} &= 1 & a_{23} &= 3 \\ c_3\,a_{31} &= -5 & a_{32} &= -4 & a_{33} &= 1. \end{aligned}\right\} \tag{3}$$

The set of equations, eqn. (2), has three unknown quantities, p_1, p_2, and p_3. In a given test it is possible, in principle, to determine the coefficients a_{ij} and the numbers c_1, c_2, and c_3. In fact, the c_i values represent the various phase detector outputs of the instrument. The a_{ij} factors can be found by observing the effect on the c_i values by each parameter as they are caused to vary individually.

1. *Separation of Parameters*

The effects of the various parameters may be separated by applying the method of Gauss successively to eliminate variables in a set of simultaneous equations. This can be done by selecting the proper linear combinations of the c_i to eliminate the effect of p_1, p_2, and p_3. This is illustrated in the line diagram in Fig. 6. The phase detector outputs c_1, c_2, c_3 are used as inputs. Y_i are summing points where signals fed on the incoming lines are summed in the required linear combinations to eliminate parameters. For example, at Y_1 combinations of c_1 and c_2 are selected to eliminate the effect of parameters p_3. Similarly, at Y_2 combinations of c_1 and c_3 are selected to eliminate p_3. The next two summing points Y_3 and Y_4 are used to eliminate p_2. As the elimination (separation) of parameters proceeds the coefficients related to the various parameters (variables) change. The new coefficients depend upon the factors applied to c_1, c_2, and c_3 and the necessary summing operations for eliminating (separating) the parameters. For example, the coefficients b_{11} and b_{12} depend on the factors applied to c_1 and c_2 and the summing operation at Y_1 for the elimination of parameter p_3. The parameters p_1 and p_2 are separated at summing points Y_5 and Y_6, and p_3 is separated at summing point Y_7. The final outputs, Ip_1, Jp_2 and Kp_3 are proportional to the parameters p_1, p_2, and p_3. Other arrangements of line connections may be desirable depending upon the functions c_1, c_2, and c_3.

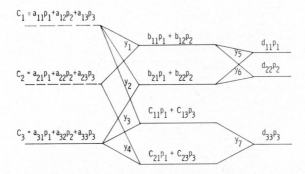

FIG. 6. Separation of parameters—algebraic concept.

2. *Complete Solution Not Required*

A complete solution of the set of equations (eqn. 2) is not required as our objective is to read out the effects of the various parameters on separate lines. Thus, it is the decoupling or separation of the parameters which is of interest here. A discussion follows in the Separation Versus Solution section.

Figure 7 shows the separation of parameters for the example given in Fig. 4, using the algebraic representation. Here the multiplying factors are given in parentheses on the appropriate line just ahead of each summing point.

C. *Vector Space Representation*

The multiparameter eddy current test may also be given a very general geometrical interpretation involving vector spaces. Figure 8 depicts the concept in simplified form. Generator A excites Test Coil Assembly B with a multifrequency or pulsed signal which is modulated by the test specimen parameters having generalized parameter P with components $p_1, p_2, ..., p_n$. The output of Receiver D produces a generalized signal C related to the input parameter P by the equation

$$C] = [A] P], \tag{4}$$

where $[A]$ is a modulation matrix which takes into account the test conditions, the test specimen, test coil and receiver characteristics.

The signal C has components $c_1, c_2, c_3, ..., c_n$ which are the coefficients of generalized Fourier series expansion. The signal is next transformed by Transformation Section E into a function, Q, proportional to the original input parameter P. The output, Q, has components $q_1, q_2, q_3, ..., q_n$ which

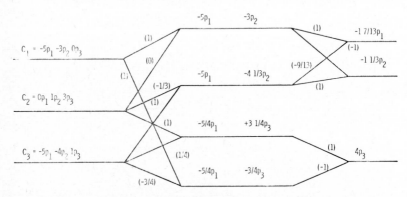

FIG. 7. Line diagram showing separation of parameters.

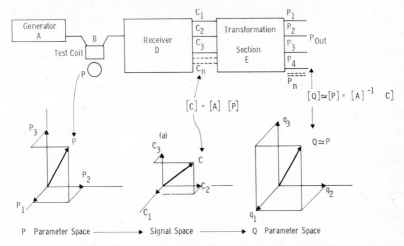

FIG. 8. Parameter space—signal space transformations.

are proportional to their input parameter counterparts $p_1, p_2, p_3, \ldots, p_n$. The equation

$$P] = [A]^{-1} C] \tag{5}$$

describes the transformation of C into output parameter signals.

Equation (4) is simply the matrix form of eqn. (2) which describes the algebraic representation. Multiplying the signal matrix by the inverse of the modulation-receiver matrix $[A]$ produces the solution (separation) described in eqn. (5). Conceptually a simple, direct set of equations, these are quite useful, although the complete solution represented is not required.

A more realistic nomenclature shows the final output being proportional to the original parameters. By denoting the output space as Q parameter space, eqn. (5) can be modified to read:

$$Q] \simeq P] = [A]^{-1} C]. \tag{6}$$

The proportionality factor can be different for each component of P. Detailed discussion appears in the section entitled Separation Versus Solution.

III. SEPARATION OF PARAMETERS

Separation of parameters refers to the adjustments or equipment functions which permit the signal response at specific instrument output terminals to be identified with specific parameters. Identification results from the variation of specific test parameters during calibration.

A parameter is defined as a test variable. The test specimen parameters of greater interest include electrical conductivity, other electrical properties, the presence of defects, location of defects and dimensions. Also of interest are the coupling between the test coil and specimen, variation in coil impedance due to environmental conditions, and instrument drift. It is desirable to select independent parameters.

Theoretically, independence refers to the parameters and components of the generalized parameter, assumed to be mutually orthogonal in parameter space.

A. Separation Versus Solution

As previously stated in the sections Algebraic Representation and Vector Space Representation, a complete solution of the basic equations, eqn. (2) and eqn. (4), is not required. This is because the essence of the multiparameter test is the separation of parameters. Once the signal effects are processed to produce read-outs for different parameters at specific output terminals, magnitude adjustments can be made separately. The sensitivity of each output channel may be set as desired using a calibration procedure in which test parameters are caused to vary by known amounts.

The difference between solution and separation can be shown by solving eqn. (4) in such a way as to give separation without complete solution. Equation (4) transforms the parameter vector P into the signal vector C:

$$C] = [A]\, P]. \tag{4}$$

To solve for P, multiply the equation by A inverse, A^{-1}, giving the complete solution

$$P] = [A]^{-1} C] = [A]^{-1} [A]\, P] = P]. \tag{7}$$

Now the output we need for separation only is a set of components $q_1, q_2, ..., q_n$, ordered in the same way as the P components, but with arbitrary amplitudes. This arbitrary amplitude may be represented by multiplying P with a diagonal matrix D_a having arbitrary values on its diagonal:

$$[D_a]\, P] = [D_a]\, [A]^{-1} C]. \tag{8}$$

Using the identity:

$$A^{-1} = \frac{1}{|A|}\, Ad_j\, A \tag{9}$$

where $|A|$ = Determinant of A

$$Ad_j\, A = \text{Adjoint of } A$$

equation (7) may then be written

$$|A| [D_a] P] = [D_a] [Ad_j A] C].$$ (10)

Since $|A|$, the determinant of A, is a scalar and D_a has arbitrary components, $|A|$ may be absorbed into D_a. However, the equation in its present form shows directly the effect of A becoming singular, that is, as $|A|$ approaches zero. This equation shows that the separating or decoupling capability is obtained through the operation of the adjoint. The role of the arbitrary or calibrating factor D is also shown, and indicates that the requirements for separation are less restrictive than for complete solution.

Equation (10) is basic in the design and development of transformation circuits for separating parameters.

B. Number of Test Frequencies Required

In general, test parameter variations in the eddy current test cause curved signal loci in two spaces or in higher spaces. This curvature, or nonlinearity, is a natural result of the spatial distribution of the electromagnetic fields and currents and the mutual interaction of the currents. For purposes of simplifying the analysis an analytical approach involving linear approximations is used.

Small variations in parameters result in small signal changes which vary approximately linearly as the parameters. This makes it feasible to apply the powerful methods of linear algebra. Fortunately, many eddy current tests involve the detection of small flaws, thus the application of the principle of linear approximation is of practical value. The number of test frequencies required for separating a given number of parameters in the linear case will be discussed first, followed by a few comments on the nonlinear case.

1. Linear approximation

The analysis of an idealized eddy current test shows that the number of parameters which can be identified equals twice the number of test frequencies used. The idealized eddy current test is represented in Fig. 9 by a lossy transmission line having infinite length with a load Z_F representing a defect at distance (depth) x from the sending end. The effect of coupling circuits between generator and line (test specimen) is not represented, not being pertinent to the immediate problem.

It is assumed that fixed a.c. currents, I_n, at various frequencies are applied to the sending end of the transmission line. The instrument reads the signal

$$E(\omega) = I_n(\omega) Z_n(\omega).$$ (11)

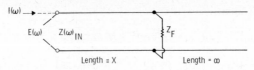

Length = X   Length = ∞

Characteristic Impedance, Both Sections: Z_0

FIG. 9. Idealized eddy current test represented by electric transmission lines.

Thus, the signal observed is proportional to the line input impedance, $Z(\omega)$. By standard methods (Everitt and Anner, 1956) the input impedance Z_{IN} of a transmission line having a characteristic impedance Z_0, a propagation constant γ, and length x is

$$Z_{IN} = Z_0 \left[\frac{1 + \dfrac{(Z_R - Z_0)}{(Z_R + Z_0)} \exp\left(-2\gamma x\right)}{1 - \dfrac{(Z_R - Z_0)}{(Z_R + Z_0)} \exp\left(-2\gamma x\right)} \right]. \tag{12}$$

As we are simulating the metal of the test specimen with the transmission line, we will let

$$x\gamma = x\,(ZY)^{\frac{1}{2}} = x(j\omega\mu\sigma)^{\frac{1}{2}} = \frac{x(1+j)}{\delta}, \tag{13}$$

where Z = series impedance

  Y = shunt admittance

  μ = magnetic permeability of free space

  σ = metal conductivity

  δ = depth of penetration, plane wave case,

  $= \dfrac{1}{(\pi f \mu \sigma)^{\frac{1}{2}}}$.

Equation (12) may be written

$$Z_{IN} = Z_0 \left[\frac{1+y}{1-y} \right], \tag{14}$$

where

$$y = \frac{Z_R - Z_0}{Z_R + Z_0} \exp\left(-2\gamma x\right). \tag{15}$$

For small values of $y(y < 1)$ eqn. (14) may be written

$$Z_{IN} = Z_0[1 + 2(y + y^2 + y^3 + y^4 \ldots)]. \tag{16}$$

The higher order terms may be omitted with less than 2% error for values of $y < 0\cdot1$. Hence,

$$Z_{IN} \cong Z_0(1 + 2y) = Z_0 + 2Z_0 y, \tag{17}$$

which gives, in terms of a new signal $Z_{IN} - Z_0$,

$$Z_{IN} - Z_0 \cong 2Z_0 y. \tag{18}$$

Referring to Fig. 9, it can be seen that the load Z_F, which represents the defect at depth x, is connected in parallel with the input impedance, Z_0, of the line section (having infinite length) beyond Z_F. The quantity Z_R in eqn. (12) is then equal to

$$Z_R = \frac{Z_F Z_0}{Z_F + Z_0}. \tag{19}$$

A simple manipulation results in

$$\frac{Z_R - Z_0}{Z_R + Z_0} = \frac{\dfrac{Z_F Z_0}{Z_F + Z_0} - Z_0}{\dfrac{Z_F Z_0}{Z_F + Z_0} + Z_0} = -\frac{Z_0}{2Z_F + Z_0}. \tag{20}$$

Substituting eqn. (20) into eqn. (15) and that result into eqn. (18) gives

$$Z_{IN} - Z_0 \cong -\frac{2Z_0{}^2}{Z_0 + 2Z_F} \exp(-2\gamma x). \tag{21}$$

Further, substituting eqn. (13) into eqn. (21) results in

$$Z_{IN} - Z_0 \cong -\frac{2Z_0{}^2}{Z_0 + 2Z_F} \exp[-(2x/\delta)] \exp[-j(2x/\delta)]. \tag{22}$$

This reduces to the following for small defect conditions, $(Z_F \gg Z_0)$

$$Z_{IN} - Z_0 \cong \frac{1}{Z_F} Z_0{}^2 \exp[-(2x/\delta)] \exp[-j(2x/\delta)]. \tag{23}$$

The impedance difference equation, (23), is now converted to a signal voltage equation by multiplying it by a fixed current $I_n(\omega)$ as in eqn. (11).

$I(\omega)$ and Z_0^2/Z_F are absorbed in a new factor, $P_n \exp(j\omega_k t)$, where the subscripts n and k identify a particular defect and a particular test frequency. The resulting equation is:

$$E_{nk} = P_n \exp(j\omega_k t) \exp[-(2x_n/\delta_k)] \exp[-j(2x_n/\delta_k)] \qquad (24)$$

where E_{nk} is the signal caused by defect n at frequency ω_k.

     ω_k is the test frequency in radians per second.

     x_n is the depth of the defect from the surface adjacent to the test coil.

     δ_k is the depth of penetration of eddy currents in the metal at test frequency ω_k.

     P_n is a factor proportional to the value or " strength " of the defect.

(a) Simplifying Assumptions: It is emphasized that eqn. (24) is approximate. Its formation takes into account only the first reflection from the defect under assumed plane wave conditions. Also, the factor P_n—which appears as such a neat linear factor in eqn. (24)—has its origin in the complex eqn. (20). Actually, P_n can also vary with the test frequency. However, it is necessary to ignore these important details in the interest of simplicity. This is justified by noting that we limit our discussion to the effects of small defect signals which we assume can be treated using linear algebra. These simplifying assumptions should not affect the validity of the conclusions since a more accurate formulation should increase the complexity of the different E_{nk}, making the new E_{nk} even more independent rather than less.

(b) Fourier Series Expansion: In the multiparameter test the signals E_{nk} are expanded into their Fourier series components

$$a_{nk} \sin \omega_k t \quad \text{and} \quad b_{nk} \cos \omega_k t$$

by applying the signals E_{nk} to phase detectors. The outputs of the phase detectors are proportional to the Fourier coefficients a_{nk} and b_{nk}. Thus, there are available two coefficients at each test frequency. These are, from eqn. (24),

$$P_n \exp[-(2x_n/\delta_k)] \sin 2x_n/\delta_k, \qquad (25)$$

and
$$P_n \exp[-(2x_n/\delta_k)] \cos 2x_n/\delta_k. \qquad (26)$$

For simplicity, miscellaneous phase shift angles at the different test frequencies are omitted without loss of generalization.

In the general case many defects or other test conditions contribute signals, yet only two Fourier coefficients are available at each test frequency. It is convenient to handle the resulting array of equations by use of matrix algebra. The phase detector output signals may now be assembled in the matrix equation

$$[a_{mn}] p_n] = y_n]$$ (27)

where $[a_{mn}]$ is a square matrix whose elements are the outputs of the various phase detectors for unit parameter value at each frequency. The new subscript m is required to identify the phase detector output signals as there are two of these for each frequency or k subscript.

      $p_n]$ is a column matrix whose elements are proportional to the parameters, $p_1, p_2, p_3, \ldots, p_n$.

      $y_n]$ is a column matrix whose elements are proportional to the outputs of the phase detectors for the existing (multiparameter valued) condition.

The matrix $[a_{mn}]$ can now be written (Table 1) and examined in detail to determine how many independent parameters, or variables, may be determined for a given number of test frequencies (see Table 1).

To determine the number of parameters, p_n, which may be separated (identified) per test frequency, it is necessary to show the number of independent equations represented by the matrix $[a_{mn}]$ as a function of frequency. It is also necessary to determine the conditions under which the set of equations represented by the matrix $[a_{mn}]$ has a solution. The matrix $[a_{mn}]$ is a square matrix; thus, for n columns it will have an equal number of rows. As there are two different rows of the matrix resulting from each set of two phase detectors operating at each test frequency, the number of variables, n, represented by the square matrix, equals twice the number of test frequencies. Examination of the individual matrix elements shows sufficient independence to solve for the n parameters. We will examine the case for $n = 2$. For this two parameter case, selecting arbitrary defect depths x_1 and x_2 will result in a matrix $[a_{mn}]$ having four elements. It is found that this matrix is nonsingular. Thus, the equations it represents can be solved for the two variables. For example, assuming $x_1/\delta_1 = 0.5$ and $x_2/\delta_1 = 0.7$, the matrix $[a_{mn}]$ is

$$[a_{mn}] = \begin{bmatrix} 0.31 & 0.24 \\ 0.20 & 0.04 \end{bmatrix},$$ (28)

TABLE I

Matrix Elements of $[A_{mn}]$. Idealized Case. Subsurface Flaws Represented by Discontinuities in an Electric Transmission Line

k		$n=1$	$n=2$	$n=3$	$n=4$	$\cdots$	n_n
$k=1$	$m=1$	$\exp\left[-2(x_1/\delta_1)\right]\sin\dfrac{2x_1}{\delta_1}$	$\exp\left[-2(x_2/\delta_1)\right]\sin\dfrac{2x_2}{\delta_1}$	$\exp\left[-2(x_3/\delta_1)\right]\sin\dfrac{2x_3}{\delta_1}$	$\exp\left[-2(x_4/\delta_1)\right]\sin\dfrac{2x_4}{\delta_1}$	$\cdots$	$\bullet\ \bullet\ \bullet$
	$m=2$	$\exp\left[-2(x_1/\delta_1)\right]\sin\left(\dfrac{2x_1}{\delta_1}+\dfrac{\pi}{2}\right)$	$\exp\left[-2(x_2/\delta_1)\right]\sin\left(\dfrac{2x_2}{\delta_1}+\dfrac{\pi}{2}\right)$	$\exp\left[-2(x_3/\delta_1)\right]\sin\left(\dfrac{2x_3}{\delta_1}+\dfrac{\pi}{2}\right)$	$\exp\left[-2(x_4/\delta_1)\right]\sin\left(\dfrac{2x_4}{\delta_1}+\dfrac{\pi}{2}\right)$		$\bullet\ \bullet\ \bullet$
$k=2$	$m=3$	$\exp\left[-2(x_1/\delta_2)\right]\sin\dfrac{2x_1}{\delta_2}$	$\exp\left[-2(x_2/\delta_2)\right]\sin\dfrac{2x_2}{\delta_2}$	$\exp\left[-2(x_3/\delta_2)\right]\sin\dfrac{2x_3}{\delta_2}$	$\exp\left[-2(x_4/\delta_2)\right]\sin\dfrac{2x_4}{\delta_2}$		$\bullet\ \bullet\ \bullet$
	$m=4$	$\exp\left[-2(x_1/\delta_2)\right]\sin\left(\dfrac{2x_1}{\delta_2}+\dfrac{\pi}{2}\right)$	$\exp\left[-2(x_2/\delta_2)\right]\sin\left(\dfrac{2x_2}{\delta_2}+\dfrac{\pi}{2}\right)$	$\exp\left[-2(x_3/\delta_2)\right]\sin\left(\dfrac{2x_3}{\delta_2}+\dfrac{\pi}{2}\right)$	$\exp\left[-2(x_4/\delta_2)\right]\sin\left(\dfrac{2x_4}{\delta_2}+\dfrac{\pi}{2}\right)$		$\bullet\ \bullet\ \bullet$
k_m		$\bullet\bullet\bullet$	$\bullet\bullet\bullet$	$\bullet\bullet\bullet$	$\bullet\bullet\bullet$		

$$a_{mn} = \exp\left[-2(x_{nl}/\delta_{[(m/2)+\frac{1}{4}(1-(-1)^m)]})\right]\sin\left[\frac{2x_n}{\delta_{[(m/2)+\frac{1}{4}(1-(-1)^m)]}} + \frac{\pi}{4}(1+(-1)^m)\right]$$

where

k = Frequency number

m = Row number (also phase detector output number)

n = Column number (also parameter number)

δ_k = Skin depth or penetration depth at frequency k

x_n = Distance from surface to nth defect

a_{mn} = Matrix element in mth row and nth column

with a determinant $|A|$ equal to -0.0356. Thus, the system of equations

$$\begin{bmatrix} y_1 \\ y_2 \end{bmatrix} = \begin{bmatrix} 0.31 & 0.24 \\ 0.20 & 0.04 \end{bmatrix} \begin{bmatrix} p_1 \\ p_2 \end{bmatrix} \tag{29}$$

has a solution.

If only one phase detector is used for each test frequency instead of two, the number of rows in the matrix $[a_{mn}]$ will be halved. With only half the rows available, the number of parameters which can be separated or identified equals the number of test frequencies.

2. Nonlinear General Case

The more general case involving the curved signal loci in signal space is all inclusive. Even the small signals, treated in this paper by linear approximation, give curved loci in signal space, however small the curvature. Thus, there is no definite delineation between the "linear" and "nonlinear" regions. However, the effects resulting from nonlinear signals are very apparent in the operation of the multiparameter testers. As the signals extend further into the nonlinear region, separation of parameters becomes more difficult and cross-channel interference occurs. For equivalent separation, fewer parameters may be separated for a given number of test frequencies than for the case of linear signals.

C. Pulse Methods

Applications of the pulse techniques for multiple parameter extraction are not as far advanced as the multifrequency methods and are not treated in detail here (cf. Chapter 12, p. 383). However, much theoretical work is available from other related fields. For example, the fields of signal analysis, system identification and pattern recognition are rich with theory and techniques pertinent to multiparameter problems. Signal analysis theory and techniques as presented by Huggins (1957) also apply. Some of his work has been reviewed (Libby and Cox, 1961) with regard to its applicability in eddy current nondestructive testing. Litman and Huggins (1963) have investigated the use of growing exponentials in the identification of multiparameter systems.

IV. INSTRUMENTATION

Instrumentation for the multiparameter test differs from that of the conventional eddy current test mainly as a result of simultaneous operation at several or many test frequencies, the demodulation of the resulting multiplicity of test signals, and the summation of the demodulated components to give the separation of parameters.

The main functions required of eddy current multiparameter test instrumentation are shown in Fig. 10. These are:

1. Excite the test coil assembly with a multidimensional driving function. This may be a multifrequency or pulsed current.

2. Provide means to bring the test coil assembly output signal to the desired state of a.c. null. This is done by compensating circuits or an a.c. bridge system.

3. Amplify, filter (as needed) and demodulate the test signal. Demodulation is performed by synchronous detectors. This expands the signal on a selected basis giving the signal vector components c_k. In the general case, the c_k are proportional to the coefficients of the terms in a generalized Fourier expansion of the signal on the selected basis. When the analysis is performed in the frequency domain and the test coil is excited by multifrequency currents, the basis is a group of sinusoidal functions and the c_k are proportional to the coefficients of the Fourier series of the test signal.

4. Extract individual test parameters by transformation of the signal vector into parameter vector components. This is accomplished by constructing linear combinations of the signal vector components c_k.

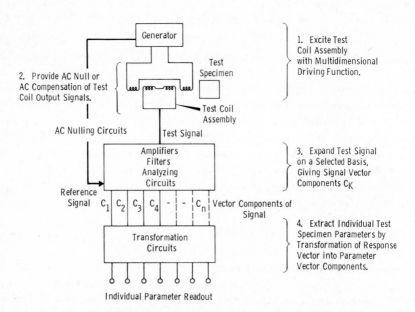

FIG. 10. Instrumentation functions of the multiparameter test.

A. Excitation

The principle of superposition applies if the test specimen is nonferrous. Thus, test coil excitation currents containing many frequency components can be applied simultaneously without cross-modulation effects. A great latitude exists in the choice of the test coil driving functions. However, the coil driving function does influence the design of the receiver equipment and choice of demodulation circuits. Synchronized or non-synchronized carriers can be used. The synchronized carriers can be generated by repetitive pulse generation circuits or by summation of a set of individual carriers which have been generated in synchronism. Single pulse excitation can also be used, as in principle all the necessary information is contained in the response of the test arrangement to an impulsive excitation.

B. Detection

The instrument designer may choose to design the equipment to perform the signal analysis in either the frequency domain or time domain. The former seems more appropriate for use in systems using excitation at several discrete carrier frequencies. In this case, the various carriers can be separated using wave filters and can then be applied to separate amplitude-phase detectors or other circuits which provide the Fourier series coefficients of the signal. In the case of repetitive pulse excitation, analysis can be made either in the frequency domain or in the time domain using pulse sampling techniques. The pulse signals may also be analyzed using an orthogonal spectrum analyzer (Huggins, 1957; Libby and Cox, 1961; Mishkin and Braun, 1961).

C. Transformation or Separation Circuits

The required transformation function is obtained by constructing linear combinations of the detector outputs c_k. The basic process is the same whether one follows either the algebraic solution concept or the signal space concept. There are, however, differences in circuit connections and adjustment techniques may vary widely. As the combinations are linear, the essential transformation circuit requirements must provide:

1. Sign changing.

2. Amplitude adjustment.

3. Summing.

The portion of the circuit here called transformation circuit has also been called separation circuit, analyzer circuit, or computer circuit.

In practical circuits, it is convenient to use operational amplifiers to provide isolation, sign changing and summing functions. Potentiometers in the summing amplifier input circuits provide amplitude adjustment of the summed signals.

Two types of transformation circuits will be described; one based on the matrix solution and a second based on an algebraic solution.

1. *General Transformation Circuit*

Figure 11 depicts this circuit, based on the solution of the matrix (eqns. (4) or (10)). This configuration provides separation of up to four parameters. It accepts the outputs of four phase detectors and delivers the selected linear combinations of these inputs to four output channels. The proper adjustment of the potentiometers U is of critical importance in the operation of the devices. Gain controls (not shown) are usually included in the output circuits for adjustment of the individual parameter channel gains.

(a) Adjustment Techniques: Two main techniques used to adjust the overall transformation circuit will next be discussed. In one of these, the first step is to determine the matrix elements of the modulation matrix $[A]$, eqn. (4). This is done by causing each parameter to vary individually and by measuring the resulting change in each of the c_k. For a four-parameter

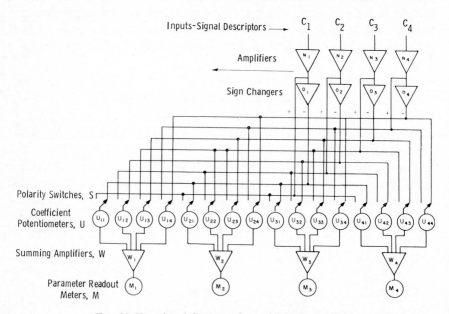

Fig. 11. Functional diagram of transformation section.

system, an array of sixteen matrix elements results. Next, one computes $[A]$ inverse, $[A]^{-1}$.

The matrix element values of $[A]$ inverse then determine the settings of the coefficient potentiometers U. Doing this, one usually finds that due to unavoidable errors and some nonlinearity of signals, optimum separation of parameters is not obtained. To obtain better separation, this leads to trial and error adjustment of the potentiometers U, a time-consuming process.

The other main overall transformation circuit adjustment technique eliminates computing the inverse elements of the modulation matrix and separates the parameters by direct adjustment of the potentiometers U. For example, in a four-parameter system, one of a group of four potentiometers U can be set at an arbitrary value. The other three are then adjusted simultaneously to eliminate the effect of three parameters on the output of the channel fed by these potentiometers and the associated summing amplifier. Although not a simple adjustment to carry out, it is flexible, thus permitting partial compensation of nonlinear effects associated with large signals. These four adjustments produce read-out, on that particular output channel, of the signal caused by only one parameter. Repeating the process for the next group of four potentiometers results in separation of the second parameter. The process is then repeated for the remaining two groups of four potentiometers.

(b) Aids in Potentiometer Adjustment: The search for the desired adjustment in either the trimming operation or search separation method may be facilitated by convenient arrangement of the potentiometer knobs so that several of them may be simultaneously adjusted with one hand.

2. Transformation Circuit Based on Algebraic Solution

Figure 12 depicts the second transformation circuit to be described. This circuit, one associated with the algebraic solution concept, applies the principle of successive elimination of parameters (variables). It follows the principles previously discussed and illustrated in Figs. 6 and 7. This particular circuit is arranged to read out two parameters and eliminate or minimize the effects of two other parameters. Four parameter read-out could be obtained by adding additional components and two more output channels.

The outputs c_1, c_2, c_3, and c_4 of phase detectors are inputs to the transformation section, or analyzer, although the detector outputs are not necessarily always connected to the correspondingly labeled input terminals of the analyzer section. The summing circuits are arranged in Fig. 12 in three columns corresponding to the positions of the summing potentiometer

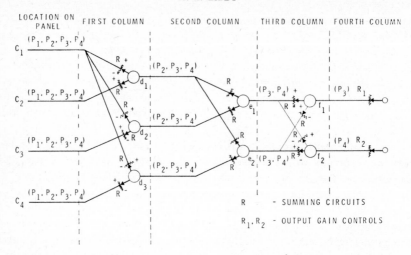

FIG. 12. Transformation circuit for successive elimination of parameters.

controls on the instrument panel. Each summing potentiometer R is driven by two operational amplifiers, not shown in Fig. 12, so that both positive and negative signals are available as desired for the summing operation. The effects of four selected parameters, p_1, p_2, p_3, and p_4 to be separated are generally available in the signals c_1, c_2, c_3 and c_4 as indicated by the p_1, p_2, p_3, and p_4 designation on the signal leads in the diagram. Since this instrument reads out two parameters, for example p_3 and p_4, the effects of parameters p_1 and p_2 are eliminated or minimized. The effects of p_1, probe wobble for example, are minimized by sequential adjustment of the potentiometers in the first column, so that signals d_1, d_2, d_3, in sequence, show a minimum variation when p_1 is varied. The effects of the second parameter p_2 are then minimized by adjustment of potentiometers in the second column so that a minimum effect of varying parameter 2 can be observed in signals e_1 and e_2. The last two parameters are similarly separated by adjustment of potentiometers in the third column to successively eliminate or minimize the effects of p_4 in signal f_1 and p_3 in signal f_2, thus separating the effects of parameters p_3 and p_4.

The two potentiometers in the fourth column are gain controls in the two output channels f_1 and f_2. The advantage of this circuit arrangement is that it eliminates the effect of each parameter in succession rather than almost simultaneously as in the general transformation circuit. The successive elimination technique is carried out faster with much more convenience in use. This is especially noted when the parameters and test specimens are such that all parameters cannot readily be presented in the rapid succession required by the technique used in adjusting the general transformation circuit.

V. Test Results

This section demonstrates some of the potential of the multiparameter eddy current test method. Three test systems, a two-frequency tubing test system, a four-frequency thickness test system, and a two-frequency system with a tubing cross section display feature are described and test results given.

A. Two-frequency Tubing Tester

A diagram of a two-frequency multiparameter tester called a multichannel tubing tester (Libby and Wandling, 1968) is shown in Fig. 13. It uses two harmonically-related test frequencies, 100 kHz and 300 kHz. Synchronous phase detectors produce the Fourier series coefficients c_1, c_2, c_3, and c_4. The transformation circuit used here is the one discussed in the preceding section and shown in Fig. 12.

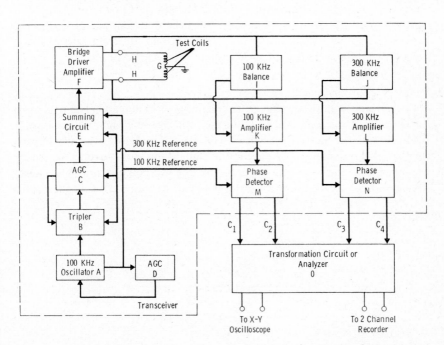

Fig. 13. Multichannel (multiparameter) eddy current tubing tester block diagram.

This eddy current tubing tester reduces the effect of test probe wobble signals on both output channels. It is flexible in operation and can be operated using either of its two test frequencies singly or both simultaneously. When operated in its single frequency modes, two test parameters may be separated, for example, probe wobble, and the presence of defects giving signals having components in quadrature with signals caused by probe wobble.

When operated in its two-frequency mode the effects of one test parameter can be minimized and the effects of two other test parameters may be read.

Comparative results were obtained operating the multichannel tubing tester in its two single-frequency modes and its single two-frequency mode using the test specimen whose cross section sketch is shown in Fig. 14. This is type 304 stainless steel tubing having 1·27 mm (0·050 in) wall and 12·7 mm ($\frac{1}{2}$ in) inside diameter. Irregularities of interest in this specimen, from right to left (in the direction of movement of the test coil assembly) are a 3·18 mm ($\frac{1}{8}$ in) long notch, approximately 0·381 mm (0·015 in) deep by 0·0762 mm (0·003 in) wide opening on the inner surface, a kink or dent in the tubing, a suspected region of intergranular corrosion on the inner wall of a tubing, and a 3·18 mm ($\frac{1}{8}$ in) long notch approximately 0·585 mm (0·023 in) deep by 0·0762 mm (0·003 in) wide opening onto the outer surface of the tubing. The results of runs made using this specimen and an inside coil assembly having two differentially connected test coils are shown in Fig. 15.

1. Single-frequency Modes

For the single-frequency runs, 100 kHz, Fig. 15(a) and 300 kHz, Fig. 15(b), the instrument was adjusted to minimize probe wobble signal on channel 1, the lower trace of each chart record. The test coil assembly was pulled nearly through the specimen tube, resulting in signal indications due to the tubing irregularities. Near the end of the tubing, the test coil assembly was caused to wobble severely, resulting in the large signals near the left end of the upper

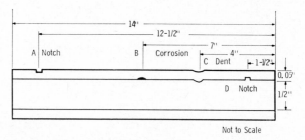

Fig. 14. Test specimen used for Fig. 15 test.

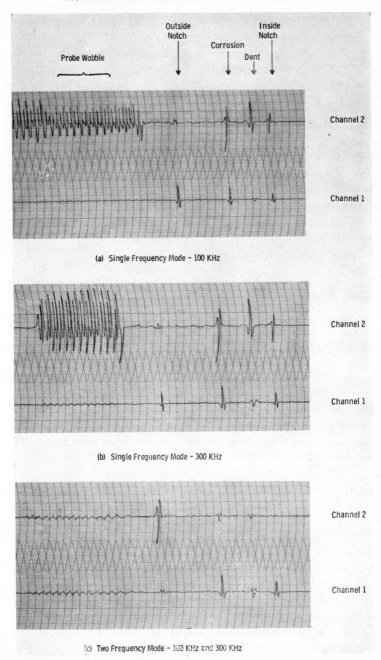

FIG. 15. Test results for tube in Fig. 14 showing comparison of single-frequency and two frequency modes of operation of a two-frequency multiparameter eddy current tester

trace of the chart records shown in Figs. 15(a) and 15(b). In these two single-frequency runs, the division of the defect signals between the two channels is very much predetermined by the phase relationship between the defect and probe wobble signals.

In the conventional eddy current test this division is not under control of the operator.

2. Two-frequency Mode

In contrast, performance of the new instrument operated in its two-frequency mode is shown in Fig. 15(c). Here the instrument has been adjusted to minimize the effect of probe wobble on both output channels, and to place the signal due to the inside notch mainly on channel 1 and the signal due to the outside notch mainly on channel 2. The probe wobble signal has been greatly reduced on channel 2 making signals of interest more useable than in the case of the single-frequency test. Some loss of signal-to-noise ratio has occurred on channel 1.

B. Four-frequency Thickness Test System

A prototype eddy current tester (Libby and Atwood, 1963) using five test frequencies, 6 kHz, 22 kHz, 70 kHz, 250 kHz and 3 MHz, is shown in the block diagram of Fig. 16. The 3 MHz test frequency was included for use in a probe lift-off compensating system but was not used in the tests described here.

These five excitation carriers were generated by crystal oscillators, not harmonically related, and thus were not phase locked. A probe-type test coil was used for the flat metal sheet test specimens.

Separate a.c. nulling circuits were used to bring the input signals at the frequency separating section to a desired degree of compensation. Off-null or incomplete compensation was required because amplitude detectors were used instead of the more common amplitude-phase or synchronous detector.

Amplitude detectors when fed with the sum of a large fixed signal and a small varying signal will give an output whose variations are approximately proportional to that component of the small signal in phase with the large fixed signal. Band-pass filters were used ahead of the detectors to separate the different carrier frequencies. The transformation section shown in Fig. 11 was used and the output read on d.c. current meters. However, an oscilloscope or strip chart could also be used. Because of the type of detection circuit used, a large d.c. component appeared in the output. Use of balanced synchronous detectors would eliminate this disadvantage, or the d.c. component could be compensated by adjustable d.c. signals. As only one signal component per test frequency is used in this instrument, it has a theoretical four parameter separating capability. However, the approximate

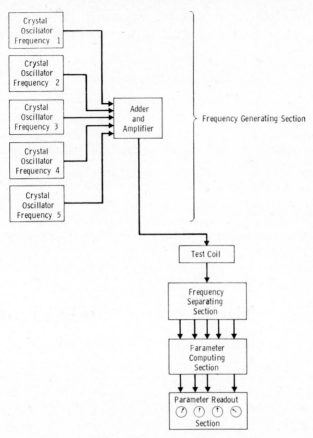

FIG. 16. Block diagram of a prototype tester.

nature or " quasi " status of the phase detection method used probably limits the operation.

1. *Test Probe and Test Specimen Arrangement*

The test probe and test specimen arrangement used for the tests to be discussed is shown in Fig. 17. The test specimen comprised two metal sheets, a top sheet of brass and a lower sheet of carbon steel. This gave a three-parameter system. These parameters were varied over the following listed ranges:

p_1 Probe spacing 0·0 to 0·762 mm (0·030 in).

p_2 Brass thickness (0·0254 to 0·152 mm (0·001 to 0·006 in).

p_3 Carbon steel thickness 0·0254 to 0·152 mm (0·001 to 0·006 in).

FIG. 17. Test probe—test specimen arrangement multiparameter thickness test.

For the tests to be described, three input signals (c_1, c_2, c_3) to the transformation section were formed by adding pairs of the outputs of four detectors as follows:

c_1 6 kHz and 22 kHz.

c_2 22 kHz and 70 kHz.

c_3 70 kHz and 250 kHz.

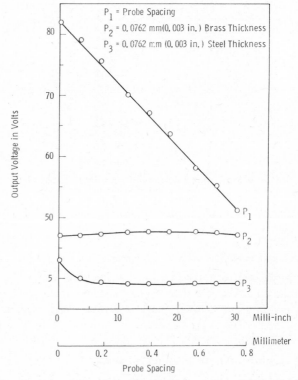

FIG. 18. Parameter separation when four test signals are used to separate three parameters (p_1 variable).

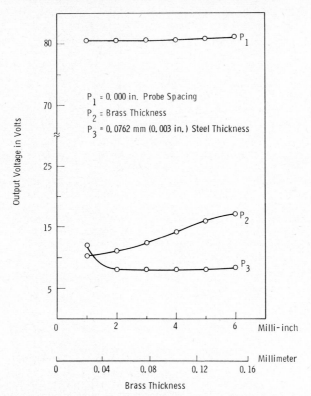

FIG. 19. Parameter separation when four test signals are used to separate three parameters (p_2 variable).

This made it possible to use information from four test frequencies to be used in the separation of three parameters.

Matrix elements of the modulation matrix (eqn. (4)) were determined by measurement, varying each parameter in turn. The functions were curved, so average slopes were used. Calculation of the inverse matrix of the modulation matrix permitted the setting of the transformation section potentiometers. Trimming these adjustments gives optimum results.

2. Results of Tests

The results of this test are shown in Figs. 18, 19, and 20. Good parameter separation is obtained over the following ranges:

p_1 Probe spacing—0·127 to 0·762 mm (0·005 to 0·030 in).

p_2 Brass thickness—0·0508 to 0·152 mm (0·002 to 0·006 in).

p_3 Carbon steel thickness—0·0254 to 0·0767 mm (0·001 to 0·003 in).

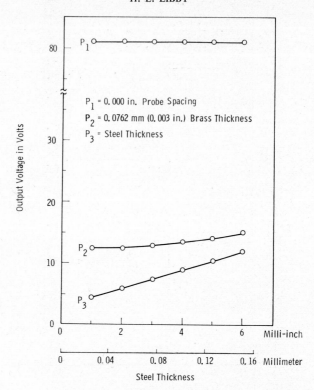

FIG. 20. Parameter separation when four test signals are used to separate three parameters (p_3 variable).

C. Tubing Tester with Cross Section Display Feature

Figure 21, describing the test system, and Fig. 22, showing typical test results, demonstrate further potential of the multiparameter test. Here a section of tubing having gross fabricated defects is inspected using a two-frequency multiparameter eddy current tester. The output of the multiparameter tester, together with tubing radial position signals obtained from a resolver, produce a simulated pattern of the tubing being inspected. The oscilloscope display shows the wall thickness or relative location and severity of flaws opening on the outer or inner walls.

1. Formation of Display Pattern

This pattern is generated by applying rotating radial sweep deflection voltages into the oscilloscope vertical and horizontal inputs, and by unblanking the cathode ray beam with defect or thickness signals derived from

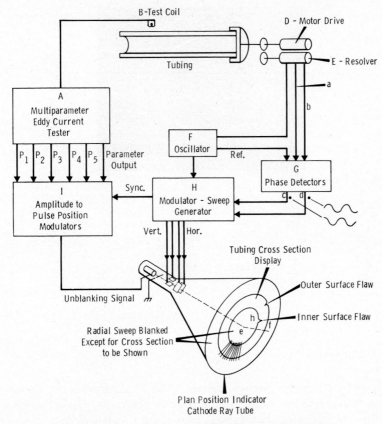

FIG. 21. Eddy current test with tubing cross-section display.

the multiparameter tester. The general system for obtaining the rotating radial sweep is patterned after the plan position indicator used in radar display systems.

2. *Description of Circuit Functions*

Referring to Fig. 21, the system comprises a multiparameter eddy current tester A, test coil B, a section of tubing C being rotated by motor D, cathode ray tube display and other associated electronic circuits. The multiparameter tester operates at two frequencies, 70 kHz and 250 kHz. Demodulation of the eddy current signals is done by synchronous detectors (amplitude-phase detectors). A $\frac{1}{2}$-inch diameter external tangent test coil is used.

Resolver E, also rotated by motor D and excited from the 5 kHz oscillator F, drives Phase Detectors G which provide the Modulator-Sweep Generator

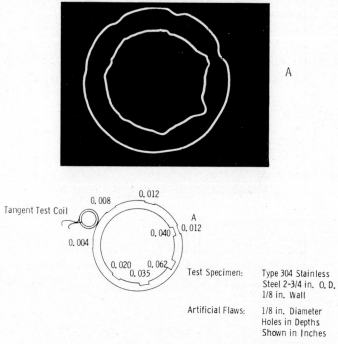

FIG. 22. Dynamic display of tube cross-section two frequency multiparameter test.

H with tubing angular position signals. The Modulator-Sweep Generator *H* produces a series of vertical and horizontal deflection voltages modulated with the sine–cosine tubing angular position information (c, d) to produce a rotating radial sweep of the cathode ray beam. The Amplitude-to-Pulse Position Modulators, *I*, convert the parameter information from the multiparameter tester into a multiplexed pulse position signal to unblank the cathode ray tube beam. These modulators are synchronized with a signal from Sweep Generator *H* to assure the proper orientation of flaw signals on the display. Amplitude-phase detectors were used making available four demodulated signals for inputs to the transformation section. Figure 11 describes the transformation system used.

3. *Dynamic Display Results*

The dynamic display of the cross section of a piece of type 304 stainless steel tubing obtained with this system is shown in Fig. 22. Similar orientation of the test specimen showing the fabricated defects and the photograph of the resulting dynamic display permit easy comparison. The tube diameter was 6·99 cm ($2\frac{3}{4}$ in) with 3·18 mm ($\frac{1}{8}$ in) wall and had artificial defects made by drilling 3·18 mm ($\frac{1}{8}$ in) diameter holes to the depths shown ranging from 0·305 to 1·57 mm (0·012 to 0·062 in). The transformation section of the tester was adjusted to give read-outs of outer surface defects and inner surface defects essentially independent of probe wobble on separate channels. In this test, probe wobble information was not used. A tangent-type test coil was used on the rotating tube. The rotating radial sweep was generated as previously described with tube position information being obtained from a sine–cosine resolver.

The inner and outer surface defect signals from the multiparameter tester were fed to the dynamic display circuitry on the parameter p_1 and parameter p_2 input circuits, respectively, shown in Fig. 21. These signals are d.c. coupled from the detector circuits to the display; thus the pattern shown is obtained at very slow to medium sample rotation speeds. Also, the d.c. connection permits wall thickness variations to be shown, even when they occur over large areas.

The simultaneous detection (and display) of the 0·305 mm (0·012 in) deep outer surface defect and the 1·02 mm (0·040 in) deep inner surface defect at location A in Fig. 22 are of special significance. This demonstrates the capability of the multiparameter tester to identify the signals due to these two types of defects even when both defects are present in the field of the coil at the same time. This type of performance was predicted in the description of the theory and original tests of the multiparameter test principle.

The adjustment of the summing networks in the transformation section of the multiparameter tester to give the desired read-out for particular test specimen standards is equivalent to a sort of instrument calibration for those standards. If, after such an adjustment, the tester is presented with a test specimen which has a new and different " uncalibrated for " parameter, such parameters will give an indication on more than one channel and should thus be readily identified as something new.

VI. Future Potential

It has been shown that application of present research study results gives test information in a form not available using conventional techniques. Thus, there is great potential just in the application of the presently demonstrated principles and in the refinement of instrumentation to implement them.

However, the greatest potential lies in the extension of the method, in a practical way, to handle an increased number of parameters and in the application of adaptive control techniques for automatic adjustment of the test equipment.

It is not expected that this new approach to eddy current test problems will replace present single frequency tests. The equipment is complex and adjustment is complicated in its present stage of development. It is expected that it will be in the area of special problems, not solvable with conventional tests, where using the more general approaches will be justified.

The multiparameter research has already given a better understanding of the single frequency tests and promises to give further clarification to aspects of conventional tests as the study is extended to more parameters or variables.

VIII. ACKNOWLEDGEMENTS

This paper reports on work performed under Contract AT(45–1)–1830 between the U.S. Atomic Energy Commission and Battelle Memorial Institute and on earlier phases of the work performed under Contract AT(45–1)–1350 between the U.S. Atomic Energy Commission and General Electric Company.

The contributions of Professor C. W. Cox and Dr. K. W. Atwood in early parts of the work involving theory, measurements and instrument design are acknowledged. The author is also grateful for many stimulating discussions with Dr. W. H. Huggins and Dr. S. Litman in the area of system identification and signal analysis.

The assistance of C. R. Wandling in the development of the cross section display device and the two-frequency multichannel tubing tester is also acknowledged.

REFERENCES

Everitt, W. L. and Anner, G. E. (1956). "Communication Engineering." 3rd edn, pp. 329–339. McGraw-Hill, New York.
Foerster, F. (1954). *Z. Metallk.* **45**, 180–187.
Huggins, W. H. (1957). "Representation and Analysis of Signals," Part 1 (ASTIA No. AD 133741).
Libby, H. L. (1958). Broadband electromagnetic testing methods, Part 1. *Nucl. Sci. Abstr.* **13**:12587, HW–59614.
Libby, H. L. (1966). U.S. Patent No. 3,229,198.
Libby, H. L. and Atwood, K. W. (1963). Broadband electromagnetic testing methods, Part 3. *Nucl. Sci. Abstr.* **18**:14055, HW–79553.
Libby, H. L. and Cox, C. W. (1961). Broadband electromagnetic testing methods, Part 2. *Nucl. Sci. Abstr.* **17**:8381, HW–67639.
Libby, H. L. and Wandling, C. R. (1968). Multichannel eddy current tubing tester. *Nucl. Sci. Abstr.* **22**:37626, BNWL–765.
Litman, S. and Huggins, W. H. (1963). *Proc. IEEE* **51** (6), 917–923.
McMaster, R. C. (ed.) (1959). "Nondestructive Testing Handbook," Vol. II. Ronald Press, New York.
Mishkin, E. and Braun, Jr., L. (1961). "Adaptive Control Systems," pp. 270–322. McGraw-Hill, New York.

CHAPTER 12

Pulsed Eddy Currents

D. L. WAIDELICH

*Department of Electrical Engineering, University of
Columbia, Missouri, U.S.A.*

I. INTRODUCTION

Possibly the earliest use of eddy currents in the determination of the physical properties of conducting materials was in 1879 when a simple apparatus was employed to test various metals (Hughes, 1879). The apparatus consisted of a clock which, with a microphone, acted as the generator to drive the two primary coils of two similar transformers connected in series. The two secondary coils were connected in series such that the induced voltage of the first coil tended to cancel the induced voltage of the second coil and a galvanometer in series showed no deflection if the coils were balanced. If two dissimilar metal samples were placed within the coils a deflection resulted whose amount depended on the degree of dissimilarity of the metals (Roberts, 1879). This apparatus has the essentials of the more recent equipment used in the nondestructive testing of metals using sinusoidal (Hochschild, 1958; Lewis, 1951; Wilcox, 1967; Russell *et al.*, 1962; Cheng, 1965; Libby, 1967) and pulsed eddy currents (Renken, 1965a; McGonnagle, 1960; McGonnagle, 1961; McGonnagle, 1966; McGonnagle, 1965; Electromagnetic Testing, 1965; Grubinskas and Merhib, 1965; McLain, 1958; Ross, 1961). Somewhat similar pulsed nondestructive testing, but with the use of thermal energy, has appeared recently (Schultz, 1966; Green, 1966 and 1967). The present work is concerned primarily with the use of pulsed eddy currents in testing metals nondestructively.

II. Early Work

Difficulties encountered in the use of sinusoidal eddy currents in nondestructive measurements led to the consideration of pulsed eddy currents (Waidelich, 1954) in making such measurements. These difficulties included the effects of low sensitivity and the presence of high harmonic content. The experimental circuit used a thyratron pulser to drive a bridge circuit containing two probes as nearly identical as possible. One probe was placed on a standard metal specimen and the other on the unknown metal specimen to be studied. The output of the bridge circuit was observed on the screen of an oscilloscope. The equipment was used to observe the thickness of one metal clad on another metal, and the measurement was made by determining the horizontal shift of the crossing point of the oscilloscope trace as the depth of the clad metal varied. The lift-off effect, i.e. the variation of the distance between the probe and the surface of the metal, was found experimentally to change the slope of the curve on the oscilloscope screen. Thus, if the slope of the CRO trace was made the same each time, the lift-off effect would be neutralized.

An analysis of the wave traveling from the air to the metal was also made. The wave was assumed to be a plane wave and the electric and magnetic field directions were assumed parallel to the surface of the metal specimen. The response to a step wave assumed in the air had a number of terms. The first term was that reflected from the air and metal boundary and was not important for the determination of the clad depth. The second term was that concerned with the reflection from the metal-to-metal boundary and should have the clad depth information included in it. This wave form is shown in Fig. 1. The objective of the bridge circuit was to balance out the first terms from both the standard and the unknown probes so that the effect of the second terms could be observed. The probes used consisted of a solenoid wound on a ferrite rod with a cylindrical ferrite shell surrounding the solenoid. The center ferrite rod was made in two pieces with an air gap in between the pieces. The second piece could be moved and this aided in the balancing of the bridge circuit.

Further details of the circuit used have been presented (Waidelich, 1955) along with a picture of the experimental equipment and a view of the probe. Additional information about the construction of the probe was also given.

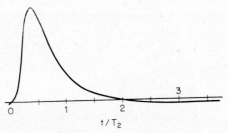

FIG. 1. Calculated pulse waveform reflected from metal-to-metal boundary.

Photographs of the output waveform as well as the expanded part of the waveform around the crossing point were shown. A comparison of the results obtained by nondestructive measurements using the pulsed eddy current instrument and by destructive measurements using optical methods (Waidelich, 1955 and 1956) was presented.

It was felt that an analysis of the reaction of the fields in the metal upon the pulsed currents in the probe would be very helpful. A start in this direction was made (Cheng, 1964 and 1965; Russell et al., 1962; Waidelich and Renken, 1956; Renken, Jr., 1961; Renken, Jr. and Waidelich, 1958), using sinusoidal currents initially. The development of masked probes changed the direction in which this analysis was channeled and further work on the problem was delayed until relatively recently.

More details on both the experimental and theoretical work has been presented (Waidelich et al., 1958). In addition, much work on obtaining information from the oscilloscope traces and on reducing the lift-off effect (Waidelich, 1958) was done. A method of using photocells to automatically record the thickness of cladding was described. This consisted of mounting two photocells on the screen of the cathode-ray oscilloscope and then by use of a differential circuit the cladding thickness could be recorded automatically. To overcome the lift-off effect a further study was made of the crossing points. In theory, it was shown that there would be an infinite number of the crossing points except that only two or three were visible experimentally before the amplitude became so small that the points could not be observed. Two crossing points are visible in Fig. 2. The change in position of the crossing points with the thickness in cladding is shown in Fig. 2 also. The middle trace of Fig. 2 is with thin cladding while the top trace has cladding twice as thick and the bottom trace four times as thick. The various waves shown in each trace are caused by varying the lift-off distance. A very brief mention of this work is available elsewhere (Libby, 1956a and b, 1958).

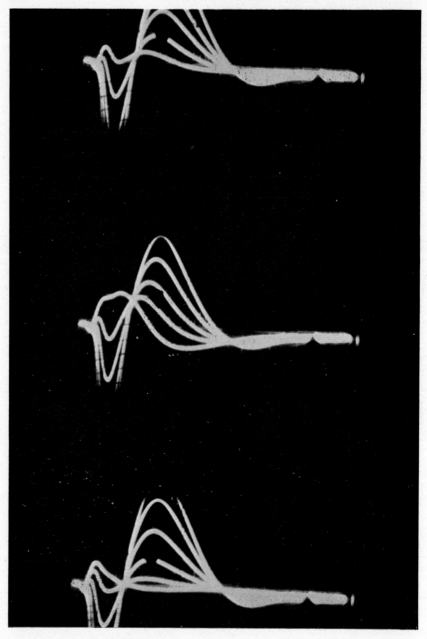

FIG. 2. Variation of the crossing points with cladding thickness.

It was desirable to eliminate the oscilloscope entirely and yet have the output recorded directly. An attempt (Meyers and Waidelich, 1958; Renken, Jr. *et al.*, 1958) to do this involved the use of a sampling pulse superimposed on the output pulse from the bridge circuit. The sampling pulse was generated by discharging a short length of transmission line through a thyratron. The resulting pulse was added to the output pulse at the crossing point where the lift-off effect was a minimum. The sum of the two pulses was measured by a peak-reading voltmeter for an indication of the plating thickness. The output voltage was presented as a function of the thickness of stainless steel on a brass base, and it was found that the results were almost independent of the lift-off distance. The advantage of this method is that it eliminated completely the use of the cathode-ray oscilloscope.

A double-frequency sinusoidal eddy current system had been developed earlier (Renken and McGonnagle, 1958; Renken, Jr. *et al.*, 1958; Beck *et al.*, 1959) with the higher frequency used to determine the probe-to-specimen spacing, i.e. the lift-off distance, and the lower frequency used to obtain information on the defects inside the metal specimen. This led to the development of a system employing two pulses (Renken and Meyers, 1959 and 1960). One pulse was short in duration and was used to obtain information on the probe-to-specimen spacing independently of the metal specimen conductivity, while the other pulse, longer in duration, penetrated into the

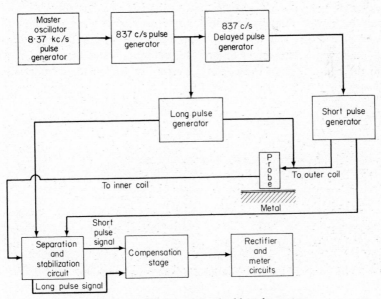

FIG. 3. Block diagram of the double-pulse system.

metal and provided information about the defects in the metal. A block diagram of the instrument is shown in Fig. 3. The circuit was so arranged that the response of the shorter pulse automatically compensated for the lift-off effect in the response of the longer pulse.

A system that combines some of the advantages and disadvantages of both the sinusoidal and pulsed systems uses a swept-frequency device (Hanysz, 1958). A sawtooth sweep generator develops a voltage which changes the frequency of a reactance oscillator linearly. Since the depth of penetration of the eddy currents varies inversely as the square root of the frequency, the reflected load of the metal and probe on the oscillator changes with frequency. The oscillator may be adjusted so that oscillations cease when the load becomes too heavy and thus the time the oscillator is oscillating during one period of the sawtooth voltage can be used as a measure of the overlay thickness. Examples of the use of this instrument are to determine overlays of lead-tin or lead-indium babbitt on bronze and silver with thicknesses of 0·3 to 1·5 mils* and of aluminum on stainless steel with thicknesses of 3 to 6 mils.

A. Masked Probes

One of the major improvements in the use of pulsed eddy currents was the idea and development of masked probes (Annual Report for 1959, Metallurgy Division, 1960; Renken, 1960a; Renken, Jr., 1965). A cross-sectional view of such a probe is shown in Fig. 4. The masking material should be a good electrical conductor such as copper. The driving coil was wound on a ferrite rod, and the peak pulse power was in the neighbourhood of one kilowatt. The center line of the driving coil was somewhat removed from the center line of the conical aperture in the masking material. The

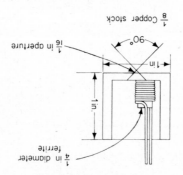

FIG. 4. Cross-sectional view of a masked probe.

*One mil is 0·001 inch or 0·0254 millimetre.

mask should be thick enough so that practically no field leaks through it during the observable length of the pulse. In this way the field is restricted to about the size of the aperture, leading to much better resolution than would be possible from a probe consisting of a coil wound on a ferrite rod or one wound on the center core of a ferrite shell.

One use of the masked probes was in testing metals using a through-transmission technique (Annual Report for 1959, Metallurgy Division, 1960; Renken, 1960a; Renken, Jr., 1965). In this application the probe is placed near one side of the metal sheet and the pick-up coil near the opposite side. The metal both attenuates and delays the pulsed field as it passes through the metal and so the magnitude or the delay or both may be used as an indicator of the conditions inside the metal. Observations were made of a plate of 16 gauge stainless steel (Renken, 1960a) with various diameter holes drilled in it to different depths. One, for example, was a 13·5 mil diameter hole drilled completely through the plate while another was a 5·9 mil diameter hole drilled to a depth of 20 mils. This method was used to test tubing with the pick-up coil placed inside the tubing.

Additional results on the through-transmission systems (Renken, 1960b) are shown in Fig. 5 which indicates the attenuation and delay caused by the passage of the pulse through the metal. It should be mentioned that an analysis (Waidelich, 1954) using the assumption that the applied field was a plane wave will give approximately the right delay as shown in Fig. 5, but the attenuation predicted by the analysis is larger by about two orders of magnitude than that measured.

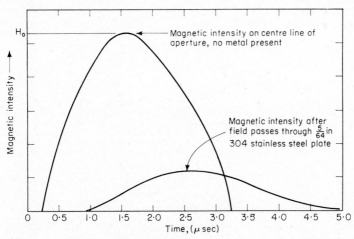

FIG. 5. Magnetic intensity at the pick-up coil showing the attenuation and delay caused by passage through the metal.

Work was also being done in the reflection technique (Renken, 1960b), i.e. using both the masked probe and the pick-up coil on the same side of the metal. The masked probes used had apertures varying from 60 to 20 mils in diameter. Figure 6 shows both the surface reflection waveform and the metal-to-metal interface reflection for a one-mil aluminum shim on a thick stainless steel base. The reflection technique has the advantage of requiring probes on only one side of the specimen being tested (Annual Report for 1960, Metallurgy Division, 1961) in contrast to the through-transmission method which requires circuit elements on both sides of the metal. For long tubes of small diameter the through-transmission method becomes impractical.

Additional results obtained from work on the through-transmission system (Renken, 1962a) showed that the magnetic flux came out one side of the aperture in a masked probe and returned on the other side of the aperture. This is demonstrated by curve A in Fig. 7. The addition of a conducting shim across part of the aperture reduces the flux on the one side of the aperture and the flux on the other side of the aperture spreads out somewhat as shown in curve B in Fig. 7. The through-transmission system was used to

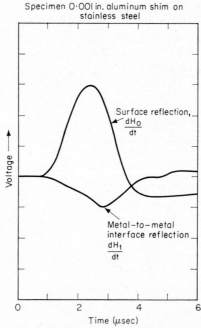

FIG. 6. Surface and metal-to-metal reflections from a 1-mil aluminum shim on a stainless steel base.

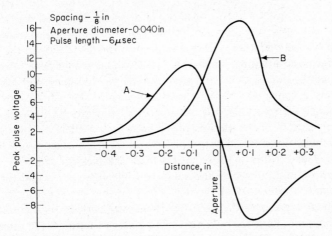

FIG. 7. Pulse voltages induced under an aperture in a masked probe.   (A) Represents the curve without a blocking shim and (B) that with a shim.

test some aluminum alloy tubes with 22 mil wall thickness.   The pick-up was on a rod passed through the inside of the tube.

Further work on the reflected pulse system was reported (Renken, 1962a; Annual Report for 1961, Metallurgy Division, 1962; Rice, 1961; Renken, Jr., 1966).   A soft iron flux guide about the pick-up coil was added to the masked probe as shown in Fig. 8. The driven field inside the mask was produced by a coil with a ferrite core although a spark discharge could also be used.   One problem with the use of a spark discharge is the jitter

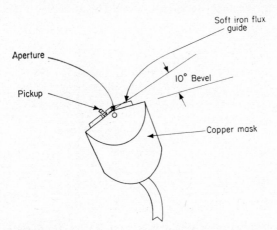

FIG. 8. Masked probe incorporating a soft iron flux guide.

that normally is present. The positioning of the pick-up coil was such that almost no voltage across the coil appeared when no test specimen was present, but when a specimen approached the aperture the fields around the pick-up coil changed and a voltage appeared. This voltage would have information in it about the conductivity and permeability of the metal and also about any defects which might be present in the metal. The soft iron flux guide would reduce the reluctance for the magnetic fields around the pick-up coil, but not all designs of the masked probe need such a guide.

An example of the pulsed output waveform from a masked probe is shown in Fig. 9. Curve (a) is the total signal from the pick-up coil with a 10 mil thick sheet of aluminum over the aperture while curve (b) is that for a very thick sheet of aluminum. It should be noticed that the negative part of the pulse, which represents the reflection from the surface of the metal, has exactly the same shape for both curves. In fact, substantially the same results would be obtained if the driven pulses had a longer or shorter length than that used, so the results depend very little upon the length of the driven pulses. If a gating circuit were used at approximately the point in time where curve (a) of Fig. 9 had its maximum, the output of the gate could be used as a measurement of the thickness of the aluminum sheet. This led to another system which incorporated the idea of using gates in a similar manner to that used in the double pulse method described earlier (Renken and Meyers, 1959).

B. Other Pulse Methods

One method of some interest was that of determining the resistivity of metals by setting up a magnetic field about the metal using a primary coil and then letting the field decay with time (Bean et al., 1959; Clarke and

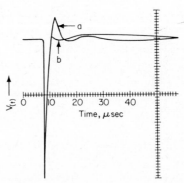

FIG. 9. Output waveform using a masked probe with curve (a) showing a 10 mil thick sheet of aluminum over the aperture and curve (b) a very thick sheet of aluminum.

Mordike, 1966; LePage *et al.*, 1968). The changing of the field in the secondary coil about the metal induces a voltage which decreases with time as viewed in a cathode-ray oscilloscope. From this voltage variation it is possible to determine the resistivity of the metal. A circuit (Clarke and Mordike, 1966) using repeated pulses makes the measurements easier and quicker.

Another method involved the application of a rectangular pulse to a ferromagnetic bar (Shinohara and Otori, 1961). A beam power tube is used to apply a pulse to the two primary coils in series. Two secondary coils are also connected in series opposing so that the difference voltage may be observed in an oscilloscope. One set of coils would encircle a standard metal specimen while the other would encircle a test specimen. The device was employed in testing for the properties of materials, in determining the thickness of the surface hardened layer of a steel cylinder and in searching for cracks.

A pulsed system which had many of the characteristics of sinusoidal systems was used to determine the sodium levels in reactor fuel rods (Ono and McGonnagle, 1961). Fig. 10 shows a cross-sectional view of the three coil assembly used for producing the pulses and the inner two coils used for picking up the response from the fuel element. The output voltage of the detecting circuit is also shown with typical deflections caused by motion of the fuel rod as the uranium of lower electrical conductivity is replaced by the sodium of larger electrical conductivity. The driving pulse was a damped

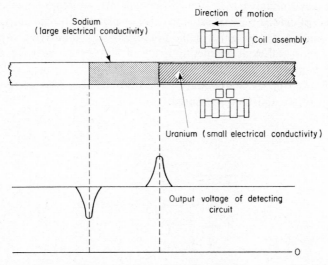

Fig. 10. System employed to detect the sodium level in reactor fuel rods.

sinusoidal wave and this was the reason for some of the sinusoidal characteristics of the instrument.

This type of pulsed system has been developed further (Ono, 1964a, 1966 and 1967) by using the double frequency idea (Renken, Jr. *et al.*, 1958) utilizing both low and high frequency pulse generators. One pair of frequencies used was 100 and 17·5 kilohertz while another pair was 45 and 8·5 kilohertz. A gate circuit is used to separate the two frequencies as obtained from the pick-up coil so that two oscillograph traces, one from each frequency, may be made. Tests were made on both monel and steel tubes, and it was found possible to discriminate between defects on the outside and on the inside of the tubing. Also the approximate size of the defects could be determined. An evaluation (Ono, 1964b) of the coils used in this method of pulsed eddy current testing has been made.

A somewhat similar pulsed nondestructive testing device (Uozumi, 1964 and 1966) using eddy currents was employed in identifying the composition of alloys, in measuring the thickness of the hardened depth of steel, and in detecting flaws. The basic pulse generator circuit is shown in Fig. 11 where the pulse is started by use of a negative gate. The resistor R_1 is used to produce a more constant amplitude of the oscillations. For output indication a trace on a cathode ray tube was used or alternately a selector pulse and a peak voltmeter gave a reading.

A study (Philippe, 1963) was made of the relationship between the time duration of the current pulse in a primary coil and relative output voltage of a secondary coil with a thin sheet of metal in between the two coils. The sheet of metal was simulated by a pool of mercury whose depth could be varied. The effects of defects were investigated also by placing a thread of nylon in the mercury pool. A double pulse method of testing was employed in which both long and short pulses at a repetition rate of 800 per second were generated and applied through the primary of a probe to a metal surface and the output was picked up by a secondary. The output pulses were separated with the aid of filters and then measured.

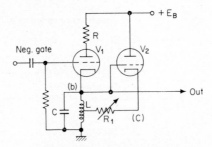

FIG. 11. Pulse generator circuit.

The idea of using pulses has also been considered in an investigation of broadband electromagnetic testing methods (Libby, 1959 and 1966; Libby and Cox, 1961; Libby and Atwood, 1963). With the use of the conventional one-frequency sinusoidal eddy current equipment not enough information may be obtained to distinguish between the various defect signals. It was proposed then that multiple frequencies should be used in the form of non-sinusoidal waves or pulses. This involves a rather formidable problem in separating the desired information. One method of achieving this employs an orthonormal exponential filter with a time domain sampler.

A model of a metal plate was constructed using Wood's alloy heated above its melting point (Atwood and Libby, 1963; Leatherman, 1965) and the driving coil was located about the surface of the alloy with the pick-up coil somewhere in the liquid alloy. The driving function was a square wave with a period long enough so that the transients in the alloy had time to disappear before the next transient came along. One interesting result of this study was the comparison between the measured current densities using the above method of measurement and the calculated current densities using the plane wave analysis as shown in Fig. 12. Notice that the decay rate is larger for the measured values than for the calculated values. Also the current densities

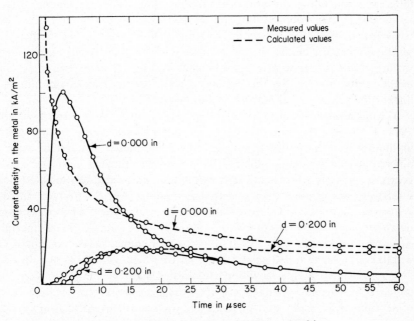

FIG. 12. Measured and calculated current densities.

for different depths in the alloy approach each other much sooner in time than predicted by calculation.

Another pulsed system (Nikolayenko, 1965) utilizes a rectangular pulse generator driving one tube which in turn drives two probes. Under one probe is placed the metal to be tested while under the other probe is the standard metal. The outputs of the two probes are compared in a second tube and the comparison voltage is amplified and is projected as an image on the screen of a cathode-ray tube. For measurements the signal is differentiated, amplified and the peak value is measured by a vacuum tube amplifier.

C. Reflection Systems

The initial work on a pulsed-field reflection system (Renken, 1962b) was outlined. The mask-aperture assemblies were driven by pulses 0·5 to 50 microseconds long, and the field which passed through the aperture had a very small cross-section leading to improved resolution and discrimination. The field was reflected by the metal specimen and detected by a pick-up coil near the aperture. The system was used in testing refractory alloy tubing.

A special pulsed system (Selner et al., 1963; Leatherman, 1964) was developed to test for cracks and other defects in Zircaloy tubing which could not be detected by radiographs or by sinusoidal eddy-current equipment. The differential encircling coil used had three windings, a center winding driven by pulses six microseconds long and shaped like a sine loop and two other windings which picked up the information. The instrument was sufficiently sensitive to detect cracks of approximately two mils in depth on the inner surface of the tubing with a 15 mil thickness. Fig. 13 shows wall inclusions in the tubing detected by this system. The inspection was carried out at a linear tube velocity of about two inches per second.

A considerable variety (Annual Report for 1963, Metallurgy Division, 1963) of alloys in the form of tubing with wall thicknesses of 0·508 millimetre or thinner were tested electromagnetically. The pulsed test system using a differential coil was later replaced by a system employing a coaxial mask-aperture assembly which had better resolution and needed less maintenance. The axial velocity used was 3·6 metres per minute. The electromagnetic test systems using pulsed fields with masks are especially useful when resolution, reliability and high sensitivity are important. A coaxial mask-aperture assembly (Renken, Jr. and Sather, 1968) with a copper mask was employed with the tube being tested encircled by a driven field coil and a pick-up coil.

A particularly useful narration (Renken, 1964a) of the work done on pulsed eddy current systems up to 1964 has been presented. This includes some theoretical results of a pulsed plane wave normally incident on a clad metal above a metal base, the use of the through transmission technique for

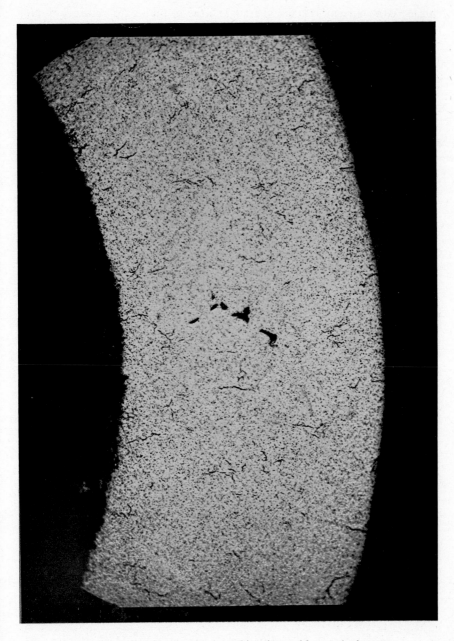

FIG. 13. Wall inclusions in Zircaloy tubing detected by a pulsed system.

pulsed eddy current systems, the advantages of masks for better resolution and a description of a double-pulse eddy current system. The pulsed field reflection system using mask-aperture assemblies is discussed.

The application of a pulsed electromagnetic test system for inspecting thin-walled tubing (Renken, 1964b) used in the nuclear field is presented. Tubing less than 5·8 millimetres in diameter and with walls less than 0·64 millimetre were tested electromagnetically using a pulsed system. A coaxial mask-aperture assembly was shown with the field coil driven by a one-microsecond current pulse of ten amperes peak value. The voltage of the pick·up coil had a maximum value of four volts and no balancing or bridge circuits were used. If two pick-up coils are used, one on each side of the field coil and connected differentially, i.e. in apposition, this helps in suppressing some effects of dimensional interference in the tubing. Views of a mask-aperture assembly are shown in Fig. 14 including a cross-sectional view and details of the aperture used. In the system employed sampling was used to extract the desired information, and the compensation for the lift-off effect was achieved as indicated previously (Renken, 1962a). The slow changes of the lift-off distance may be eliminated from the signal by a filter, but the

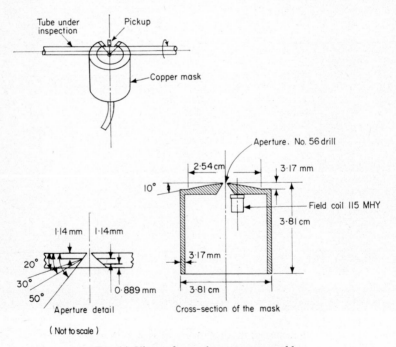

Fig. 14. Views of a mask-aperture assembly.

rapid changes such as those of vibration require a much more carefully designed compensation system so that the demodulated signals from the sampling process add algebraically while the variations caused by vibrations tend to cancel. An automatic gain control is also required to correct for the change in signal amplitude caused by the changes in lift-off distance. One test system used had a pulse repetition rate ranging from one to two thousand pulses per second and with linear inspection rates up to six metres per minute. This is slower than some commercial sinusoidal eddy-current systems but the reject levels used in the pulsed systems for the nuclear applications are also lower.

Some experiences with the pulsed eddy-current systems in the inspection of tubing have been described (Annual Progress for 1964, Metallurgy Division, 1964). One application was in the inspection of the ten per cent defect level of type 304 stainless steel jacket tubing using an encircling mask-aperture assembly. The tubing, 3·96 millimetres inside diameter and 0·229 millimetre wall thickness, was to be used in jacketing the fuel elements in the EBR-II Fuel Cycle Facility. The inspection speed was 3·66 metres per minute and 18,600 metres of the tubing were inspected. Point-type mask-aperture assemblies were also used to test type 304 stainless steel tubing 0·508 millimetre in wall thickness. As a ten per cent standard defect, an Elox notch, 0·508 millimetre long and ten per cent of the wall thickness in depth, was cut on the inside surface of the tubing. It was found that the noise and interference caused by the tubing was less than twenty per cent of the signal from the standard notch. When interference problems are severe point-type mask-aperture assemblies with two distinct apertures about 1·5 millimetres apart have been tried with success. The fields are of the same amplitude but opposite in polarity so that any condition which appears under both apertures at the same time will not be detected in the pick-up coil. These double aperture assemblies have been especially useful in vanadium-titanium alloy tubing development.

Another example (Renken, 1965a) of the possibilities of amplitude sampling in the time domain is shown in Fig. 15. The Hastelloy-X tubing has an outer diameter of 4·98 millimetres and a wall thickness of 0·38 millimetres. The effect of a 0·1 millimetre coating of tungsten on the inside of the tubing is shown. To measure the thickness of the tungsten coating an amplitude sampling in the vicinity of point A of Fig. 15 would be best. Again the relatively slow rate of the diffusion process means that the information on the conditions deeper in the metal appears later. Three disadvantages of the broadband electromagnetic test methods are mentioned. These are, first, the increase in noise voltages as a result of the wide bandwidths employed; second, the loss in inspection speed if the time between pulses is long compared

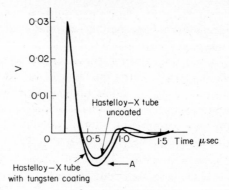

FIG. 15. Effect of a tungsten coating on the inside of a Hastelloy-X tube.

to the length of the pulses; and third, the fact that the idea of the impedance plane, so useful in sinusoidal eddy current testing, is of little help in pulsed eddy current testing.

The peak value of the pulse as determined in the pick-up coil (Renken, 1965b) is shown in Fig. 16 as a function of the aperture-to-metal distance. The parameter for the various curves is the resistivity of the metal being tested. The length of the pulses used is not at all critical since very much the same results are obtained when the pulses are much shorter, i.e. about one microsecond long instead of the ten microsecond pulses used in obtaining Fig. 16.

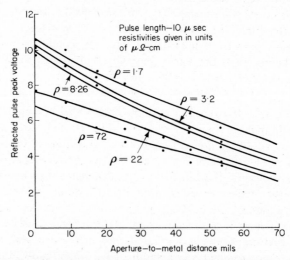

FIG. 16. The peak pulse voltage in the pick-up coil as a function of aperture-to-metal distance for various metal resistivities.

An interesting comparison of the measurement of the thickness of the wall of a zircaloy tube has been carried out by the use of three different methods (Destribats *et al.*, 1965), pulsed eddy current, sinusoidal eddy current and ultrasonics. The results obtained were illustrated by the use of four traces. One trace showed the measured results, a second trace the ultrasonics measurements, a third trace the sinusoidal eddy current results and the fourth trace the pulsed eddy current results. The wall of the tubing was slightly less than one millimetre in thickness. The measuring head was unusual in that provision was made for either eddy current or ultrasonic probes. Another study (Bonnet and Jansen, 1965) comparing the various methods of testing has been made.

The use of pulsed eddy current equipment in testing approximately 136,000 feet of EBR-II fuel-jacket tubing has been presented (Busse *et al.*, 1967; Leatherman, 1968) and the equipment employed is described in some detail. A multivibrator drives two thyratrons which pass large pulses of current through the differential-type encircling probe. The output from the pick-up coil of the probe is amplified and then rectified and monitored. The change in the rectified level is also amplified and recorded. Further amplification activates an alarm relay which turns on a loudspeaker powered by an audio oscillator. The driving coil is in the center while the pick-up coils are on the ends. Resistors and capacitors aid in balancing the coils while resistors are also used to damp the coils to reduce ringing. A recording showing two flaws in a tube is also shown with the flaw indications standing out above the ever-present tube noise.

Further information on the use of the differential coils in the inspection of tubing has been given (Leatherman, 1967; Renken *et al.*, 1966). Any slowly varying change in the tubing will affect both coils of the probe in the same way and will cancel in the output. This also means that no bridge circuit is necessary in balancing out the output of the pick-up coils when no defects are present.

D. *Correlation and Filtering Methods*

The advantages and disadvantages of the various nondestructive methods are discussed (Renken, 1965) for the testing of tubing. Electromagnetic methods are especially useful in detecting cracks which terminate on the tube surface at a very shallow angle, but, on the other hand, deep but short defects are detected better by radiographic methods. Ultrasonic and electromagnetic methods are both particularly adapted to the testing in production of tubing in great quantities, but the electromagnetic method has a considerable advantage in the speed of testing. The problem of unwanted signals for reasons other than defects has been overcome by decreasing the scanning area to that of the same magnitude of the area of the defects themselves. Also, a system of sampling in time aids in reducing the unwanted signals. These signals directly increase the cost of inspection. The use of masked probes in pulsed eddy-current testing decreases the scanned area but also decreases the speed of inspection so that the ultrasonic and eddy-current methods become comparable in the speed of testing. Then the ultrasonic method would be used for wall thicknesses of tubing greater than 0·75 millimetre and the eddy current method for wall thicknesses less than 0·5 millimetre thick with a range in between where either method might be suitable.

As the electromagnetic field diffuses into the metal a delay may be observed between reflections from a defect near the surface of the metal and another defect deeper in the metal. Hence sampling the output signal at various times will discriminate against unwanted signals. Fig. 17 shows a recent sampling system as a block diagram. If two sampling pulses are employed, the first one may be used early in reflected pulse from the pick-up coil so the information is primarily the spacing between the aperture and the metal, and this information may be used to increase the amplification of the system auto-

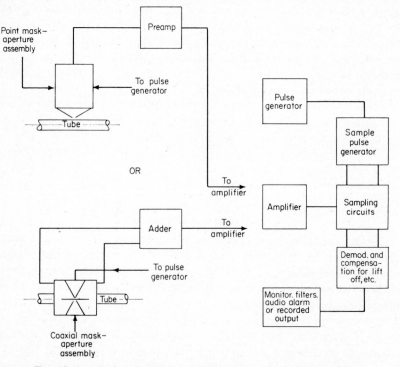

FIG. 17. A pulsed testing system showing the provision for sampling.

matically as the spacing increases. The second sampling pulse may be positioned at the best point in time for the observation of the particular defect desired. The output of the second sampling pulse is filtered and then read out or recorded as desired.

A large quantity of stainless steel with a 0·508 millimetre wall thickness was inspected this way at a rate of four metres per minute using one mask-aperture assembly. The positioning of the sampling pulse was such that both inside and outside notches were recorded with the same sensitivity. About one defect per 100 metres of tubing were found with labor costs of about $0·03 per metre for tubing cost of $1·50 per metre.

The noise in an electromagnetic testing system (Annual Progress Report for 1965, Metallurgy Division, 1965) may be divided into two main categories of noise, i.e. system noise and test specimen noise. System noise is the noise arising within the testing system itself and may include shot, partition, flicker, Johnson, jitter and similar noises. These noises require a wide band of frequencies for transmission and so a narrow-band electromagnetic testing system has a distinct advantage in discriminating against this type of noise. The test specimen noise would comprise changes in geometry, conductivity and permeability and vibration and would not necessarily be random. Methods useful in suppressing test specimen noise may actually work to increase the system noise. A wideband system such as the pulsed eddy current system will minimize the test specimen noise but the wideband width increases the system noise. It would help then to reduce the system noise first.

The noise is a system such as that of Fig. 17 has three main sources, one of which is preamplifier noise, which in turn is largely of thermal and shot origin. Low-noise preamplifiers have been constructed using type 7586 nuvistors and also type 2N3833 N channel field effect transistors. These reduced the system noise figure about 2·5 decibels. Another important noise source is the noise resulting from time jitter of the sampling circuits. More stable sampling circuits would reduce this source. The third source of noise is the pulse generator noise, and little has been done to reduce this noise as yet. The effect of system noise may be reduced by increasing the signal strength such as might result from a better design of the mask-aperture assembly. The reduction of the pulse generator noise would depend possibly on the development of a better method of effecting the separation of the information from the output pulses.

A comparison of the inspection of metals with high melting points using both ultrasonic and pulsed eddy current methods is presented (Renken, 1966). Electromachined (Elox) notches are probably the best artificial standards available at the present time for pulsed eddy current testing. Sometimes it is possible to use a standard defect on a tube of one metal to produce the nearly equivalent defect indication on a tube of different metal.

In a discussion (Lyke, 1967) of the statistical properties of the demodulated output of the sampling circuits, it is considered that the output is both stationary and ergodic. The outputs of two sampling points are selected and probability density functions, auto and cross correlation functions and power and cross spectral density functions are shown. The power density functions for the two sampling time functions are shown in Fig. 18. This shows that simple filtering in the bands outside the spectrum of the signal would help in increasing the signal-to-noise ratio. Unfortunately, the spectra of the noise and the desired signals overlap. Another method of processing the signals is to subtract one from the other. The noise in both channels has a high correlation factor so that subtraction may help a great deal.

The method of generalized filtering using the idea of representing a signal by a series of orthonormal functions has been suggested. One possible set of

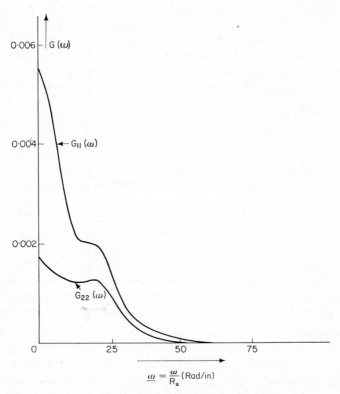

$$\underline{\omega} = \frac{\omega}{R_s} \,(\text{Rad/in})$$

Fig. 18. Power spectral density functions for two sampling channels.

orthonormal functions would be a set of exponential functions. Another method would be to match the filters to the given signal and the accompanying noise in such a way that the signal-to-noise ratio is optimized. The use of decision theory was suggested, and a possible system for the detection of various classes of faults was outlined. Adaptive filters might also be employed.

The liquid-metal fast breeder reactor in its fuel elements requires a jacket tube that has to pass some difficult nondestructive tests. One method of testing (Renken, 1968b) would be that of electromagnetic pulses. The concept of an " impulse defect " is suggested as a small element of current equal in magnitude and opposite in direction to the normal element of current at the location of the test specimen in question. Then the time response of the whole testing system to the impulse defect and its frequency spectrum by use of the Fourier transform may be employed as comparisons of the resolutions of two electromagnetic testing systems. Much of the noise that originates in the test specimen is nearly periodic. The spectrum that results consists mainly of some discrete lines. These may be filtered out by the usual low pass filter but much of the information is lost doing this, as indicated in Fig. 18. A differential transducer has much the same effect since it has a zero at zero frequency. This creates difficulties in detecting, for example, a long longitudinal crack since much of the output will be at the low frequencies.

It would be possible, but probably not very practical, to take the multi-dimensional signal from a pulse, compare it with similar signals from various defects stored in the memory of a computer and read out the type of defect. It probably would be more practical to process the pulse itself (Libby and Atwood, 1963) or to sample the pulse, demodulate the samples and then filter, cross-correlate or combine them.

A newer transducer used in testing tubing has one driving coil and a number of apertures at equal angular intervals around the circumference. The tubing passes longitudinally along the axis. The system is more complex but has the advantages of larger inspection speeds and better resolution. Also, the outputs of two or more pick-up coils may be combined to reduce interference. The signals may be produced by an artificial defect, i.e. a notch, on the inside surface of a tube, and a great deal of the noise and interference may be made to disappear by subtracting, for example, the number four sampling point output of one aperture from the same output of another aperture removed twelve degrees from the first in a circumferential direction.

In the liquid-metal fast breeder program, sodium is used to transmit the heat from the fuel to the jacket. To test this type of fuel element, thicknesses up to three millimetres in depth are tested using very long pulses. One important change in the masks of the probes is that they must be made of a high permeability material such as mu metal. Figure 19 shows the voltage

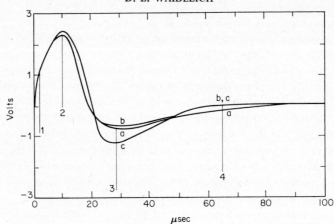

FIG. 19. The voltage pulses used in testing sodium capsules.

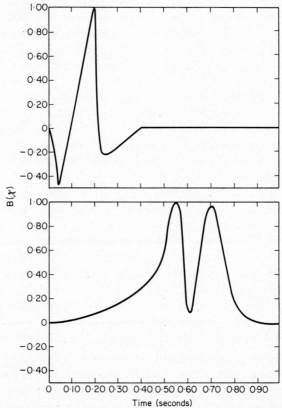

FIG. 20. Response curve of a mask-aperture assembly (upper curve) and of a small solenoid (lower curve).

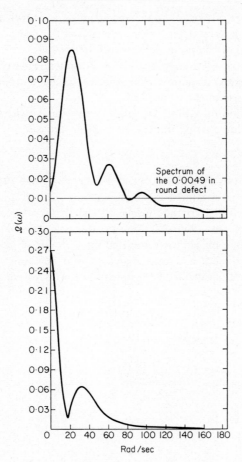

FIG. 21. Frequency spectra of a mask-aperture assembly (upper curve) and of a small solenoid (lower curve).

pulses used in testing a sodium capsule. Trace (b) in the figure is a normal section of the capsule while trace (a) shows a void about 1·5 millimetres deep in the sodium and trace (c) shows a similar void about 0·75 millimetre deep. Note that sample point 3 would be used for the void 0·75 millimetre deep while point 4 would be used for the void 1·5 millimetres deep.

The response curve (Renken, 1968a) of a mask-aperture assembly is shown in the upper part of Fig. 20 and that of a small solenoidal probe in the lower part of the figure. The mask-aperture assembly would have a much better resolution than that of the small solenoid. The corresponding frequency spectra as obtained from a computer is shown in Fig. 21. The straight line in the upper part of Fig. 21 is the spectrum of a 0·0049 inch round defect which simulates a point defect. The spectrum of the mask-aperture assembly fits the spectrum of the point defect much better than that of the solenoidal probe, but there is still a great deal of improvement possible, particularly in the high frequency response. Filtering could be employed to level the frequency response of the transducer.

Further work on the use of the pulsed eddy current method (Russkevich, 1968) in measuring coating thickness has been reported.

E. Field Analysis

An early analysis (Waidelich, 1954; Waidelich *et al.*, 1958) of the pulsed fields in the metal specimen assumed that the electric and magnetic fields were directed parallel to the surface of the metal. Experiments (Annual Progress Report for 1965, Metallurgy Division, 1965) showed that the attenuation when a pulsed wave traveled through a plate was much less than the analysis predicted, and thus indicated that an analysis assuming that the fields were not parallel to the metal surface probably should be tried.

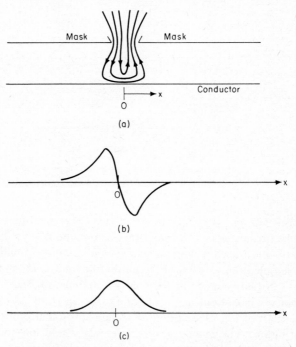

Fig. 22. The magnetic field lines near an aperture.   (a) Sketch of field lines; (b) vertical field variation; (c) horizontal field variation.

Further work (Kadochikov, 1965) using the magnetic field parallel to the metal surface was presented. The applied field to the infinitely deep metal was assumed first a ramp and then a step function of time. Then the applied field in the form of the same functions of time was applied to both surfaces of a metal plate of finite thickness. In all of these cases the magnetic field at any point in the metal as a function of time was derived. Lastly, three different saw-tooth waves repeated periodically were applied to both sides of the metal plate. The resulting calculated magnetic field at a point inside the metal plate was presented.

A relaxation method and a solution in the form of definite integrals were obtained (Leatherman, 1966; Dodd, 1967) for a coil with a rectangular cross-section carrying sinusoidal currents near a plane conductor or encircling a conducting rod. For a pulsed wave it is possible to use a time sequential relaxation method in determining the fields and currents in the conductor and in the space surrounding the conductor.

A method of using the Laplace transform in obtaining the response in a metal or in a coil near the metal is given (Nikolaenko, 1966). The first example given involves an applied field in the form of a step function in time and also parallel to the surface of a plane conducting half-space. The solution for a steady-state sinusoidal field is employed to find the response function and use of the inverse transform then will give the transient field desired. A second example is concerned with an impulsive magnetic field applied parallel to the axis of a very long conductor with a rectangular cross-section.

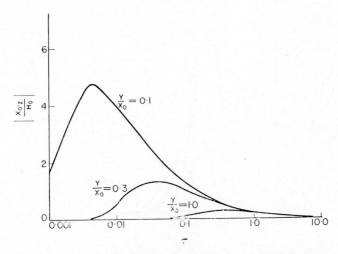

FIG. 23. Variation of the current density with time and the depth into the conductor.

The output is the transient voltage induced in a coil wound on the conductor and again the steady state sinusoidal response is used in obtaining the output.

Experimental tests (Waidelich, 1967) on a mask-aperture probe indicated that the magnetic field projecting through the aperture had a shape resembling that of Fig. 22. A mathematical analysis in two dimensions was made of the fields and current densities in the metal assuming that the impulsive horizontal field intensity at the surface of the metal decreased exponentially with distance from the center of the aperture. Fig. 23 shows the variation of the current density with time as a function of depth into the metal. Note that for a small depth into the metal the current density is large and its peak occurs early in time. For a large depth into the metal the current density is small and its peak occurs later in time.

An extension of the analysis (Chan, 1968; Chan and Waidelich, 1968) but in three dimensions was made with the vertical magnetic field at the surface of the metal assumed to decrease exponentially from the center of the aperture. Fig. 24 shows the current density variation as a function of time for three different conductivities of the metal. The lowest conductivity is that of 304 stainless steel, the middle one that of zirconium, while the largest one is that of lead. The curve for the lowest conductivity has the largest peak value and the peak occurs earliest while that of lead has the smallest peak and the peak occurs latest. The current density for all three curves reverses direction at larger values of time.

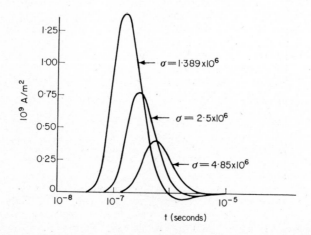

FIG. 24. Current density variation with time for three different conductivities.

III. CONCLUSIONS

An outline of much of the work done in the research effort on and in the development of pulsed eddy current systems has been presented here. The work being done at the present time is divided into three main parts; first, the effort, largely experimental, to improve resolution and lower the noise level by the use of better designs of equipment. Secondly, there is w ork being pursued on the use of correlation methods and electronic computers to obtain more reliable information from the raw pulsed data at the output of the systems. Thirdly, a mathematical analysis of the systems is continuing in the effort to develop a better understanding of how the systems operate with the expectation that better designs will result from the work. It is expected also that these systems will find a considerable number of applications in the burgeoning field of nuclear energy in the next ten to twenty years. It is believed that many applications in nondestructive testing exist in other fie lds, such as the aerospace field, in which the use of pulsed eddy-current systems would be highly advantageous and it is hoped that the present exposition will help in increasing their use in these other fields.

REFERENCES

" Annual Report for 1959, Metallurgy Division " (1960). Argonne National Laboratory Report ANL-6099, pp 55–56.

" Annual Report for 1960, Metallurgy Division " (1961). Argonne National Laboratory Report ANL-6330, pp. 130–131.

" Annual Report for 1961, Metallurgy Division " (1962). Argonne National Laboratory Report ANL-6516, pp. 67–68 and 185–190.

" Annual Report for 1963, Metallurgy Division " (1963). Argonne National Laboratory Report ANL-6868, pp. 105–108 and 227–228.

" Annual Progress for 1964, Metallurgy Division " (1964). Argonne National Laboratory Report ANL-7000, pp. 58–59.

" Annual Progress Report for 1965, Metallurgy Division " (1965). Argonne National Laboratory Report ANL-7155, pp. 163–164.

Atwood, K.W. and Libby, H. L. (1963). " Diffusion of Eddy Currents ", General Electric Company, Hanford Atomic Products Operation, Report No. HW-79844.

Bean, C. P., DeBlois, R. W. and Nesbitt, L. B. (1959). Eddy-current method for measuring the resistivity of metals. *J. appl. Phys.* **30**, 1976–1980.

Beck, W. N., Renken, C. J., Meyers, R. G. and McGonnagle, W. J. (1959). " Nondestructive Testing of EBR-1 Mark III Fuel Elements and Components ", Argonne National Laboratory Report ANL-5893.

Bonnet, P. and Jansen, J. (1965). " The Application of Various Non-Destructive Testing Methods to Fuel Elements of the Orgel Type ", *in* " Non-Destructive Testing in Nuclear Technology ", Proceedings of the Symposium on Non-Destructive Testing in Nuclear Technology, International Atomic Energy Agency, Bucharest, Romania, May 17–21, 1965, Vol. 1, pp. 205–231.

Busse, T. H., Wood, D. R. and Beyer, N. S. (1967). " Nondescructive Inspection of EBR-II Fuel-Jacket Tubing using Electromagnetic Techniques ", Argonne National Laboratory Report ANL-7334.

Chan, S. B. (1968). " The Penetration of Pulsed Electromagnetic Fields into a Conductor ", Ph.D. Thesis, University of Missouri.

Chan, S. B. and Waidelich, D. L. (1968). " The Penetration of Concentrated Pulsed Electromagnetic Fields into a Conductor ", Twelfth Metallurgical Congress on Nondestructive Testing and Control in the Field of Nuclear Metallurgy and Technology, Saclay, France, June 24–26, 1968

Cheng, D. H. S. (1964). " The Reflected Impedance of a Circular Coil in the Proximity of a Semi-Infinite Medium ", Ph.D. Thesis, University of Missouri, Columbia, January 1964.

Cheng, D. H. S. (1965). The reflected impedance of a circular coil in the proximity of a semi-infinite medium. *Trans. IEEE* IM-14, 107–116.

Clarke, J. M. and Mordike, B. L. (1966). Continuous contactless resistivity measurement. *Appl. Mater. Res.* 5, 181–186.

Destribats, M.-T., Allain, C., Prot, A. and Thome, P. (1965). " Control Methods used in the Department of Metallurgy for Structure and Fuel Elements ", *in* " Non-Destructive Testing in Nuclear Technology ", Proceedings of the Symposium on Non-Destructive Testing in Nuclear Technology, International Atomic Energy Agency, Bucharest, Romania, May 17–21, 1965, Vol. 1, pp. 167–189.

Dodd, C. V. (1967). " Solutions to Electromagnetic Induction Problems ", Oak Ridge National Laboratory Report ORNL-TM-1842.

" Electromagnetic Testing " (1965). Handbook H54, 15 October 1965, Superintendent of Documents, U.S. Government Printing Office, Washington, D.C.

Green, D. R. (1966). Thermal surface impedance for plane heat waves in layered materials. *J. appl. Phys.* 37, 3095–3099.

Green, D. R. (1967). Thermal surface impedance method for non-destructive testing. *Mater. Evaluation* 25, 231–236.

Grubinskas, R. C. and Merhib, C. P. (1965). " A Report Guide to Literature in the Field of Electromagnetic Testing ", DDC Report AD-615346, AMRA-MS-65-03.

Hanysz, E. A. (1958). Swept frequency eddy-current device to measure overlay thickness. *Rev. scient. Instrum.* 29, 411–415.

Hochschild, R. (1958). *In* " Progress in Non-Destructive Testing " (Stanford, E. G. and Fearon, J. H., eds.), Vol. 1, pp. 59–109. Heywood, London.

Hughes, D. E. (1879). Induction-balance and experimental researches therewith. *Phil. Mag.* Series 5, 8, 50–56.

Kadochikov, A. I. (1965). Electromagnetic processes in a conducting plate subjected to pulsed magnetic fields. *Defectoscopy*, No. 1, 9–12.

Leatherman, A. F. (1964). Nondestructive testing. *Reactor Core Mater.* 7, 207–209,

Leatherman, A. F. (1965). Nondestructive testing. *Reactor Core Mater.* 8, 49–51.

Leatherman, A. F. (1966). Nondestructive testing. *Reactor Core Mater.* 9, 70–71.

Leatherman, A. F. (1967). Nondestructive testing. *Reactor Core Mater.* 10, 131–133.

Leatherman, A. F. (1968). Nondestructive testing. *Reactor Core Mater.* 11 141–143.

LePage, J., Bernalte, A. and Lindholm, D. A. (1968). Analysis of resistivity measurements by the eddy current decay method. *Rev. scient. Instrum.* **39**, 1019–1026.

Lewis, D. M. (1951). " Magnetic and Electrical Methods of Non-Destructive Testing ". Allen and Unwin Ltd., London.

Libby, H. L. (1956a). " Basic Principles and Techniques of Eddy Current Testing ", General Electric Company, Hanford Atomic Products Operation, Report No. HW-44751.

Libby, H. L. (1956b). Basic principles and techniques of eddy current testing. *Mater. Evaluation* **14**, 12–18, 27.

Libby, H. L. (1958). " Basic Principles and Techniques of Eddy Current Testing ", American Society for Testing Materials, Special Technical Publication No. 223, " Symposium on Nondestructive Tests in the Field of Nuclear Energy ", pp. 13–26.

Libby, H. L. (1959). " Broadband Electromagnetic Testing Methods, Part I— Analytical Basis ", General Electric Company, Hanford Atomic Products Operation, Report No. HW-59614.

Libby, H. L. (1966). " Eddy Current Nondestructive Testing Device for Measuring Multiple Parameter Variables of a Metal Sample ", U.S. Patent No. 3,229,198, Jan. 11, 1966.

Libby, H. L. (1967). *In* " Progress in Applied Materials Research " (Stanford, E. G., Fearon, J. H. and McGonnagle, W. J., eds.), Vol. 8, pp. 121–173. Iliffe Books, London.

Libby, H. L. and Atwood, K. W. (1963). " Broadband Electromagnetic Testing Methods, Part III—Parameter Extraction ", General Electric Company, Hanford Atomic Products Operation, Report No. HW-79553.

Libby, H. L. and Cox, C. W. (1961). " Broadband Electromagnetic Testing Methods, Part II—Signal Analysis ", General Electric Company, Hanford Atomic Products Operation, Report No. HW-67639.

Lyke, R. F. (1967). " Pulsed Electromagnetic Testing Systems—The Nature of the Signals and Proposed Methods for Increasing the Information Yield ", M.S. Thesis, South Dakota School of Mines and Technology.

McGonnagle, W. J. (1960). " New Techniques for Testing Nuclear Reactor Components ", Proceedings of the Third International Conference on Non-Destructive Testing, Tokyo, Japan, pp. 432 to 441, March, 1960.

McGonnagle, W. J. (1961). " New Developments in Non-Destructive Testing at Argonne National Laboratory ", Proceedings of the Bureau of Naval Weapons Missiles and Rockets Symposium, Concord, California, pp. 51–65 April, 1961.

McGonnagle, W. J. (1966). " Non-Destructive Testing ", 2nd Edn., Chs. 12 and 14. Gordon Breach, New York.

McGonnagle, W. J. (1965). " Non-Destructive Testing of Nuclear Reactor Components ", Proceedings of the Third International Conference on the Peaceful Uses of Atomic Energy, Geneva, Switzerland. *Reactor Core Mater.* **9**, 570–576.

McLain, S. (1958). " Non-Destructive Testing in the Nuclear Energy Field ", American Society for Testing Materials, Special Technical Publication No. 223, " Symposium on Non-Destructive Tests in the Field of Nuclear Energy ", pp. 3–9.

Meyers, R. G. and Waidelich, D. L. (1958). A system for gauging plating thickness. *Trans. Am. Inst. elec. Engrs* **77** (I), 770–778.

Nikolayenko, A. T. (1966). Frequency-domain analysis of transient electromagnetic fields in conductors. *Defectososcopy* **2**, 94–96.

Nikolayenko, A. T. (1965). " Pulse Method in the Nondesctuctive Control of Metal Products ", *in* Proceedings of the Fourth Conference on Automatic Control and Methods of Electric Measurement, Novosibirsk, 1964. National Aeronautics and Space Administration Report TTF-384, pp. 147–151.

Ono, K. (1964a). Double frequency eddy current instrument for nuclear fuel clad tubing inspection. *Nippon Genshiryoku Gakkaishi* **6**, 447–454. (See *Nucl. Sci. Abstr.* No. 41628, **18**, 1964.)

Ono, K. (1964b). An approach to the evaluation of eddy current testing coils. *Jap. J. Non-Destructive Inspection* **13**, 203–209.

Ono, K. (1966). The use of double frequency eddy instruments for the detection of defects in metal tubing. *Jap. J. Non-Destructive Inspection* **15**, 85–90.

Ono, K. (1967). " High-Frequency and Low-Frequency Nondestructive Testers Utilizing Eddy Currents to Test for Surface and Sub-Surface Defects ", U.S. Patent 3,340,466, September 5, 1967.

Ono, K. and McGonnagle, W. J. (1961). " Pulsed Eddy Current Instrument for Measuring Sodium Levels of EBR-II Fuel Rods ", Argonne National Laboratory Report ANL-6278.

Philippe, A. (1963). " Non-destructive Testing by an Instrument with Double Pulsed Eddy Currents ", Euratom Report EUR 356f.

Renken, C. J. (1960a). A through transmission system using pulsed eddy current fields. *Mater. Evaluation* **18**, 234–236.

Renken, C. J. (1960b). " The Diffusion of Pulsed Current Fields in Good Conductors ", Symposium on Physics and Nondestructive Testing, October 1960, Argonne National Laboratory Report ANL-6346, pp. 127–138.

Renken, C. J. (1962a). " Progress Report on Nondestructive Testing by Electromagnetic Methods ", Argonne National Laboratory Report ANL-6414.

Renken, C. J. (1962b). " Eddy Current Techniques ", Argonne National Laboratory Report ANL-6677, pp. 182–183.

Renken, C. J. (1964a). " Theory and Some Applications of Pulsed Current Fields to the Problems of Non-Destructive Testing ", *in* " Progress in Applied Materials Research ", Vol. 6, pp. 239–261. Heywood Press, London.

Renken, C. J. (1964b). " A Pulsed Electromagnetic Test System Applied to the Inspection of Thin-Walled Tubing ", Argonne National Laboratory Report ANL-6728.

Renken, C. J. (1965a). " Broadband Electromagnetic Testing Methods ", Encyclopedic Dictionary of Physics Annual Supplementary Volume, pp. 19–21. Pergamon Press, Oxford.

Renken, C. J. (1965b). A pulsed eddy current test system using reflected field. *Mater. Evaluation* **23**, 622–627.

Renken, C. J. (1966). Refractory metal tubing inspection using ultrasonic and pulsed eddy current methods. *Mater. Evaluation* **24**, 257–262.

Renken, C. J. (1965). " The Economical Application of Non-Destructive Testing to Reactor Components, Expecially Jacket Tubing ", *in* " Non-Destructive Testing in Nuclear Technology ", Proceedings of the Symposium on Non-Destructive Testing in Nuclear Technology, International Atomic Energy Agency, Bucharest, Romania, May 17–21, 1965, Vol. 2, pp. 125–137.

Renken, C. J. (1968a). The role of high resolution fields and filtering in eddy current testing. *Mater. Evaluation* **26**, 191–195.

Renken, C. J. (1968b). " The Application of Pulsed-Electromagnetic Field Methods to Liquid-Metal Fast Reactor Nondestructive Test Problems ", presented at the Twelfth Metallurgical Congress on Nondestructive Testing and Control in the Field of Nuclear Metallurgy and Technology, June 24–26, 1968, Saclay, France.

Renken, C. J. and McGonnagle, W. J. (1958). " Eddy Current Techniques for Testing Liquid Metal Bonding ", American Society for Testing Materials, Special Publication No. 223, 273–277.

Renken, C. J. and Meyers, R. G. (1959). " A Double Pulse Eddy Current Testing System ", Argonne National Laboratory Report ANL-5935.

Renken, C. J. and Meyers, R. G. (1960). " Metal Resistivity Measuring Device ", U.S. Patent No. 2,965,840, December 20, 1960.

Renken, C. J., Lapinski, N. P., Cox, C. W. and Selner, R. H. (1966). " Nondestructive Testing in the Nuclear Rocket Program ", Argonne National Laboratory Report ANL-7169 (Classified Report).

Renken, Jr., C. J. (1961). " Nondestructive Eddy Current Testing ", U.S. Patent No. 2,985,824, May 23, 1961.

Renken, Jr., C. J. (1965). " Device for Testing Metal Sheets by Measuring the Time Required for Electromagnetic Pulses to Pass Therethrough ", U.S. Patent No. 3,189,817, June 15, 1965.

Renken, Jr., C. J. (1966). " Pulsed Electromagnetic Field System for Nondestructive Testing ", U.S. Patent No. 3,229,197, Jan. 11, 1966.

Renken, Jr., C. J. and Sather, A. (1968). " Pulsed Nondestructive Eddy Current Testing Device using Shielded Specimen Encircling Coils ", U.S. Patent No. 3,361,960, January 2, 1968.

Renken, Jr., C. J. and Waidelich, D. L. (1958). " Minimizing the Effect of Probe-to-Metal Spacing in Eddy Current Testing ", American Society for Testing Materials, Special Technical Publication No. 223, " Symposium on Nondestructive Tests in the Field of Nuclear Energy ", pp. 181–190.

Renken, Jr., C. J., Meyers, R. G. and McGonnagle, W. J. (1958). " Status Report in Eddy Current Theory and Application ", Argonne National Laboratory Report ANL-5861.

Rice, W. L. R. (1961). " Nuclear Fuels and Materials Development ", U.S. Atomic Energy Commission Report TID-11295, First Edition, pp. 161–162, February 1961. Also First Edition Supplement, p. 46, February 1961.

Roberts, W. C. (1879). Note on the examination of certain alloys by the aid of the induction-balance. *Phil. Mag.* Series 5, **8**, 57–60.

Ross, J. D. (1961). " Electromagnetic Testing of Reactor Components ", *in* Proceedings of the Symposium on Non-Destructive Testing Trends in the AEC Reactor Program, Germantown, Maryland, May 20, 1960. U.S.A.E.C. Office of Technical Information Report TID-7600, pp. 65–80.

Russell, T. J., Schuster, V. E. and Waidelich, D. L. (1962). The impedance of a coil placed on a conducting plane. *Trans. Am. Inst. elect. Engrs* **81**, (I), 232–237.

Russkevich, Y. N. (1968). Use of the pulsed eddy-current method for monitoring coating thickness. *Defektoskopiya*, No. 1, 74–78.

Schultz, A. W. (1966). " The Feasibility of Using a Novel Infrared Non-Destructive Testing Technique for Determining the Thermal Conductivity of Solids (With Emphasis on Graphites) ", Avco Corporation Report R-270-10-66-214.

Selner, R. H., Renken, C. J., Perry, R. B. and Balaramamoorthy K. (1963). " Nondestructive Tests of Components of EBR-I, Core IV ", Argonne National Laboratory Report ANL-6632.

Shinohara, U. and Otori, S. (1961). " Nondestructive Test by Pulsed Electric Current ", *in* Proceedings of the Third International Conference on Non-destructive Testing, Tokyo, March 1960, pp. 696–700. Pan-Pacific, Tokyo.

Uozumi, S. (1964). " Non-Destructive Testing by the Pulsed Electromagnetic System ", *in* Proceedings of the Fourth International Conference on Non-destructive Testing, London, 1963, pp. 295–303. Butterworths, London.

Uozumi, S. (1966). " Non-Destructive Testing of Materials by Pulsed Electro-magnetic Waves ", U.S. Patent No. 3,235,795, February 15, 1966.

Waidelich, D. L. (1954). Coating thickness measurements using pulsed eddy currents. *Proc. nat. Electron. Conf.* **10**, 500–507.

Waidelich, D. L. (1955). Pulsed eddy currents gauge plating thickness. *Electronics* **28** (11), 146–147.

Waidelich, D. L. (1956). Measurement of coating thickness by use of pulsed eddy currents. *Mater. Evaluation* **14**, 14–16.

Waidelich, D. L. (1958). " Reduction of Probe-Spacing Effect in Pulsed Eddy Current Testing ", American Society for Testing Materials, Special Technical Publication No. 223, " Symposium on Nondestructive Tests in the Field of Nuclear Energy ", pp. 191–200.

Waidelich, D. L. (1967). " An Analysis of the Pulsed Magnetic Field under a Masked Eddy-Current Probe ", *in* Proceedings, Fourth Congress on Material Testing, Budapest, Hungary, October 11–14, 1967, Vol. 3, pp. 879–892. An abstract appears in *Mater. Evaluation*, **26**, 26A, February 1968.

Waidelich, D. L. and Renken, Jr., C. J. (1956). The impedance of a coil near a conductor. *Proc. natn. Electron. Conf.* **12**, 188–196.

Waidelich, D. L., DeShong, J. A. and McGonnagle, W. J. (1958). " A Pulsed Eddy Current Technique for Measuring Clad Thickness ", Argonne National Laboratory Report No. 5614.

Wilcox, Jr., L. C. (1967). Prerequisites for quantitative eddy current testing. *Mater. Evaluation* **25**, 237–244.

CHAPTER 13

Microwave Techniques

D. S. Dean and L. A. Kerridge

Rocket Propulsion Establishment,
Westcott, Aylesbury, Buckinghamshire, England

I. Introduction

The use of microwaves for nondestructive testing is relatively recent, the first papers describing such techniques appearing in the early 1950's but the bulk of papers being published after 1960. Before this time equipment was not generally available for the generation and measurement of such short electromagnetic waves and it is likely that the exploitation of their full potential in this field will have to await the development of cheap robust generators at the higher microwave frequencies. The microwave region, although not rigidly defined, is taken to be between 1 and 1000 mm wavelength in free space, that is between 300 and 0·3 gigacycles per second. The region is bounded at the long wave end by radio waves into which category fall eddy current testing devices, and at the short wave end by the infra-red region in which techniques are now being developed. The electromagnetic spectrum is shown in Table I classified by wavelength, frequency and photon energy; three ways of describing the same phenomenon, the convention changing for different parts of the spectrum.

Most inspection techniques have, at the time of writing, employed wavelengths in the region between 5 and 200 mm. Since individual defects much smaller than a wavelength cannot be detected there are obvious advantages in

TABLE I

The Electromagnetic Spectrum

Wavelength (m)	Frequency (Hz)	Photon Energy (eV)	Usual division of radiation
10^{-14}	3.10^{22}	10^8	Cosmic radiation
10^{-13}	3.10^{21}	10^7	
10^{-12}	3.10^{20}	10^6	
10^{-11}	3.10^{19}	10^5	X and Gamma radiation
10^{-10}	3.10^{18}	10^4	
10^{-9}	3.10^{17}	10^3	
10^{-8}	3.10^{16}	10^2	Ultra-violet
10^{-7}	3.10^{15}	10	
10^{-6}	3.10^{14}	1	Visible light
10^{-5}	3.10^{13}	10^{-1}	Infra-red
10^{-4}	3.10^{12}	10^{-2}	
10^{-3}	3.10^{11}	10^{-3}	
10^{-2}	3.10^{10}	10^{-4}	Microwaves
10^{-1}	3.10^{9}	10^{-5}	
1	3.10^{8}	10^{-6}	Radiowaves

using shorter wavelength radiation. Most work in this field has been applied to military research so that many examples quoted are aimed at the testing of devices for use in space applications, although similar applications in industry are obvious.

Since conducting materials are opaque to radiation in this frequency band, microwaves can only be used to examine the interior of non-conducting materials or the surface of conductors. Unlike X-rays, at the powers generally used for testing purposes, they present no hazard to health and are partially reflected at a change of dielectric constant or completely at a conducting surface.

Although they are different in nature from the acoustic waves used in ultrasonic testing, the defect detection of the two systems can be compared on the basis of wavelength in the specimen.

II. Propagation of Microwaves

A. General Equations

It is not possible in this chapter to give a complete coverage of the theory of microwave propagation and the treatment will be limited to a statement of the more important equations required to understand the application of microwaves to nondestructive testing. The derivation of the equations, based on Maxwell's equations of the electromagnetic field, is explained in any textbook on microwaves.

Let us consider uniform plane waves of angular frequency, ω, radiated into a medium of conductivity σ, permeability μ and permittivity ε. The distribution of the wave in the direction of propagation, z, in relation to mutually perpendicular axes x, y and z can be described in terms of the electric field vector E_x, and the perpendicular magnetic field vector, H_y, as follows

$$E_x = E \exp(-\gamma z),$$

where $\gamma = (j\omega\mu\sigma - \omega^2 \varepsilon\mu)^{\frac{1}{2}}$ and is called the propagation constant. This is made up of the attenuation constant which describes the reduction in signal strength and the phase constant which describes the phase of the wave. The corresponding magnetic vector is given by

$$H_y = -\frac{\gamma E_x}{j\omega\mu}.$$

A useful concept is that of the wave impedance, Z, given by E_x/H_y such that

$$Z = \left(\frac{j\omega\mu}{\sigma+j\omega}\right)^{\frac{1}{2}}.$$

If the conductivity of the medium is zero, as in free space, this reduces to

$$Z = \left(\frac{\mu}{\varepsilon}\right)^{\frac{1}{2}}$$

and its value in free space is 377 ohms, so that the impedance of any radiating device in air should be matched to this. The velocity of the wave, c, in free space is given by

$$c = \frac{1}{(\mu_0\varepsilon_0)^{\frac{1}{2}}}$$

where the subscript 0 denotes values in free space.

This has the value $2\cdot998.10^8$ m sec^{-1} and the presence of the earth's atmosphere alters this by only three parts in 10^4.

B. Propagation in Materials

Provided that we are considering transmission in a material of low conductivity and permeability, the ratio μ to μ_0 will be very close to 1 and the velocity of the wave in the medium will be given by $v = c/(\varepsilon)^{\frac{1}{2}}$.

ε for the majority of materials which may be tested by microwaves lies between 1 and 10 so that the velocities with which we are concerned lie between the free space velocity and about $0\cdot3c$. The wavelength, λ_m, in the material will thus be given by

$$\lambda_m = \frac{\lambda_0}{(\varepsilon)^{\frac{1}{2}}}.$$

If the conductivity is not zero we may need to use an expression involving the loss tangent, $\tan\delta$, as follows

$$\lambda_m = \frac{\lambda_0}{(\varepsilon)^{\frac{1}{2}}}\left(1 - \frac{(\tan\delta)^2}{8}\right) + \text{terms of the order of } (\tan\delta)^4.$$

This is not usually necessary, as can be seen if we insert a value for $\tan \delta$ of 0·1, which represents a fairly lossy material, when the error involved in using the simpler equation will be only 0·05%.

The amplitude of the signal in a material will diminish with distance from the source for many reasons. There will usually be a square law reduction due to divergence of the beam and an exponential reduction due to interaction with the material. The interaction can be by generation of circulating currents in a partially conducting medium or by excitation of electrons in the atoms into oscillation, the final result being to raise the temperature of the material. If the material is not homogeneous and the inhomogeneities are of comparable dimensions to a wavelength of the radiation, they can cause scattering which will again reduce the transmitted signal.

If the radiation is incident normally on an interface between two non-conducting materials of different dielectric constants, ε_1 and ε_2, some will be reflected and some transmitted. The power reflection coefficient r is given by

$$r = \left(\frac{(\varepsilon_1)^{\frac{1}{2}} - (\varepsilon_2)^{\frac{1}{2}}}{(\varepsilon_1)^{\frac{1}{2}} + (\varepsilon_2)^{\frac{1}{2}}} \right)^2$$

and the transmission coefficient, t, by

$$t = 1 - r = \frac{4(\varepsilon_1 \varepsilon_2)^{\frac{1}{2}}}{[(\varepsilon_1)^{\frac{1}{2}} + (\varepsilon_2)^{\frac{1}{2}}]^2} .$$

It is interesting to compare the similarity in form of these equations with those governing transmission of acoustic energy across an interface.

If the radiation is incident on the surface at an angle θ, it will be refracted at an angle θ_2 given by Snell's Law

$$\frac{\sin \theta_1}{\sin \theta_2} = \frac{\varepsilon_2}{\varepsilon_1}$$

and there will exist a critical angle at which all radiation will be reflected as in the case of light.

Some further theoretical treatment will be required but this will be dealt with at the appropriate points in the text.

III. MICROWAVE COMPONENTS

In order to understand inspection techniques using microwaves it is necessary to have some knowledge of how the individual components operate, so these will be described before the techniques.

A. Generators

1. Klystron

Various types of generators have been used in the microwave region, but for other than high-power applications the reflex klystron shown in Fig. 1 is now the most commonly used. Frequency is controlled by the dimensions of a resonant cavity shaped like a flat cylinder with holes in the centre of each end closure. The cavity may be tuned by screwing a plunger into or out of it.

A beam of electrons is projected along the axis of the cylinder through the holes, accelerated by a potential difference between cathode and cavity of 1–2 kV. If we assume that the cavity is oscillating, the edges of these holes will become alternately positive and negative so that the electrons between the lips of the holes will experience an accelerating field during one half cycle and a decelerating field during the next. The net result is that the average energy in the cavity due to this effect will remain constant as equal amounts are transferred to or removed from the electrons at each half cycle.

After leaving the cavity the electrons travel towards a reflector held at a negative potential which reverses their direction of travel. Those with the greatest energy travel furthest towards the reflector and the ultimate effect of this differential energy is that they arrive back at the cavity in bunches.

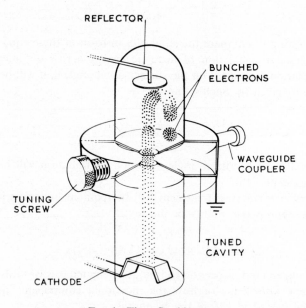

FIG. 1. The reflex klystron.

If all potentials in the klystron are correctly adjusted the bunches will pass through the cavity when the phase of the oscillation is such that they will give up some of their energy to it. There is thus a net transfer of energy from the electrons to the cavity and oscillation is maintained. The tunable range of these devices is small and they are more fragile than any of the other components used in a microwave circuit. Their output for most NDT purposes is up to 250 mW, although higher powers are available.

2. Gunn Diode

A solid state generator known as the Gunn diode is, at the time of writing, coming into limited use at the lower frequency end of the microwave band. It seems reasonable to expect that it will eventually be available at all frequencies in the band. Q band diodes are operating experimentally but their output is limited to about 10 mW. They are robust and operate with a supply of a few volts so that they should be better suited than klystrons to industrial applications.

The diode consists of a thin layer of GaAs separating two contacts between which a potential is applied. The operation depends upon the fact that at electric fields above $300 \, Vmm^{-1}$ conduction electrons are transferred to higher energy states at which they have a lower mobility. This condition usually arises near to the cathode and the electrons at the anode side of the region move more rapidly towards the anode leaving a depletion layer followed by an excess of electrons on the cathode side. The electric dipole so formed drifts through the GaAs layer at a velocity of $10^5 \, m \, sec^{-1}$. As soon as it leaves the anode a further domain forms and travels through in a similar manner. This causes a series of pulses in the current flowing through the diode, their recurrence frequency depending on the thickness of the GaAs layer, and being the fundamental frequency of the diode. The output can be rendered more nearly sinusoidal by connecting a resonant tuned circuit across the diode, usually a cavity at microwave frequencies. Other modes of oscillation are possible (Carroll, 1967; Foxell and Wilson, 1966; Robson and Mahrous, 1965).

B. Waveguide

At lower microwave frequencies co-axial conductors may be used to connect components, but at the frequencies most used for materials examination it is necessary to use waveguides. These are circular, or more commonly, rectangular metal tubes, accurately dimensioned and with a high internal finish. More than one mode of transmission in rectangular waveguide is possible but that most commonly used is such that the electric field is at a maximum in the centre of the guide falling to zero at the two shortest sides. It also varies sinusoidally along the axis of the guide with the electric vector parallel to the shortest sides. Associated with this is a magnetic field pattern as shown in Fig. 2.

The wavelength of the wave in the guide, λ_g, is not the same as that in free space, λ_0, and is given by

$$\lambda_g = \frac{\lambda_0}{[1 - (\lambda_0/\lambda_c)^2]^{\frac{1}{2}}}.$$

λ_c is a critical wavelength above which any wave in the guide is rapidly attenuated given by

$$\lambda_c = 2a \quad (a \text{ is the maximum width of the guide}).$$

If a material of dielectric ε and permeability μ fills the guide

$$\lambda_g = \frac{\lambda_0}{[(\varepsilon\mu/\varepsilon_0\,\varepsilon_0) - (\lambda_0/\lambda_c)^2]^{\frac{1}{2}}}.$$

To reduce losses to a minimum, the internal surface of the guide should be highly conducting, and this is achieved by plating or constructing the guide of a good conductor such as silver.

Any discontinuity in the guide, such as a change in the internal section, will result in a change of impedance and partial reflection of the wave at the discontinuity. Guides and components are joined by coupling flanges with suitable machined surfaces for accurate location of the guide internal section.

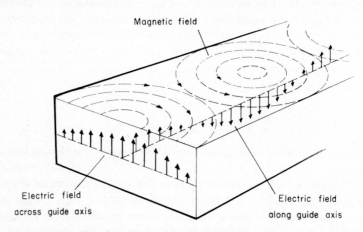

FIG. 2. Distribution of fields in waveguide.

C. Isolator

To avoid any effect on the klystron by the external circuit it is usual to insert an isolator between them which will allow transmission in one direction but attenuate heavily in the other. To accomplish this a section of waveguide within the isolator is twisted to rotate the plane of polarization through 45°. The wave then passes into a ferritic material which, when subjected to the field from a permanent magnet, rotates the plane back through 45° and thence into the exit waveguide. The plane of any returning wave is also rotated through 45° in such a sense that it reaches the other end of the ferrite at right-angles to the plane necessary for it to enter the guide.

D. Attenuators

Attenuation of the wave in a guide can be accomplished by inserting a conducting plate along the axis of the guide parallel to the shortest side, as in Fig. 3, so that it lies in the plane of the electric field and hence has currents induced in it which tend to destroy the field. The degree of attenuation can be controlled by moving the plate across the guide until attenuation will be reduced almost to zero when it lies flush with the side wall in a region of zero electric field.

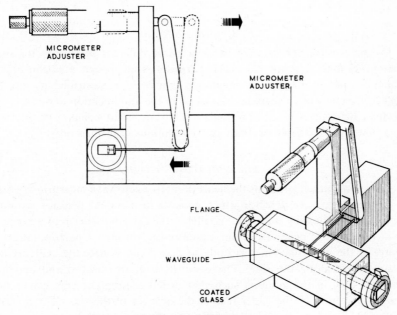

FIG. 3. Attenuator.

To avoid reflections from the plate the ends are tapered so that its effect on the wave is gradual. Movement of the plate is effected by a micrometer if a calibrated attenuator is required.

E. Directional Couplers

It is often necessary to divide a signal into two separate waveguides and control the direction in which waves may travel. Devices performing this function, known as directional couplers, may take different forms but a typical example is constructed as follows. Two waveguides are brazed together along the broader side of the guides and connected by a row of small holes between the guides spaced at quarter wavelength intervals. Consider a wave travelling along one of the guides and reaching the first of the holes. Some of the energy will radiate through the hole and spread in both directions in the second guide. The wave travelling in the same direction as that in the first guide will arrive in phase with it at the second and subsequent holes so that it will be reinforced each time. The division of energy can be controlled by the hole size and the number of holes. The backward travelling wave will, however, arrive at each successive hole out of phase with the energy radiating through it at that time, resulting in destructive interference and very little energy travelling backwards from the holes.

F. Crystal Detector

The microwave energy may be detected in different ways, but the most commonly used detector for NDT purposes is at present a silicon crystal which can rectify microwave signals fed to it by a probe projecting into the field in a section of waveguide. To assist in the amplification and resolution of small signals it is usual to modulate the transmitted signal at the klystron at a few kiloHertz rather than use continuous transmission.

G. Standing Wave Detector

A component which forms the heart of many microwave measuring systems is the standing wave detector which consists of a crystal detector mounted on a carriage allowing it to slide parallel to the axis of a section of waveguide as in Fig. 4. The probe from the crystal passes through a narrow slit in the centre of the wider face of the guide so that it can sample the field strength at any point along the slit. The disposition of the electric and magnetic lines of force in the guide is such that no circulating currents would flow across the centre of the wider face in the absence of the slit. The presence of the slit, therefore, has negligible effect on the field distribution in the guide.

The instrument is used to compare two waves when these are made to travel in opposite directions in the detector, thus setting up a standing wave pattern with successive maxima spaced at intervals of $\lambda/2$. The ratio of maxima to minima is a measure of the ratio of the amplitudes of the two waves. The crystal detector is moved by a micrometer to allow the wavelength of the radiation to be measured accurately.

H. Wavemeter

An alternative method of measuring the wavelength is to use a wavemeter consisting of a tunable calibrated cavity coupled to the waveguide, resonance being indicated by the output of a crystal detector.

I. Matched Load

To avoid reflections from the end of a waveguide section it can be terminated in a matched load which is similar to an attenuator in which the wedge-shaped plate is made of a strongly absorbing material which dissipates the energy as heat.

IV. APPLICATIONS

A. Classes of Techniques

Since microwave radiation in free space is travelling at about 3.10^8 m sec^{-1} a 10 nanosecond pulse will occupy 3 m in space, so that pulse reflection

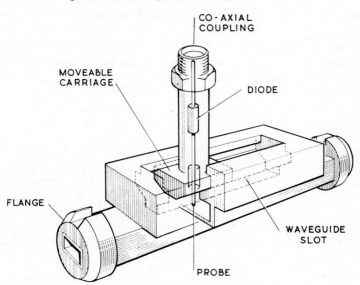

FIG. 4. The standing wave detector.

techniques, similar to those used in ultrasonic testing, are not yet a practical proposition. All techniques, therefore, employ continuous radiation modulated at a few kHz to assist in amplification and detection. The techniques may be divided into three classes.

(a) Transmission, where the internal condition of a sample is measured by its effect on the signal passing through it. The thickness of non-conducting materials may be measured this way, and the presence of flaws detected or the state of the material determined.

(b) Reflection, where the signal reflected from the surface or internals of a specimen is monitored to give similar information to that just mentioned. In addition, the thickness of conducting materials can be measured and the state of the surface or surface coating of a conductor may be found.

(c) Scattering, which is a particular case of reflection where the reflecting medium consists of many small discontinuities in a non-conducting medium. The detection of porosity in such a medium is a particularly effective use of microwaves.

Some techniques are not easily categorized as above and may be a cross between the techniques.

B. Transmission

The basic transmission technique is shown in block diagram form in Fig. 5. The output of a klystron is fed through an isolator to a waveguide horn and thence into the sample. A second horn receives the energy which is measured with a crystal detector and meter. An attenuator can be used in various ways to measure the loss in the sample and when connected as in the figure it is adjusted to keep the received signal constant whilst the level of the transmitted signal is monitored. Lavelle *et al.* (1967) describe such

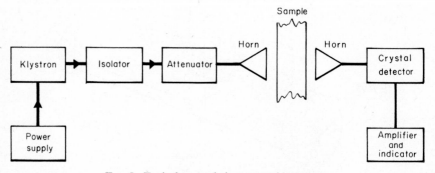

FIG. 5. Typical transmission measuring system.

a technique to measure the state of cure of resin samples placed between the horns and measured an increase in transmitted power of twenty times the initial value over a curing period of 75 minutes. The frequency used was 9·6 GHz. The author has measured a change in attenuation of casein glue during curing of 5·5 dB in a sample 2 mm thick. A change was just detectable in a layer 25 μm thick.

The method has been used to measure the state of cure of a rocket propellent and good correlation between the measured loss and the elastic properties of the propellent has been reported.

An instrument to measure the water content of walls has been developed by the Building Research Establishment (Watson, 1963) which entails placing a transmitting horn on one side of the wall and a receiving horn on the other, the water content being indicated by the strength of the received signal.

This apparatus has been produced commercially by AEI Ltd., and has been adapted to measure the water content of a variety of materials such as soap, sand and grain. It has been refined to minimize transmitter power variations by including a motor-driven attenuator to reduce the received signal to the same value as a reference signal from the transmitter. A correction can also be applied for variations in sample temperature sensed by a suitable transducer. The final signal, proportional to attenuator position and corrected for temperature, can be recorded or indicated to an accuracy equivalent to a variation of 0·2% water content.

So far the effects of the mismatch at the surfaces of the sample have been ignored, but these will cause standing waves in the apparatus on both sides of the sample and within it. The thickness of the sample will thus have a marked effect on the energy distribution, maximum energy being transmitted when the sample thickness is $\lambda/2$. The ratio of power transmitted to incident power, T, for any thickness of material, d, being given by

$$T = \frac{1 + 2r + r^2}{1 + 2r \cos(4\pi d/\lambda_m) + r^2}.$$

If loss in the material is high the signal reflected back into the specimen will be small and interference effects may be insignificant, but the other surface reflections cannot be ignored. The problem has been considerably reduced by Lavelle *et al.* (1967), who matched into the specimen on both sides by a tapered wedge of material of similar dielectric constant placed with its base against the surface of the sample. If the wedges are made of the same material as the sample the interface effects can be eliminated.

C. Reflection

The thickness of a conducting plate may be measured by reflecting signals from both sides and comparing amplitude or phase. Beyer *et al.* (1960) describe early examples of both techniques, the arrangement of components in the amplitude measuring version being shown in Fig. 6. A common signal source was used to feed antennae on each side of the specimen. Each antenna was constructed so that the reflected signal varied linearly in amplitude with movement of the specimen and slot and dielectric antennae are described which possess this property over short distances. The sum of the outputs from the two antennae was inversely proportional to the distance between the horns less the specimen thickness, so that the latter could be determined. Since the crystal detectors used had the usual square law characteristic it was necessary to feed the output to a series resistor and solid state diode with a similar characteristic, the voltage across the diode thus being corrected to provide an output proportional to antenna signal.

The system was stable and unaffected by translation of the specimen between the antennae over a distance of 1·5 mm, and a thickness change of 0·025 mm was measured with a strip specimen moving at $0·5\ \mathrm{m\ sec^{-1}}$, whilst smaller variations would appear to be detectable.

The same authors also describe a similar system which depended on the change of phase of the reflected signals from both surfaces. Each signal was mixed with a reference signal derived from the signal source and detected by

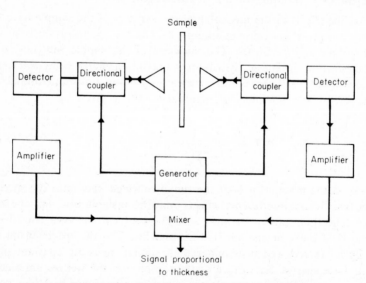

Fig. 6. Thickness measurement by amplitude of reflected signals.

the usual crystal. The authors found this system to be less satisfactory since the phase sensitive detectors were also sensitive to amplitude changes and greater drift effects were observed. This is a fundamental difficulty with phase measuring systems which would otherwise seem preferable.

A high-resolution system to measure burning rates in solid propellent rocket motors is described by Dean and Green (1967) using 37·5 GHz radiation fed via an isolator and standing wave detector to a directional coupler which split the signal into two channels, one leading to a waveguide horn and the other to a tuner and termination as shown in Fig. 7. The signal from the horn passed through a window in the motor, transparent to microwave radiation, through the propellent and was reflected from the burning surface. This acted almost as a metallic reflector due to the high concentration of free electrons, the signal returning to the horn being a combination of this reflected signal and any standing reflections from other surfaces. The combined signal was mixed in the directional coupler with that from the termination which was varied in amplitude by the tuner until it nearly cancelled the standing reflections from the motor before ignition.

After ignition the additional signal from the burning surface mixed with the transmitted signal in the standing wave detector to form a standing wave which moved along the guide as the burning surface moved. The signal at the fixed detector varied through one cycle from maximum to minimum and back to maximum as the burning surface moved through a distance equal to a half wavelength of the radiation in the propellent. Since the dielectric constant of propellent was about 5 this wavelength was about 4 mm, so that a resolution of 0·2 mm was easily obtained, corresponding to 0·1 cycle change of the standing wave. Higher reading resolution than this was complicated by the fact that the amplitude of the standing wave increased as the burning progressed and attenuation in the propellent decreased. Absolute accuracy was limited by the accuracy with which wavelength in the propellent was known. This was determined by inserting small specimens in a waveguide using the method due to Von Hippell, but there remains some doubt about how representative the samples were of the propellent in the motor.

A similar technique can be used to monitor the velocity of the shock wave due to initial stages of burning and final detonation in samples of explosive materials (Cawsey et al., 1958). In this case the progress of the shock into and out of the explosive can be seen by a change of velocity, or trigger devices can be inserted in the explosive, so that the velocity of the wave in the explosive need not be known beforehand.

Rockowitz and McGuire (1966) were able to detect defects with dimensions as small as $\frac{1}{8}$ inch in one direction in 2-inch-thick glass fibre honeycomb

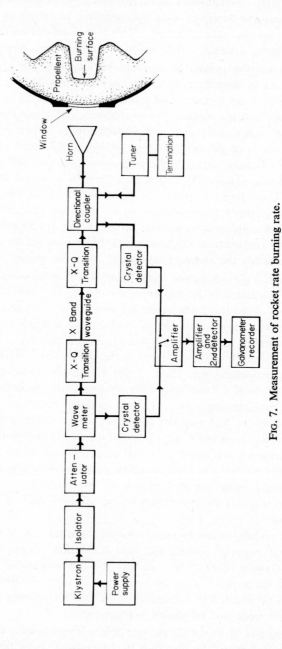

FIG. 7. Measurement of rocket rate burning rate.

filled with ablative material for heat shields. They used 69 GHz radiation directed at an angle to the surface of the specimen and picked up the radiation reflected from the back surface in two closely spaced horns, these signals being displayed on two traces of an oscilloscope. The presence of a void was indicated by a simultaneous reduction in signal in both horns caused by the signal being totally reflected from the interface between the fibreglass honeycomb and the void and hence being scattered away from the receiver horns. Two horns were necessary as it was found that the refractive index of the material might vary, causing refraction of the beam and hence variation of signal in a single horn which could be interpreted as a void.

The surface conditions of metals can be measured by a microwave reflection technique and Owston (1968) describes such a method for examining the interface between ceramic coatings and metals to detect the onset of corrosion and hence failure of the coating. The method relies on the fact that the wave will penetrate the surface of a conductor to a distance δ given by

$$\frac{1}{\delta} = \left(\frac{\pi \mu_0 f \mu}{\rho}\right)^{\frac{1}{2}},$$

where f is the frequency of the wave and ρ is the resistivity.

Variation of the electrical properties of this layer will, therefore, affect the reflected signal and the resistivity of any layer formed by corrosion will be considerably higher than that of the parent metal.

Owston made the specimen to be examined one wall of a resonant cavity, and it can be shown that the change of resistivity of the cavity wall is related to the change of Q of the cavity by the equation

$$\delta\rho = -\frac{4L^3}{Q\lambda_0^2} \left(\frac{2\pi f \mu_0 \rho}{\mu}\right)^{\frac{1}{2}} \cdot \frac{\delta Q}{Q}.$$

The smaller the cavity length, L, the more sensitive the system, so the specimen was made one of the larger flat sides of a short cylindrical cavity which was fed with a 25 GHz signal from a klystron. The output from the cavity was detected and indicated on a meter, the reading of which was maintained constant by adjusting the power input, E, to the cavity. The Q of the cavity was given by

$$Q \propto \frac{1}{E}$$

and $\rho = (AE+B)^2$ where A and B are constants.

The Q of the cavity varied from 5000 with a copper plate fitted, to 50 with $1\,\Omega\,\mathrm{cm}$ semiconductor layer. One of the biggest changes observed was from about $100\,\mu\Omega\,\mathrm{cm}$ to $5000\,\mu\Omega\,\mathrm{cm}$ for a sample of disilicide-coated niobium when subjected to a temperature of $1250\,^\circ\mathrm{C}$ for 32 hours. The results were most promising, but the author suggests that considerably more work is required.

A further cavity technique, which is difficult to classify, is that due to Brown and Billeter (1967), who used a resonant cavity to measure the water content of helium gas intended as a cooling medium for a nuclear reactor. They used a reference cavity tuned to the frequency of the measuring cavity when filled with dry helium to control the frequency of the microwave generator feeding energy to the measuring cavity. The deviation of the measuring cavity centre frequency from that of the generator output was detected by the second harmonic signal generated by driving the cavity slightly off frequency.

The frequency of the cavity, f, is related to the dielectric constant of the contents, ε, by the relationship

$$f^2 = \frac{k}{\varepsilon}$$

and it can be shown that the change of frequency, Δf, is a function of the frequency, the polarizability of helium, α_2, and of water vapour, α_1, and the number of water molecules, n, as follows

$$\Delta f = 2\pi f (\alpha_1 - \alpha_2)\, n.$$

Detection sensitivity at ambient temperature was 85 ppm change in concentrations of 400 ppm water vapour in helium, but the system was intended for use with high-temperature gas and the polarizability varies with temperature as follows

$$\alpha = \alpha_0 + \frac{\mu^2}{3kT}$$

where α_0 is the optical polarizability which does not charge and μ is the dipole moment.

A temperature rise of $1000\,^\circ\mathrm{C}$ results in a reduction in polarizability of about four times and will result in fewer molecules being present in a given volume, so to test this effect the system was operated with gas at $436\,^\circ\mathrm{C}$ and the sensitivity was reduced to 875 ppm.

D. Scattering

The scattered field from an object at any point is strictly the field measured at that point less the component of the incident energy, so that all microwave techniques can be said to rely on scattered radiation. In this chapter, however, the term scattering techniques applies to those where the scattered field predominates over the incident field and where the linear dimensions of the scattering object are small compared with, or comparable to, a wavelength of the radiation.

A useful concept is the scattering cross section of an object, σ, given by

$$\sigma = \lim_{D \to \infty} 4\pi D^2 \frac{S_s}{S_i}$$

where S_i is the incident power flux density and S_s is the scattered power flux density at a distance D, so that σ is proportional to the power scattered into unit solid angle divided by the incident power per unit area. Lord Rayleigh has shown that the scatter from a conducting sphere with radius, r, small compared to a wavelength is given by

$$\frac{\sigma}{\pi r^2} = 1 \cdot 403 \times 10^4 \left(\frac{r}{\lambda} \right)^4 .$$

This relationship holds approximately up to $r/\lambda = 0\cdot 15$ and as the sphere becomes larger $\sigma/\pi r^2$ varies in a cyclic manner, the amplitude of the variation decreasing with increasing sphere diameter to approach a constant value.

Since the scattered radiation can be regarded as emanating from an aperture comparable with the dimensions of the scattering object, the angle, β, at which radiation can be detected will increase with decreasing object diameter, d, according to the relationship

$$\beta = 2 \sin^{-1} \left(\frac{\lambda}{d} \right) \text{ radians.}$$

It can be seen, therefore, that individual defects in a material much smaller than a wavelength are difficult to detect, but that if there are a large number of small defects, as in a porous material, radiation will be scattered over a wide angle and the total scatter from all defects may be detectable.

Lavelle et al. (1967) describe experiments to measure the scattered radiation from hollow glass beads embedded in epoxy resin blocks and obtained results of the correct order of magnitude, but measurement was complicated by internal reflections on the block giving a larger scatter signal than theory would predict.

Dean and Green (1967) describe a scatter technique for the examination of bonded rubber engine mountings to detect porosity in the rubber. 35 GHz radiation was directed at one face and a receiver horn picked up scattered radiation at right-angles to the incident beam. No scatter was detectable with unflawed samples, but porosity could be readily seen over a wide range of pore sizes. Defective areas of 10 mm cube containing 0·5 mm diameter pores at about 2 mm spacing could be detected with simple equipment. Penetration of greater thicknesses of rubber was possible by microwaves than would have been the case if ultrasonic radiation had been used of a wavelength giving similar defect resolution, typical figures for attenuation being 5 dB in^{-1} for microwaves and 20 dB in^{-1} for ultrasonic radiation. In addition, no coupling material was required and inspection could be carried out more rapidly than samples could be moved mechanically.

V. Microwave Visualization Techniques

Unless one requires a simple form of microwave examination in which the amplitude or phase of the signal can be monitored and rejection levels set, interpretation of the signal in terms of flaw type and size is difficult. With most forms of NDT it is common to attempt to display the information in pictorial form, such as by X-ray film or an ultrasonic " plan position " plot of some function of the signal, and such a presentation would be highly desirable when using microwaves.

A. Field Scanning

A direct equivalent of the ultrasonic plot can be obtained by scanning a horn through the microwave field, the motion of the horn synchronized with that of a plotter. If the plotter is of a type which can vary the density of the plot according to signal strength, a cross section of the field can be displayed which may allow specimen defects to be seen.

B. Cholesteric Liquid Crystals

A much less tedious method and one which allows real time displays of information is the use of cholesteric liquid crystals which change colour over a narrow temperature range. These substances, usually cholesteryl ester mixtures, flow in the manner of a liquid but have a regular crystal structure similar to a solid. When white unpolarized light falls on these crystals it is split into two components with oppositely rotating vectors, one of which is transmitted and the other reflected. The reflected component is thus coloured and the colour depends on the relative positions of the crystals which change with temperature. Various types are available which undergo a full range of colour changes for temperature ranges between 1 °C and 10 °C, the process usually being reversible and the crystal life indefinite, although some loss of brilliance occurs with time. This can be restored by lightly brushing the surface exposed to light.

For microwave applications a thin layer of crystals is sprayed or otherwise deposited on a thin substrate which has ideally been painted matt black to ensure that all reflected light is from the crystals. This is then placed in the microwave field and is differentially heated, depending on the microwave energy absorbed in different parts of the layer. A colour pattern is thus produced mapping the variations of microwave energy.

Augustine (1968) describes such a system used for mapping energy in microwave equipment in which a thin metal coated mylar film was used as the substrate for the crystals. He reports that when this was placed 1 inch in front of a radiating X band waveguide distinct concentric bands of colour were observed when the output rose above 20 milliwatts. Optimum conditions can be achieved with a metal coating of 377 ohms per square, which matches the impedance of free space, and a metal plate placed $\frac{1}{4}$ wavelength from the substrate.

C. Photographic Techniques

Another method of mapping microwave fields also relies upon their heating effect, but in this case the detector is a photographic film. This is first exposed to even illumination which would result in moderate blackening of the film if it were developed normally. The film, shielded from light, is then given a thin film of developer and placed in the microwave field. The rate of development of the film depends upon its temperature, so that parts exposed to more intense radiation develop more rapidly. If development is arrested before it is complete the areas in more intense microwave fields will appear darker than the rest.

The method is described by Iizuka (1968), who also employed colour film to give an enhanced effect. He exposed the cyan dye layer, since this is at the bottom of the colour layers and the developer takes longer to reach it, the diffusion constant, D, being given by

$$D \simeq Ae^{-E/kt}$$

where A is a constant, E is the activation energy, k is Boltzmann's constant and t is the temperature of the film. It can be seen that D is strongly affected by temperature and that maximum sensitivity is achieved if the film is first cooled. He also reports satisfactory results using Polaroid packet film, and quotes the minimum energy required to produce a clear image as $0.06 \ W \ in^{-2}$.

D. Holographic Techniques

By the use of microwave lenses and any of the above detectors it should be possible to produce microwave images of defective areas in samples, but a three-dimensional display is a possibility using holographic techniques. Since microwave radiation from a horn is coherent there is no fundamental reason why a hologram should not be produced. The fundamental theory of holography is described elsewhere in this book (cf. chapters by Ennos (p. 155) and Aldridge, p. 133), so that it will not be repeated here. Suffice it to say that the object must be illuminated by coherent microwave radiation and the resultant diffracted radiation must be mixed with a reference beam from the same source to form a diffraction pattern which can be recorded in some way, usually on a photographic film.

If this film is now illuminated by coherent light from a laser, for example, a three-dimensional representation of the object originally illuminated by microwaves will be visible. Unfortunately the wavelength of the laser light is 10^{-4} to 10^{-5} times smaller than that of the microwave radiation, so that the reconstructed image will be reduced in the same ratio. It is thus necessary to use a lens system to view the hologram.

Aoki (1967) produced holograms by scanning through the microwave field with a horn, the output of which modulated a light moving over a photographic plate in step with the movement of the horn. He was able to produce recognizable reconstructions of metal letters placed in the microwave beam and evaluated the effects of the scanning lines produced by this method on the quality of the image, since they act as a grating and may result in overlapping images. Using a scanning method it is possible to dispense with the reference microwave beam and feed in an electronic signal which must be varied in phase as the horn is moved in the same manner as the reference beam would have done.

VI. FUTURE MICROWAVE TECHNIQUES

The methods of pictorial presentation of microwave images are still very much in their infancy and it is likely that considerable development along these lines will be made in the near future. Possibilities which suggest themselves are the use of a thin mylar film, as in the cholesteric techniques, looked at with a direct viewing infra-red camera to make the temperature patterns visible. Such cameras are now available (cf. chapter by Lawson and Sabey, p. 443) with sensitivities similar to that of cholesterol, but it is possible to adjust them to operate over a narrow range anywhere in a wide temperature spectrum. A given cholesterol mixture will only operate over a fixed temperature band.

With modern methods of fabricating semiconductors it is comparatively cheap to construct a matrix of circuits in compact form which could be used with an associated matrix of microwave detectors. These could be electronically scanned to build up a picture without the need for the intermediate step of converting microwaves to heat. Such a system could be sensitive to considerably lower microwave intensities than any techniques involving heat and could be made with sufficient resolution for many applications.

Solid state power supplies and generators are likely to be available shortly giving output powers at least equal to present klystrons in the shorter wavelengths, and associated components for use in the region of 1 mm wavelength may become available at reasonable prices.

Range resolution equivalent to that possible with ultrasonic techniques would extend the uses of microwaves to a wide variety of new tasks, but with the present design of equipment this seems fundamentally impossible.

In conclusion it can be said that although microwaves are unlikely to occupy a position in the nondestructive testing field comparable with ultrasonics, radiography or conventional eddy current testing, they nevertheless carry out tasks which would be difficult by the established techniques and they are likely to grow in importance. A reference list is given at the end of this chapter which will allow the interested reader to obtain more information, although all the publications are not referred to in the text.

REFERENCES

Aoki, Y. (1967). Microwave holograms and optical reconstruction. *Appl. Opt.* **6** (11), 1943–1946.

Augustine, C. F. (1968). Field detector works in real time. *Electronics* **41**, 118–122.

Beyer, J. B., Bladel, J. V. and Peterson, H. A. (1960). Microwave thickness detector. *Rev. scient. Instrum.* **31** (3), 313–316.

Brown, D. P. and Billeter, T. R. (1967). "Microwave Detection of Moisture in Reactor Coolant Gas." Battelle Northwest Report BNWL-406.

Carroll, J. E. (1967). Mechanisms in Gunn effect microwave oscillators. *Radio electron. Engr* **34**, 17–30.

Cawsey, G. F., Farrands, J. L. and Thomas, S. (1958). Observations of detonation in solid explosive by microwave interferometry. *Proc. R. Soc.* **A248**, 499–521.

Crissey, J. T., Gordy, E., Fergason, J. L. and Lyman, R. B. (1965). A new technique for the demonstration of skin temperature patterns. *J. invest. Derm.* **43**, 89–91.

Dean, D. S. (1967). Microwaves for materials inspection—an introduction. *Non-destruct. Test.* **1** (1), 19–24.

Dean, D. S. and Green, D. T. (1967). The use of microwaves for the detection of flaws and measurement of erosion rates in materials. *J. scient. Instrum.* **44**, 699–701.

Deschamps, G. A. (1967). Some remarks on radio-frequency holography. *Proc. IEEE* **55**, 570–571.

Dickinson, A. and Dye, M. S. (1967). Principles and practice of holography. *Wireless Wld.* **73** (2), 56–61.

Dowden, W. A. (1967). Cholesteric liquid crystals. *Non-destruct. Test.* **1** (2), 99–102.

Fergason, J. L. (1964). Liquid crystals. *Scient. Am.* **211**, August, 76–85.

Fowler, K. A. and Hatch, H. P. (1966). "Detection of Voids and Inhomogeneities in Fibre Glass Reinforced Plastics by Microwave and Beta-Ray Back Scatter Techniques." Springfield Armory Report No. SA–TR19–1519.

Foxell, C. A. P. and Wilson, K. (1966). Gallium arsenide varactor diodes. *Radio electron. Engr* **31**, 245–255.

Gates, J. W. C. (1968). Holography with scatter plates. *J. scient. Instrum.* Series 2, **1**, 989–994.

Giangrande, R. V. (1965). Microwave Inspection Techniques for Determining Ablative Shield Thickness and Ceramic Materials Properties. U.S. Army Materials Research Agency Technical Report AMRA TR 65–31.

Groble, K. K., Gruetzmacher, R. W., McClurg, G. O. and Retsky, M. W. (1964). Corona and microwave methods for the detection of voids in glass-epoxy structures. *Mater. Evaluation* **22**, 311–314.

Hochschild, R. (1963). Applications of microwaves in non-destructive testing. *Non-destruct. Test.* **21**, 115–120.

Hochschild, R. (1968). Microwave non-destructive testing in one (not-so-easy) lesson. *Mater. Evaluation* **26**, 35A–42A.

Iizuka, K. (1968). Mapping of electromagnetic fields by photochemical reaction. *Electronics Lett.* **4** (4), 68–69.

Kovalev, V. P. and Aleksandrov, Ku. B. (1967). Microradio waves as a means of inspecting articles made of plastic. *Defektoskopiya* No. 2, pp. 29–34.

Kovalev, V. P. and Kuznetsov, M. G. (1965). Application of radio waves in flaw detection. *Defektoskopiya* No. 5, pp. 383–387.

Lavelle, T. M. (1967) Microwaves in non-destructive testing. *Mater. Evaluation* **25**, 254–258.

Lavelle, T. M., Lehman, J. T. and Latshaw, D. (1967). "Microwave Techniques as applied to Non-destructive Testing of Non-metallic Materials." Frankford Arsenal Test Report T67–9–1.

Leith, E. N. and Upatnieks, J. (1962). Reconstructed wavefronts and communication theory. *J. opt. Soc. Am.* **52** (10), 1123–1130.

Leith, E. N. and Upatnieks, J. (1964). Wavefront reconstruction with diffused illumination and three-dimensional objects. *J. opt. Soc. Am.* **54** (11), 1295–1301.

Owston, C. N. (1968). "An Investigation of the Degradation of Ceramic Coatings on Metals Using a Microwave Technique." College of Aeronautics Memo 157.

Robson, P. N. and Mahrous, S. M. (1965). Some aspects of Gunn effect oscillators. *Radio electron. Engr* **30**, 345–352.

Rockowitz, N. and McGuire, L. (1966). A microwave technique for the detection of voids in honeycombed ablative materials. *Mater. Evaluation* **24** (2), 105–108.

Ryan, A. H. and Summers, S. D. (1954). Microwaves used to observe commutator and slip ring surfaces during operation. *Electl Engng, N.Y.* **73** (3), 251–255.

Stroke, G. W. (1968). US-Japan Seminar on Holography, Tokyo and Kyoto, 2–6 October, 1967. *Appl. Opt.* **7** (4), 622 and 631.

Watson, A. (1963). "Measurement and Control of Moisture Content by Microwave Absorption." Building Research Station Current Papers Research Series 3.

Whistlecroft, D. (1962). Microwave thickness measurement of dielectric materials. *J. Br. Instn Radio Engrs* **23**, 151–155.

Woodmansee, W. E. (1966). Cholesteric liquid crystals and their applications to thermal non-destructive testing. *Mater. Evaluation* **24**, 564–572.

Zinoviev, O. A. (1968). The properties of silver bromide emulsions irradiated by radio waves. *Zh. nauch. Prikl. Fotogr. Kinem.* **13** (5), 353–354.

CHAPTER 14

Infrared Techniques

W. D. Lawson and J. W. Sabey

Royal Radar Establishment,
Malvern, Worcestershire, England

I. Introduction

This chapter outlines the basic concepts of infrared radiation, transmission and detection, describes current technology of components and systems, and then goes on to illustrate how infrared techniques can be of value in nondestructive testing.

II. Infrared Radiation

All bodies above a temperature of absolute zero are continuously generating and emitting electromagnetic radiation by virtue of the motion of their constituent atoms and molecules. The spectrum and intensity of this radiation depends on the temperature of the body and if it is a perfect radiator or " black body " the radiation it emits is described by Planck's equation

$$W_\lambda \, d\lambda = \frac{C_1}{\lambda^5} \cdot \frac{1}{\exp(C_2/\lambda T) - 1} \, . d\lambda, \tag{1}$$

where T = absolute temperature in °K,

λ = wavelength of emitted radiation,

W_λ = Spectral radiant emittance, i.e. the radiation emitted from the black body in the wavelength interval $d\lambda$, per unit surface area; usually measured in watts per cm² per micrometre,

$C_1 = 2\pi hc^2$ where h = Planck's constant,
                                and c = velocity of light

      $= 3 \cdot 74 \times 10^{-12}$ watts . cm²,

$C_2 = hc/k$ where k = Boltzmann's constant

      $= 1 \cdot 44$ cm . degree.

From this equation the curves of Fig. 1 have been calculated and they show the spectral radiant emittance in watts . cm² . μm^{-1} against wavelength of black bodies at temperatures ranging from 300°K to 6000°K. The importance of the infrared spectrum, which extends from 0·75 μm at the lower wavelength end to about 1000 μm or 1 mm at the upper wavelength end, becomes apparent from Fig. 1 which shows that the bulk of the radiation emitted from most of these bodies, and which can be used to detect them, appears in

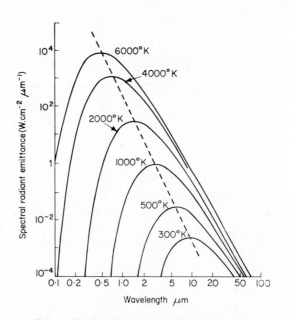

FIG. 1. Black body spectral radiant emittance.

the infrared. Bodies at temperatures above 1000°K emit sufficient radiation in the visible to be seen by the eye, but for bodies below this temperature infrared detectors are required.

The total radiation emitted by a black body over the whole spectrum is obtained by integrating W_λ with respect to $d\lambda$ between $\lambda = 0$ and $\lambda = \infty$, i.e.

$$W = \int_0^\infty W_\lambda \, d\lambda.$$

This gives the well-known Stefan–Boltzmann equation

$$W = \sigma T^4, \tag{2}$$

where $\sigma =$ the Stefan–Boltzmann constant

$$= 5 \cdot 67 \times 10^{-12} \text{ watts/cm}^2/°\text{K}^4$$

and $W =$ total radiant emittance in watts/cm^2.

Note that W, which is no longer wavelength dependent, is proportional to the fourth power of the absolute temperature.

The wavelength of maximum emission is obtained by differentiating Planck's equation with respect to λ and equating the derivative to zero. This gives Wien's law relating λ_m and T

$$\lambda_m = \frac{C}{T}, \tag{3}$$

where C is a constant $= 2898 \; \mu\text{m.}°\text{K}$ if T is in °K and λ_m in μm. Thus λ_m decreases as T increases and the locus of λ_m is shown by the dotted line in Fig. 1. Black bodies at 300°K have their peak emission at $9 \cdot 66 \; \mu$m and those at 6000°K peak in the visible at $0 \cdot 483 \; \mu$m.

The radiation curves of Fig. 1 and the value of W given by eqn. (2) will be reduced for any practical body by a factor equal to its emissivity. This emissivity factor is generally a property of the body's surface and is a measure of its efficiency as a radiator (or absorber). For black bodies the emissivity factor is unity, for others it is defined as

$$\varepsilon = \frac{\text{radiant emittance of the body}}{\text{radiant emittance of a black body at the same temperature}}.$$

The radiation emitted by a black body is always the maximum possible at that temperature, i.e. the value of ε is never greater than unity. When the emissivity is invariant with wavelength the body is termed a "grey body", the radiation curves of Fig. 1 retain the same shape at reduced levels, and the Stefan–Boltzmann equation becomes

$$W = \varepsilon \sigma T^4. \tag{4}$$

When emissivity is not constant with wavelength a corresponding factor ε_λ can be introduced into eqn. (1) and the curves of Fig. 1 take a modified shape as well as a reduction in level. The radiation curves for most opaque bodies, however, will approximate in shape to those of Fig. 1 but transparent materials, gases and flames emit predominantly in narrow bands corresponding to the frequencies of resonant vibration of their molecules, i.e. their emission consists of narrow spikes superimposed on a low level grey body radiation. The spectral characteristics of this emission are of course used with considerable advantage in chemical analysis.

The emissivity of an opaque body depends primarily on its surface condition, although it may also have some dependence on temperature. For instance, highly polished metals have emissivity factors below 0·1, while for the same metals with oxidized surfaces the factor is 0·5 or higher. Emissivities are usually calculated from measurements of the complementary property reflectivity. A list of emissivities of typical materials is given in Table I.

TABLE I

Emissivities of Typical Materials

Ag polished	0·03	Sand	0·75
Al polished	0·05	Cement	0·95
Cu polished	0·15	Brick	0·94
Fe polished	0·05	Wood	0·8–0·9
Al anodized	0·3–0·5	Paper	∼0·9
Cu oxidized	0·6		
Fe oxidized	0·6–0·9		

The modified Stefan–Boltzmann equation, (4), is utilized in radiation pyrometry but the accurate calculation of temperature from a radiation measurement is rendered difficult if ε is not accurately known. In a number of processes, particularly those concerned with heat treatment of metals, this uncertainty may arise. Returning to the curves of Fig. 1 it is evident that the effect of temperature on spectral emittance is greater at the shorter wavelengths. The dependence of emission at peak wavelength on temperature can be seen by substituting for λ_m in eqn. (1) when we get

$$W_{\lambda m} = \text{constant} \times T^5. \tag{5}$$

Thus, whereas the total radiation increases with the fourth power of wavelength, the radiant emission at peak wavelength follows a fifth power law and higher power laws are followed at wavelengths shorter than λ_m. Dreyfus (1963) has shown that within certain limits spectral radiant emittance can be written approximately as

$$W_\lambda = K T^n \tag{6}$$

where K is a constant and n approaches the value

$$n = \frac{5\lambda_m}{\lambda} \quad \text{when} \quad \frac{\lambda}{\lambda_m} \leqslant 2\cdot 5.$$

Thus at $\lambda = \lambda_m/2$, $n = 10$, and $W_\lambda = K T^{10}$. The effect on pyrometry of uncertainty in the value of ε can therefore be minimized by carrying out the measurement at as short a wavelength as possible.

The radiant emittance defined by Planck's and the Stefan–Boltzmann laws refer to emittance from unit area into a hemisphere. Where the radiating area is a perfect, plane diffuser its projected area in a direction θ from the normal varies as $\cos \theta$, and if N is the emittance from unit area in the normal direction, the emittance in direction θ is $N \cos \theta$ (Fig. 2). Integrating over the hemisphere we get that the total emittance in the hemisphere

$$W = \pi N \quad \text{or} \quad N = W/\pi. \tag{7}$$

Surfaces which are perfectly diffusing are known as Lambertian and the law they follow is Lambert's cosine law.

The radiant flux or irradiance incident on the detecting surface in Fig. 2 distant d from the Lambertian radiator of area A is then

$$\frac{W A}{\pi d^2} \cos \theta. \tag{8}$$

Not all surfaces are perfect diffusers; many have a direction-dependent emissivity, e.g. liquids, and the appropriate expression for emissivity ε_θ must then be inserted in the equations.

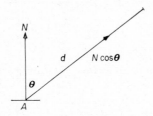

FIG. 2. Lambert's cosine law.

A number of radiation slide rules have been devised and are available commercially (Martin, 1959) for carrying out rapid calculations of the radiation quantities above.* They take account of a variety of conditions and the answers they provide are usually sufficiently accurate for most purposes.

The discussion above has dealt solely with the radiation emitted from a body by virtue of its temperature, i.e. with its self-radiation. However, the body itself is irradiated from its surroundings with which it attempts to achieve thermal equilibrium, and a fraction of this irradiance complementary to the emissivity is reflected from its surface, adding to its self-radiation. If W' in Fig. 3 is the irradiance on the body, in the same units as its self-emitted radiation, then the net radiation from the body is given by

$$\varepsilon W + (1 - \varepsilon) W' \qquad (9)$$

where W is the black body radiation.

Thus the radiation from a body is determined by its temperature, emissivity and the radiation falling on it, and it is this radiation which forms the basis for thermal detection, measurement and imaging systems. Some care is necessary in experiments to separate these components of the radiation from the object under examination, and if possible W' should be reduced to negligible proportions.

III. Atmospheric Transmission

One of the most attractive features of radiation sensing for nondestructive testing is that it can be carried out remotely without the need for physical contact with the object, and so virtually eliminates thermal disturbance

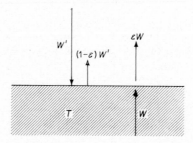

Fig. 3. Net radiation from a surface of emissivity ε.

---

* Radiation slide rules are available from General Electric Co., 1, River Road, Schenectady, New York, U.S.A., and A. G. Thornton Ltd., 93, Bridge Street, Manchester 3, England.

and possible contamination by the sensor. It provides a useful approach to the otherwise difficult problem of surface temperature measurement. It also enables widely spaced and moving objects or sheets of large area to be rapidly sampled, thereby making production-line control systems with rapid feed-back possible.

All this depends on the intervening medium having suitable transmission properties. In most practical situations the medium will be clear air, the transmission characteristics of which are shown in Fig. 4. This shows that clear air has several transparencies in the near infrared, i.e. from visible wavelengths to 2·5 μm, other "windows" between 3 and 5 μm, and a fairly wide window from 8 to 14 μm, which coincides with the broad peak of radiation from bodies at room temperature. Beyond 14 μ the only window is a partial transparency between 17 and 24 μ and all wavelengths above that in the far infrared out to its juncture with the mm wave region are fairly strongly attenuated over any practical path-length.

The main attenuating constituent of air is water vapour, with some contribution from CO_2 and other molecules, so the attenuating factor depends on the amount of precipitable water vapour present in the path between the sensor and the source, which in turn can be calculated from the air temperature and its relative humidity. For the path lengths likely to be encountered in laboratory or factory nondestructive testing, atmospheric attenuation within the windows shown in Fig. 4 can normally be ignored, but where testing is carried out in the open air over longer path lengths a factor for attenuation within the window must be included.

Figure 4 shows that radiation sensing in other than evacuated laboratory paths is confined to wavelengths within the atmospheric windows shown. Moreover, in factories where hot moist air can intervene the use of filters to prevent radiation emitted by it from falling on the sensor may be necessary.

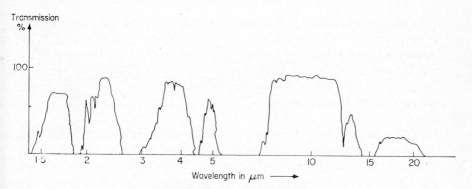

FIG. 4. Atmospheric transmission of infrared.

Transmission of infrared, particularly the longer wavelength (8–14 μm band) is markedly superior to visible transmission through smoke and clouds of condensed steam and this makes the sensing and examination of hot objects possible in conditions where they cannot be seen by eye. Some results for infrared transmission through smokes are given in Table II.

TABLE II

Infrared Penetration through Smoke

| Type of Smoke or Fuel | Attenuation Coefficients metre^{-1} | | IR Coefficient |
	Visual (0·4–0·7 μm)	Infrared (7–12 μm)	Visual Coefficient
Linseed Oil	5·7	0·38	15
Oil Cake	6·0	0·23	26
Rubber	4·3	0·41	10·5
Polystyrene	2·7	0·16	17
Timber	1·8	<0·10	>18
Textiles	1·8	<0·15	>12
Special fine particle smoke	6·0	0·04	150

IV. INFRARED DETECTION

The heart of any infrared instrument is its detector. The process of detection requires an interaction of the incident radiation with the detector material to produce some form of measurable signal output, usually and most conveniently electrical. Two types of interaction are used, giving rise to two different types of detector, the thermal, and the photon or quantum detectors.

In the first the radiation interacts with and is absorbed by the molecules of the detector or a blackened layer in close thermal contact with it, increasing its temperature and producing a second, measurable effect (Di Ward and Wormser, 1959). For instance, in the thermopile the change in temperature of the junctions creates a thermal e.m.f., in the bolometer a measurable change of resistance occurs, and in the Golay cell a volume of gas expands and moves an optical lever.

In photon detectors, photons in the incoming radiation interact with electrons in the detector material raising them to excited states and producing a conductivity change or voltage directly. If the photon energy is large enough electrons can escape from the detecting surface completely; this gives the photo-emissive detector.

Thermal detectors have the advantage that they can respond, with proper surface treatment, usually blackening, to all wavelengths uniformly, whereas the photon detector response is wavelength selective and in theory increases linearly with λ to the limit of response. Thermal detectors are, however, relatively slow having response times measured in milliseconds (*cf.* microseconds for photon detectors) and are less sensitive since two energy conversion processes are involved.

The spectral response limits of the photon detectors are properties of the materials. In photo-emissive detectors the photon should excite the absorbing electron sufficiently to overcome the surface work function. Most photo-emissive surfaces employ caesium, the element with the lowest work function, as emitter, and by suitable preparation surfaces with effective work functions as low as 1 eV can be achieved. The quantum energy is given by

$$E = \frac{hc}{\lambda} \tag{10a}$$

$$= \frac{1\cdot24}{\lambda} \tag{10b}$$

where E is in eV if λ is in μm. This means that only wavelengths up to about $1\cdot2$ μm can be detected by these surfaces (Martin, 1959).

Photo-emissive surfaces are employed in image converter tubes to provide an image of the infrared scene, up to $1\cdot2$ μm. The design of such a tube is shown in outline in Fig. 5. They can be useful for observing objects at intermediate temperatures, i.e. not quite hot enough to be clearly visible.

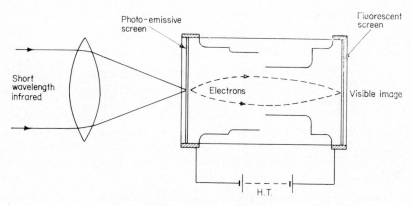

FIG. 5. Principle of infrared image converter tube.

Infrared film is also a detector of near IR wavelengths. Most infrared film is sensitive out to 0·9 μm only, but special varieties can be sensitized to respond, with sacrifice of speed, to about 1·2 μm. The chromatic properties of the camera lens should be checked before using infrared-sensitive film since the focus may be different at the longer wavelengths.

Photon detectors sensitive beyond 1·2 μm operate by excitation of electrons from their valence to conduction bands and the electrons and holes created become available for conduction. The detector materials are semiconductors, usually in the form of thin single crystals, and their sensitivities can be increased by several orders of magnitude by cooling to say 77°K. The materials can be operated in the photoconductive mode, when a charge of conductivity is effectively measured, or in the photovoltaic mode, when an internal p–n junction separates the carriers and generates a small voltage (Petritz, 1959).

The upper spectral response limit of these detectors is again given by eqn. (10) but in this case E is the energy gap separating the valence and conduction bands in the semiconductor. Thus for detection to 12·4 μm a semiconductor with an intrinsic energy gap of 0·1 eV is required.

The crucial property of an infrared detector is its noise equivalent power (NEP). This is defined as the incident power, modulated at a specific frequency, required to produce a signal equal to noise when the detector is followed by an amplifier with unit bandwidth.

Thus

$$\mathrm{NEP} = \frac{V_n}{V_s} \frac{W \cdot A}{B^{\frac{1}{2}}} \text{ watts} \cdot \mathrm{Hz}^{-\frac{1}{2}} \tag{11}$$

where V_s = RMS signal voltage,

V_n = RMS noise voltage,

B = amplifier bandwidth,

W = irradiance on the detector in watts/unit area,

A = area of detector.

It is clear that the higher the sensitivity of the detector the lower is its value of NEP, and the inverse of NEP, the detectivity, is now commonly used to describe and compare infrared detectors. This has the advantage that it increases with detector sensitivity.

The detectivity of semiconductor detectors and some others varies inversely with the square root of area, and for ease of comparison of their performance their detectivities are normalized to unit area. Thus if A is the detector area, then the normalized detectivity D^* is given by

$$D^* = \frac{A^{\frac{1}{2}}}{NEP}$$

$$= \frac{V_s}{V_n} \frac{A^{\frac{1}{2}} B^{\frac{1}{2}}}{W} \text{ cm Hz}^{\frac{1}{2}} \text{W}^{-1}. \tag{12}$$

Figure 6 shows spectral responses in terms of detectivity for a variety of detectors. Available detectors have been described in the literature (Shapiro, 1969).

A fundamental limit is set to the detectivities of detectors by the fluctuation in quanta emitted by the background which the detector sees. This limit is shown by the dotted curve in Fig. 6 which is drawn for a background at a temperature of 300°K and detector field of view of 2π steradians.

The pyroelectric detector is a significant new development. This is operated essentially as a temperature sensitive capacitor and consists of a crystal of ferroelectric material, e.g. triglycine sulphate (TGS), which changes its polarization when heated. Its internal polarization is normally neutralized by stray charges and no external field is apparent until absorption of radiation heats up the crystal. The consequent dimensional change alters the internal

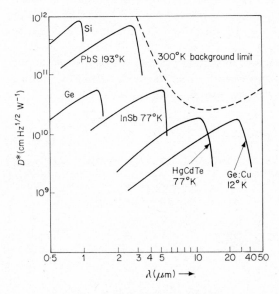

FIG. 6. Typical detector responses.

polarization of the crystal and generates a measurable change in the external field. Although it is a thermal detector its speed of response can be quite high because of its capacitive nature, and the fact that its noise decreases with increase in frequency allows it to retain high sensitivity up to quite high frequencies, e.g. 10 kHz. When ample incident power is available it can still give a useful response at 1 MHz. Its great advantage is lack of need for cooling, and in addition to its greater speed is likely to be more robust than other thermal detectors, but microphony in vibrating environments is likely to be a serious disadvantage.

V. DETECTOR COOLING (Goodenough 1959)

The highest sensitivity, however, is obtained with cooled semiconductors. A variety of simple reliable cooling techniques is available and detector cooling presents little difficulty.

Small closed cycle cooling engines, based mostly on the Stirling thermodynamic cycle, are becoming available commercially as a fallout of military demands, but they are still relatively complex and expensive and require a fair amount of maintenance.

The simplest method of cooling is to encapsulate the detector in a dewar capable of holding an adequate amount of liquid nitrogen. This, however, leads to a bulky dewar if several hours operation is required without refill, and a more convenient technique employs liquid transfer, a process whereby the liquid cryogen is transferred along uninsulated, flexible pipes to the detector dewar by the process of Liedenfrost transfer, i.e. liquid droplets are carried along on insulating cushions of their own vapour. The rate of transfer can be governed to meet the demand and the storage dewar can be placed up to 6 ft away from the detector (Fig. 7).

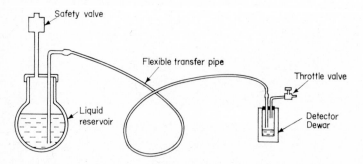

FIG. 7. Liquid transfer cooling.

Another practical, reliable cooling device is the Joule–Thomson mini-cooler designed to be inserted in the detector dewar. High-pressure air or nitrogen cools itself by expansion through a small orifice at the end of the minicooler and by regenerative cooling, liquid is eventually formed.

VI. Optical Components (Wolfe and Ballard, 1959)

A comprehensive and detailed knowledge of infrared optical materials and filters is perhaps not so essential to the NDT engineer as it is to the instrument designer, but there are likely to be occasions when special windows and filters have to be used, and this section is added for completeness to meet these needs.

Common glasses and optical glasses are based on silica as the main glass-forming constituent. Most transmit well through the visible out to wavelengths between 2 and 2·5 μm, but then a series of absorption bands renders them virtually opaque to longer wavelengths. The first strong absorption band is centred on 2·7 μm and is due to a resonant vibration of water which becomes dissolved in the glasses in small percentages during their formation. If care is taken to prevent the entry of water, e.g. by melting the glass in vacuum, good transmission can be achieved, especially in pure silica, out to 4·4 μ where the first absorption due to the Si–O band is encountered. Thereafter silica and all glasses based on it are opaque until wavelengths in the very far infrared; such wavelengths are outside the scope of this chapter.

Glasses based on germania and on calcium aluminate transmit beyond silica to between 5 and 6 μ because of the lower resonant frequencies of their heavier basic constituents, and they cover the atmospheric windows out to 5 μ. Again special glass-making techniques have to be employed to avoid solution of water vapour.

The limit of transmission of oxide glasses is about 6 μm, but glasses made of As_2S_3 and As_2Se_3 give good transmission to longer wavelengths, although their refractive indices are higher. As_2Se_3 when free from oxide and other impurities gives good transmission over the whole of the 8–14 μm atmospheric window. It is a relatively soft glass with a low softening temperature—about 150°C—but it can be hardened thermally and mechanically with additions of germanium without impairing its transmission. It is just becoming available commercially. Figure 8a shows the transmission of glasses.

Semiconductors like silicon, germanium, CdTe, etc., can be used for transmission at wavelengths beyond their intrinsic absorption edges, provided they are sufficiently pure. Their unbloomed transmission is about 50% but simple blooming can increase this to about 90%. Silicon is

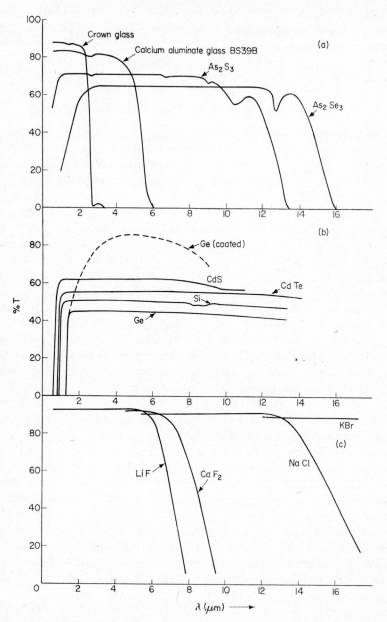

Fig. 8. Transmission of infrared optical materials. (a) glasses; (b) semiconductors; (c) alkali halides.

a useful hard window material but its manufacturing difficulty—it melts at 1420°C and is very reactive—limits its available size to some inches. Other semiconductors apart from germanium and silicon have not been extensivey. developed as window materials. Their transmission is shown in Fig. 81b

A number of alkali halides in single crystal form give good transmission through the visible and well into the infrared. Most are attacked by water to some extent and as single crystals they need careful handling.

A range of optical materials called Irtrans have become available recently by a technique developed by the Eastman–Kodak Co. of hot-pressing fine powders of crystalline materials. Thus Irtran 1, compacted MgF_2, gives good transmission to 6 μm and Irtran 2, ZnS, to 13 μm. Others covering longer wavelengths are also available, but the size of these materials is limited to a few inches, because of the process.

The optical glasses, semiconductors and crystals described above can be used as low-pass or high-pass absorption filters. For more exacting filter requirements, however, multilayer interference filters are essential and a wide range of sophisticated designs is now available to meet most requirements in the infrared.

VII. INFRARED SYSTEMS

In systems design, the final choice of components and parameters is often arrived at in a two-stage process. The first stage involves the selection of the most appropriate basic design, and the detailed parameters are then determined by an iterative process of successive approximations to the final choice. This "circular" rather than "serial" approach is the result of the fact that usually no important component can be chosen independently of the other components and no important performance parameter can be altered without affecting others. However, this brief discussion of systems will probably be clarified by the adoption of an artificially "serialized" approach, as far as is possible.

The detector must enable the system to achieve the desired radiance discrimination by a suitable matching of its spectral detectivity to that of the relevant spectral radiance or differential spectral radiance expected from the target. The spectral transmission of the air path between system and target must also be taken into account. The time-constant of the detector must be short enough for it to respond at the maximum frequency demanded. This may be set by the maximum rate of change of target emission in an unchopped radiometer, or the corresponding maximum upper sideband frequency in a chopped system. In a scanning system it is determined by the rate at which the basic angular resolution elements are traversed. The effective size and shape of the detector, together with the focal length of the optical system, will determine the angular size and shape of the received "beamwidth". Finally, the detector choice may be influenced by considerations such as the requirement to operate in a given part of the target spectral emission curve, in order to derive the appropriate output/temperature scale, or to make the temperature reading less dependent on target emissivity by "short wavelength tail" operation.

In most systems the detector or its real image is placed at the focus of a suitable telescope or microscope (Fig. 9A). The optical system together with the detector define the angular resolution of the system and its radiance resolution. For example, when a detector (or its image) of characteristic dimension L cm is placed at the focus of a telescope of focal length F cm, the beam divergence angle is given by θ where

$$\theta = L/F \text{ radians.} \tag{13}$$

It can also be shown that the minimum detectable radiance difference for such a simple system is given by

$$N_{min} = \frac{F}{A_c \theta} \cdot \frac{B^{\frac{1}{2}}}{D*} \text{ watts.cm}^{-2}.\text{sr}^{-1}, \tag{14}$$

where A_c = effective collecting area of the optics in cm^2 and N_{min} is defined for a signal to R.M.S. noise of unity.

The corresponding minimum detectable temperature difference is given by

$$\delta T_{min} = \frac{F}{A_c \theta} \cdot \frac{B^{\frac{1}{2}}}{K} \text{ °C,} \tag{15}$$

where

$$K \equiv \int_0^\infty D*(\lambda) \cdot \left(\frac{\partial N_\lambda}{\partial T}\right)_{T_0} .d\lambda \qquad \text{cm}^{-1}.\text{sr}^{-1}.\text{Hz}^{\frac{1}{2}}\text{°C}^{-1} \tag{16}$$

and the temperature T_0 refers to that of the black body on which the temperature discrimination is measured.

Now the quality of the optical system must be adequate for the required angular resolution and it can be seen that as θ is reduced, A_c must be increased and the f number reduced. This implies increased aberrations and

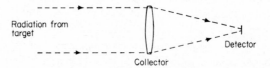

FIG. 9A. Simple telescope system.

sets one of the limits to the angular resolution achievable. Some computed aberrations for reflecting mirror systems are given in Fig. 10.

Fortunately, it is possible to partially offset the requirement for very small f numbers in detectors limited by background radiation noise with the use of radiation shielding. This takes the form of a baffle around the detector, cooled to near the detector temperature in the cryostat. If the detector background flux is restricted to the same f number as that of the optical system feeding it, it is possible to achieve an improvement in D^* of up to $2f$ times. $N_{\min}$ is then determined by the collector diameter rather than the f number.

A feature of many radiometers is the chopper (Fig. 9B), usually in the form of a rotating reticle. Choppers may be employed for various reasons. Where a high radiance sensitivity is required, it is often advantageous to employ chopping so that the detector receives radiation which alternates between that from the target and that emitted and reflected from the chopper at a high frequency. This puts the information on a carrier which can be chosen so that carrier and sidebands are all in the optimum frequency band for the detector and following amplifiers. As photoconductive detectors show excess noise of a "$1/f^n$" form below about 100–1000 Hz, it is advantageous to chop these systems at or above this band to realise the best D^* values and system performance. (As "n" tends to lie between 1 and 3, the fact that the system is not viewing the target all the time is more than compensated by the reduction in noise.)

By reflecting radiation from a standard radiation source into the detector from the reticle blades, it is possible to keep the signal calibrated absolutely. This is usually necessary if precision measurements are required on low temperature targets. In one commercial radiometer, the temperature of such a standard source (a black body cavity) is servoed from the detector

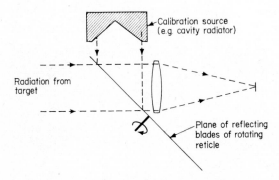

FIG. 9B System using a chopper and reference.

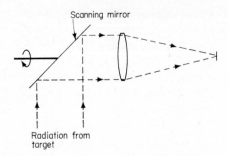

Fig. 9c. Simple line scan system.

output until a null is obtained. The cavity is then at the equivalent "radiation" temperature as the target and this can be read from a thermocouple in the cavity. The system has the advantage of linear temperature scales, but requires a few seconds for the cavity temperature to become adjusted. A radiometer employing a double reticle chopper at its input aperture, acting as both modulator and reference, has been described (Sabey, 1966). Having thus "labelled" the incoming radiation at its point of entry, the system is not affected by temperature drifting of internal components.

The radiative nature of infrared sensing makes it possible to employ scanning techniques, the advantages of which are discussed later. Scanning could be done by rotating the basic radiometer system. This is possible when only very slow scan rates are demanded. For line-scanning, a rotating or vibrating mirror may be added to determine the line of sight of the radiometer (Fig. 9c). If a frame scan is required, two such mirrors may be used. As scanning in one direction normally has to be done very rapidly, a drum with mirror facets is often used (Fig. 9d). To obtain a rapid scan

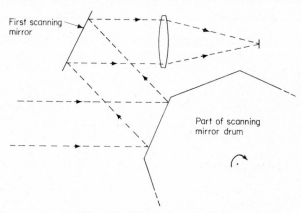

Fig. 9d. Frame scan system.

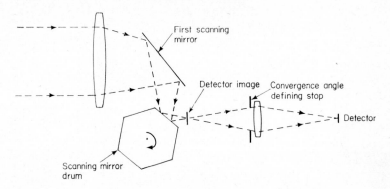

FIG. 9E. Convergent beam scanning.

rate at low drum rotation rates, it is often necessary to use several facets, with the result that the drum size can sometimes become too large for convenience.

The size of the scanning optics can be greatly reduced if the scanning is carried out in the path between the detector and the converging optics, since the facet sizes then have to deal only with a "beam" which has been partially converged (Fig. 9E). Indeed, the detector could be imaged on the facets, reducing the drum to very small dimensions. Such convergent beam scanning systems require that the optical system should be adequate for the angular resolution demanded not only on axis, but at the extreme angles of coverage. It can be seen from Fig. 10 that this can be best achieved when the angular resolution specified is low, or the scan angles small. Nipkow disc scanning is also feasible under these conditions (Jervis, 1968).

The high video frequencies characteristic of such scanning systems usually preclude the use of reticle choppers by which the radiance reference was established in the slower systems. Instead, references can be inserted into the scan sweep, and the video clamped to suitably controlled voltage level in this period.

The line and frame scan systems described often demand a much shorter time constant detector than the point sensing radiometer. Consequently, whereas the slower thermal detectors are often used in radiometers, scanning systems will usually demand quantum detectors. An imaging system using a pyroelectric detector has also recently been reported (Astheimer and Schwarz, 1968).

In addition, the larger bandwidths demanded by line and frame scanning would lead to a lowering of the temperature discrimination, as can be seen from eqn. (15). For example, radiometers generally operate in an effective

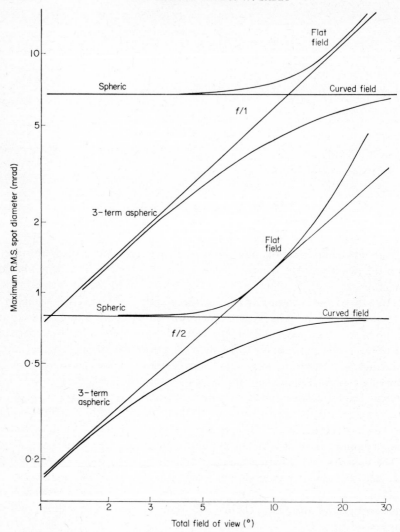

FIG. 10. Computed aberrations in mirror systems showing the effects of f/number, surface type and image plane curvature.

final bandwidth of the order of 10 Hz, whereas a frame scanner may require a bandwidth of 100 kHz. Thus the temperature discrimination would be two orders lower.

To compensate for this, the detector is usually cooled, with a resultant increase in D^* and K. With indium antimonide, for example, D^* is increased by about the requisite two orders by cooling from room temperature to 77°K. Some typical values of K are shown in Fig. 11 for various infrared detectors as functions of the mean temperature of the target. A rough indication is given of the order of K required to satisfy the sensitivity requirements of the different types of system which have been discussed.

The data rate which can be extracted from a single infrared detector is limited, and this limits the overall system performance to a large extent. There are also problems in mechanically scanning at very high speeds. Multielement arrays of detectors have now become available, and a number of high performance experimental systems have been built which employ them. The use of a linear array with sufficient elements to cover one dimension of a frame scan can avoid the requirement for a mirror drum. By

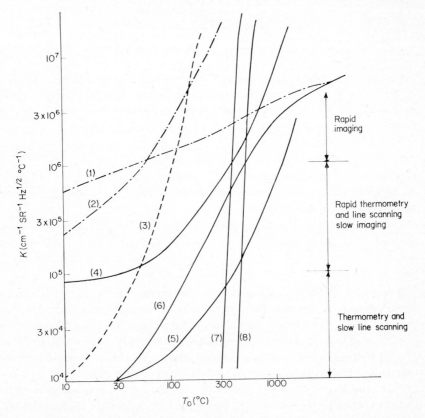

Fig. 11. $K(T_0)$ for typical detectors. (1) CdHgTe at 77°K, $D^*_{pk} = 1·2 \times 10^{10}$; (2) InSb at 77°K, $D^*_{\lambda pk} = 4·0 \times 10^{10}$; (3) InAs at 193°K, $D^*_{\lambda kp} = 1·2 \times 10^{11}$; (4) thermal detector, $D^* = 1·0 \times 10^9$; (5) thermal detector, $D^* = 1·0 \times 10^8$; (6) InSb at 300°K, filtered for 3–5·5 μm, $D^*_{\lambda pk} = 5·0 \times 10^8$; (7) Ge at 300°K, $D^*_{\lambda pk} = 4·0 \times 10^{11}$; (8) Si at 300°K, $D^*_{\lambda pk} = 2·5 \times 10^{12}$.

scanning the elements in parallel, the response speed demanded of each element is lowered and with it the bandwidth. Consequently the temperature discrimination is improved. For an N element array, this treatment is equivalent to increasing K by a factor $N^{\frac{1}{2}}$.

As will become clear later, many applications require an assessment of temperature to be made from the infrared radiance emitted from a target. Difficulties arise because this radiance is a function not only of temperature but of its emissivity, and it can be seen from eqn. (9) that this can lead to interpretational difficulties when W' is appreciable compared to W, and when ε is small. Thus, in view of the importance of accurate temperature measurement it is worthwhile devoting some space to the methods available for compensating or reducing the effects of emissivity.

If it can be assumed that in eqn. (9)

$$\varepsilon W \gg (1 - \varepsilon) W'$$

then it may be possible to compensate the reading for ε by a simple increase in gain of ε^{-1}. Values of ε for many commonly occurring surfaces have been tabulated for this purpose, and many radiometers have the facility to compensate the reading for ε under these conditions, e.g. Table I.

Some situations allow a low emissivity surface to be coated with a thin layer of high emissivity material. This method runs the risk, however, of altering the surface temperature characteristics, especially if there are small air inclusions under the coating. However, a fairly successful example of this method is quoted later on.

If it is possible to use near-contacting methods, two related techniques for emissivity compensation are available. Both depend on forcing the radiation above the area being sampled to approximate to that of black body radiation at the temperature of the surface.

In the first technique, a highly reflecting hemispherical cavity is placed over the surface. The resultant cavity radiation is largely determined by the temperature of the surface being measured (unless its emissivity is comparable with that of the hemisphere), and consequently approximates to that of a black body at the appropriate temperature. As the hemisphere is highly reflecting it is not necessary for it to be in temperature equilibrium with the surface, and hot surfaces can therefore be measured. Conduction is minimized by standing it on insulating points around the base of the hemisphere. The detector views the cavity radiation through a small hole in the hemisphere.

The second near-contacting technique (Kelsall, 1963) involves the formation of the enclosure by means of a highly absorbing cavity placed over the surface. The walls of this cavity are capable of being controlled in temperature over the required range of measurement temperatures from the detector output. The detector alternately views radiation emitted directly from the upper heater surface and radiation from inside the cavity formed between the test surface and the lower heated surface, by the use of a reflecting chopper and the heater temperature is adjusted from the detector output signal until a null is obtained. Under these conditions it can be shown that the heater surface temperature is equal to that of the test surface, and can therefore be read by a thermocouple in the heater.

This method has been applied to the measurement of low emissivity moving surfaces, such as motor commutators. Its disadvantages lie in the fact that it must be used close to the test surface, and that the servo-time of the heater can be a few seconds.

Neither of these methods is possible when working at long range, and under these conditions no perfectly satisfactory solution is possible. Methods based on the comparison of emission at two different wavelengths have been used. These "two colour" pyrometers require assumptions to be made about the ratio of emissivities at the two wavelengths in question, and are therefore open to some objections. A second method is to operate in the short wavelength "tail" of the target emission. It can be shown that equivalent fractional charges in emission due to emissivity or temperature changes are related by

$$\frac{\delta\varepsilon}{\varepsilon} = n \cdot \frac{\delta T}{T}, \tag{17}$$

where

$$n = \left[\frac{\exp\left(\dfrac{4\cdot96}{f}\right)}{\exp\left(\dfrac{4\cdot96}{f}\right) - 1} \right] \cdot \left(\frac{4\cdot96}{f}\right) \tag{18}$$

and

$$f \equiv \frac{\lambda}{\lambda_m} \quad \text{see eqn. (3).} \tag{19}$$

For values of $f < 0\cdot5$, x can be approximated by:

$$n \simeq \frac{4\cdot96}{f}. \tag{20}$$

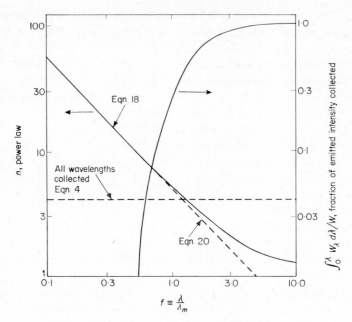

FIG. 12. Short wavelength " tail " operation.

Thus, at the expense of the available energy it is possible to reduce the temperature error to very small values. A pay-off graph is given in Fig. 12.

Problems arise when there are sources in the vicinity of the target which are at higher temperatures, since the radiation from these reflected from the target can rapidly become appreciable if small values of f are used. "Short wavelength tail" near-surface pyrometers for use under these conditions therefore often carry radiation screens. Another method of increasing emissivity when measuring some materials is to restrict the sensing to the highest emission bands. Organic materials, for example, have a narrow band at 3·4 μm due to the −CH absorption. When dealing with thin films of semi-transparent materials, it is sometimes possible to increase the apparent emissivity by placing a reflector behind them.

Before leaving systems, it would complete the picture to mention the types of display available. Pyrometer outputs are readily displayed on meters, recorder charts and oscillographs. Scanning system outputs are usually also suitable for oscillograph presentation, and A-scan or brightness-modulated displays are both frequently used.

Recently systems have become available which allow isotherms to be plotted in place of or on the normal thermographic picture (Fig. 13). An

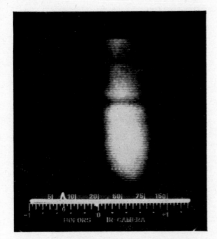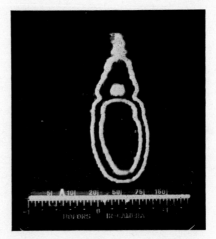

FIG. 13. Thermograph and isotherm presentations of surface temperature distribution in a plate, distorted by subsurface voids. (Crown copyright. Reproduced by permission of H.M. Stationery Office.)

alternative method which has been found useful is the multiple *A*-scope presentation (Fig. 14).

Information is presented more vividly in colour, and displays have been described (Borg, 1968; Sundstrom, 1968) in which successive isotherms have been designated different colours, and others in which different spectral regions of the infrared have been displayed in different colours and combined to form a final coloured picture (Nichols and Lamar, 1968).

VIII. CHARACTERISTICS OF INFRARED SENSING

The capabilities for nondestructive testing afforded by infrared sensing techniques arise partly from the nature of infrared and the laws governing its behaviour, and partly from the extent to which they can be exploited by current component and systems technology. It will be worthwhile considering these characteristics and their general practical implications before embarking upon a detailed discussion of the various types of nondestructive testing technique available.

Since all infrared techniques depend on the sensing of infrared radiation emanating directly or indirectly from the target to the detector, the only basic requirement for remote sensing is that a suitable optical path should exist between them. This ability to sense remotely constitutes one of the major advantages of the technique. It often confers a freedom in the positioning of the sensor, and it enables scanning or other techniques to be employed when a survey of an area is required.

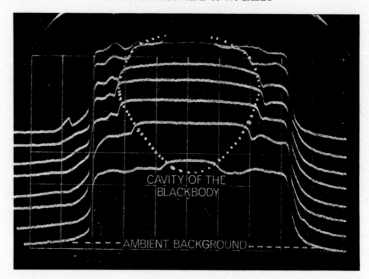

FIG. 14. Multiple *A*-scope presentation of the radiance distributions across the aperture of a black body cavity. (Successive traces were separated by 2° in elevation.) (Crown copyright. Reproduced by permission of H.M. Stationery Office.)

Because no mechanical contact between source and sensor is required, most of the undesirable ways in which target and sensor could interact are at once eliminated. Naturally, the presence of the sensing equipment should not disturb the conditions to be measured, and the contactless sensing enables this requirement to be met, even for objects of very small thermal capacity. In addition, the possibility of damage to mechanically fragile or viscous materials is avoided. A particularly important practical result of remote sensing is the ability to measure moving targets and sheets which would present difficult problems for contacting sensors.

It is frequently desirable to measure bodies which would damage or at least disturb the detecting equipment. Some targets are also located in an immediate environment which would similarly be detrimental to the detector system. Non-contacting techniques are, in principle, able to deal with such situations. Examples have been encountered in which bodies at extreme temperatures or high voltages have to be measured. The problems of sensor contamination by radioactive or chemically active sources are similarly removed.

Usually the presence of the detecting equipment creates a negligible disturbance in the radiation field around the target. It is true that some infrared techniques, to be described later, involve the use of active radiation sources. However, the sensitivity of infrared systems can generally be made high enough to enable low power densities to be used. Moreover, the particular properties being investigated by such methods would not, in general, be affected appreciably by this radiation. Consequently, it can be fairly said that infrared methods usually satisfy the conditions for a body of test techniques which are not only nondestructive but non-perturbing.

Most materials are opaque in the infrared, and consequently infrared sensing will yield information on surface conditions. Accurate surface temperature measurements are particularly difficult by contacting methods on some types of surface, so that infrared sensing can often complement bulk temperature measurements by other techniques. Bulk radiation temperature measurements can be made if a suitable cavity can be let into the target. This, although sometimes undesirable, is often necessary in employing contacting methods also. Surface measurements can be of value in assessing boundary and interface conditions.

The penetrating ability of infrared radiation in haze, dust and smoke has already been mentioned. Provided that the regions of high molecular absorption by the atmospheric components are avoided, therefore, infrared transmission can often be appreciably better than in the visible.

The other useful capabilities of infrared methods are the result of component technology, particularly that of detector fabrication. The short time constants of quantum detectors enable infrared surveying to be carried out very rapidly. This is necessary when dealing with dynamic problems such as are found with fast-moving objects or transient thermal distributions. On-line operation and real-time displays are therefore possible in principle in several important industrial applications.

It has been shown how cooling these detectors enables them to achieve high detectivities while maintaining the rapid speed of response. This results in high angular resolution and radiance (or temperature) discrimination for the sensing equipment. A high angular resolution is useful not only when small areas need to be sampled, but also when operation at long range is desirable in order to avoid adverse target environments or to cover widely separated points. Many situations in practice demand a high discrimination of radiance or temperature, particularly those in which the detection of small non-uniformities in temperature or a linked variable is required. There are also advantages in being able to restrict the wavelength band sampled. Operation in the "short wavelength tail" of the Planck distribution has been mentioned as a means of reducing the effects of emissivity in radiation temperature measurement. Another advantage lies in the ability to measure solids, liquids and gases which are only optically thick in relatively restricted wavelength bands.

Table III summarizes the characteristics of infrared sensing which have been discussed.

TABLE III

Characteristics of Infrared Sensing

Characteristic	System Capability	Advantages and Applications
Radiative nature of infrared	Remote, contact-less sensing	Freedom of sensor positioning, Ability to survey, Measurement of targets which would be disturbed or damaged by contacting methods, Measurement of targets which would disturb or damage a contacting sensor
Opaque nature of many materials in the infrared	Surface sensing	Measurement of surface and some interface conditions
Low attenuation coefficients in smoke dust and haze	Good penetration	Measurement in adverse atmospheric conditions
Rapid response in quantum detectors	High data rates Rapid scanning High resolution	Measurements in dynamic situations On-line capability Real-time display
High detectivity in cooled detectors	High resolution High radiance and temperature discrimination	Small sampling areas Long range operation Detection of small non-uniformities Operation in narrow wavelength bands

IX. INFRARED APPLICATIONS

It is now possible to consider, in greater detail, the various types of problem to which infrared techniques can be applied. There exists such a wide variety of infrared applications, that it is worthwhile attempting some kind of classification at the outset. Of the numerous ways of doing this, one method, used in Table IV, takes as the main criterion, the mechanism responsible for the infrared radiation conveying the information from subject to sensor. A secondary criterion is also used, which takes into account the number of cause–effect links which have to be traversed before the quantity of primary interest gives rise to the detectable infrared signal. Each of these links requires the establishment of a relation between the quantities representing cause and effect, and the latter may, at each stage, be modified by other variables than the one of interest. Consequently, the simplest chains tend to yield the most "natural" applications for infrared techniques, and the longest ones tend to require the greatest care in the interpretation of the infrared signal.

TABLE IV

Classification of Infrared Applications

Classification	Quantity of Primary Interest	Conversion Mechanism	Quantity Detected
1·1	Emitted radiance ─────────────────────→		Emitted radiance
1·2	↑ Temperature		
1·3	↑ Other variables		
2·1	Thermal properties→	Applied thermal gradient or pulse ──────→	Emitted radiance
2·2	↑ Other variables		
3·1	Emissivity ─────→	Surface heating ──────→	Emitted radiance
3·2	↑ Other variables		
4·1	Reflectivity ─────→	Surface irradiation ─────→	Reflected radiance
4·2	↑ Other variables		
5·1	Transmission ─────→	Transmitted radiation→	Transmitted radiance
5·2	↑ Other variables		

One broad class of problems is that in which information is conveyed to the sensor by infrared radiation which is directly emitted by the target as a result of its temperature characteristics.

Within this class, perhaps the simplest and most readily assessed problem is that in which the information required concerns some aspect of the emitted infrared radiance, such as its absolute value, its spatial distribution or its variation with time. Relatively few applications appear to be of this fundamentally straightforward type. There are applications in the measurement of the radiance distributions in radiant heaters for furnaces and driers.

An infrared scanner has been used to measure the radiance distribution across the aperture of a black body radiator which had been designed as a large uniform standard radiance source for radiometer calibration (Beardsley and Sabey, 1968). (Fig. 14)

In many applications, information is required about similar aspects of target temperature. As the emitted radiance is a function of temperature, the process is of a basic type. Difficulties of interpretation arise when the target emissivity is low, and when reflected radiation effects may have to be taken into account. In many instances the primary interest is in assessing some quality in the target closely determined by the temperature. For example, automatic steel rolling mills ideally require controlling from a measurement of steel strip rolling qualities. These are tied to the strip temperature. Another example recently examined by the authors concerned the need to measure the temperature of a thin extruded polythene sheet before laminating with cellulose in making food wrappings and packagings. The range of temperatures was critical to 10°C, too high a temperature giving oxidation and an unacceptable odour to the polythene, and too low a temperature giving poor adhesive qualities in laminating. In the annealing of glass, the correct temperature distribution must be set up in order to leave the surface layers correctly stressed. This is also important before glass is quenched to toughen it.

There are possible applications to the continuous measurement of the dimensions of moving sheets or webs of material. If the sheet temperature or radiance is significantly different from its background, a measurement of the length of the pulse generated in scanning an infrared sensor across the sheet will be a basis for width assessment. Examples have been found in width measurement of mechanically fragile or non-rigid sheets such as plaster board prior to setting, and thin organic films such as polythene.

The anomalous temperature of a solid body or even a liquid can often be used as a label, and positional or distributional information can then be obtained with infrared equipment. Thus the technique can sometimes be used to indicate the level of liquids inside opaque containers, and the movement of the surface layers of liquids have been studied, both on the laboratory and on large scale. Figure 15 shows an airborne infrared line scan picture of the hot water outfall from Aberthaw Power Station on the Welsh coastline, showing the water surface temperature distribution. The dynamic range of the picture is about 3°C. A useful medical application investigated with rapid thermal imaging equipment was in the detection of the sites of incompetent perforator veins which give rise to varicose conditions. These veins connect the inner high-pressure system with the outer lower pressure system, and have valves to prevent the flow of the higher pressure blood

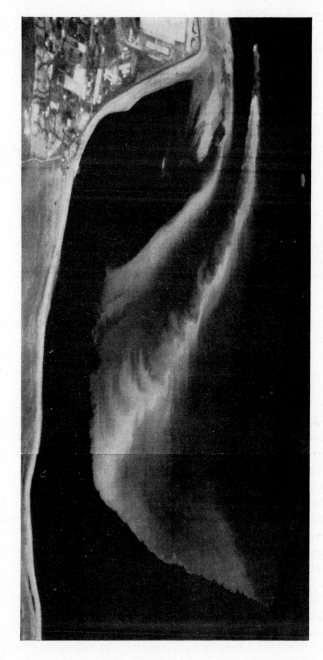

FIG. 15. Aerial infrared line scan picture of the hot water outfall from Aberthaw Power Station. (Crown Copyright. Reproduced by permission of H.M. Stationery Office.)

to the outer veins. The failure of these valves can cause the varicose condition. The technique is to apply a torniquet to the limb above the suspected area restricting the general blood flow to the limb. The skin and the blood in the surface veins is then "labelled" thermally by cooling it. On releasing the tourniquet, hotter blood from the interior of the limb passes into the surface venous system via the incompetent perforators, giving an initial hot spot.

Infrared measurements can often indicate the state of physical variables which are related to temperature either as the cause or the effect. Many applications of this type arise through the conversion of various forms of energy into heat, and infrared methods can often give a sensitive indication as to whether or not all is well. For example, faults and overloading in electrical circuitry, ranging from microcircuits (Yoder, 1968) to power lines (Borg, 1968), have been located because of the anomalous dissipation of electrical energy as heat. Infrared techniques have been used in checking for undue friction in machine bearings, locomotive axle boxes and brake blocks. The fibre rollers of super calender stacks used in greaseproof-paper manufacture are deliberately slipped relative to alternating steel rollers while under considerable pressure and at high temperature. The fibre rollers can develop hot spots, spoiling the paper or even causing bursting at the rollers, the onset of which could be detected by infrared imaging.

There are further applications in monitoring the state of exo- or endothermic chemical reactions, an extreme instance of which occurs in the detection of combustion (Coleby, 1968).

The medical uses of thermography as an indicator of unusual physiological activity are steadily becoming accepted, and there is evidence that infrared thermography will be a useful diagnostic technique in a wide range of applications (Barnes, 1968). Aerial infrared surveying has been used to assess the state of crops and trees.

Often applications may be found in the indication of evaporation rates and the uniformity of moisture distributions in drying processes such as occur in the textile and paper industries, and in moisturizing processes as in tobacco preparation.

The second type of application listed in Table IV is concerned with the detection and possible evaluation of bulk thermal properties of materials by infrared techniques. Transient methods have been described (Schultz, 1968) for the determination of thermal inertia, thermal conductivity and thermal diffusivity in solids. Many defects represent anomalies in the thermal properties of the host material. It may be possible to produce detectable surface infrared emission effects from such defects by the application of suitable thermal gradients or pulses to the material. The optimum conditions for observing such surface effects, and the order of defect size which might be observed, have been calculated by Fergason (1968). A practical limitation proposed for the depth of a void defect was given as twice the linear size of the defect. Fig. 13 shows the effect of void defects, in the form of holes drilled below the surface, on the surface thermal distribution caused by a thermal pulse applied to the other surface of the plate. The distortion of the distribution is seen in both the brightness modulated and the isotherm presentations.

Practical examples have been found in ceramic ware and in brake linings, in which air inclusions were detected fairly readily by thermal pulse techniques. In some instances, natural thermal gradients can show up such defects as the erosion of the refractory linings of glass and blast furnaces and areas of poor thermal insulation. Figure 16 shows the greater heat losses through the thin spots of a crucible than through the thicker areas.

The other types of application given in Table IV are concerned with the properties of emissivity, reflectivity and transmission. Emissivity effects may be observed thermographically by applying surface or bulk heating,

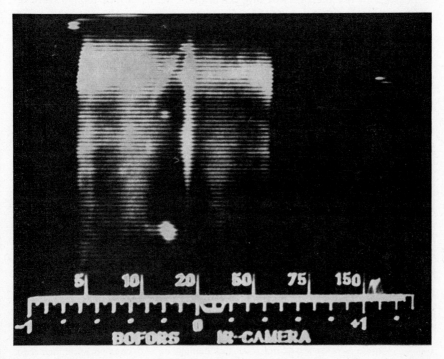

FIG. 16. Thermograph of steel crucible showing a hot spot. The vertical narrow band is a weld line. (Courtesy of Guest Electronics.)

yielding a directly emitted radiance. Information about the reflective properties of the target can be obtained by the application of a suitable infrared surface irradiation, while the transmitting properties can be investigated by passing infrared radiation through the material. Since the emissivity, reflectivity and transmission sum to unity, anything affecting one of these properties will also affect at least one other. The method chosen for testing may therefore be determined by which of these quantities shows up the required information to best advantage. In testing for defects, for example, one criterion to be applied will be that of estimating which of the three properties will be most affected by the defect. In many practical instances, however, either emissivity, reflectivity or transmission will be negligible and the complementary pair will be left.

Spot measurements of emissivity can be made with an emissivity compensated instrument, but taking the ratio of radiance readings obtained without and with compensation. For example, the reflecting hemisphere instruments can be fitted with a blackened hemisphere insert to provide the reading without compensation. If emissivity surveys are required, it may be possible to use thermal imaging techniques, if the surface temperature of the body can be independently assessed, and provided it is possible to neglect or allow for reflected background radiation. To enhance the effects of emissivity, long wavelength operation would normally be used, but this is limited by the increasing importance of reflected radiation from the lower temperature background. These are precisely the opposite conditions to the "short wavelength tail" methods.

Figure 17 shows the variations of emitted radiance from an aluminium sheet due to surface defects. The surface emissivity in this sample varied

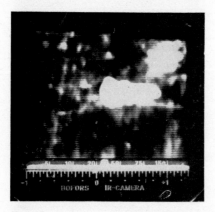 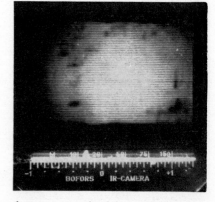

FIG. 17. Variations of emissivity due to surface damage on an aluminium sheet, and their reduction on applying a thin aquadag coating. (Crown copyright. Reproduced by permission of H.M. Stationery Office.)

from 0·2 in the undamaged regions to 0·9 in an area of severe abrasion. An application of aquadag to the surface immediately rendered it much more uniformly emitting at an emissivity of about 0·95.

Infrared reflection methods are now being used in the on-line measurement of the constituents of paper and board. The product is irradiated from a modulated infrared source such as a filament lamp. Reflected radiation is then collected and is sampled sequentially in the various wavelength bands corresponding to the characteristic absorption bands of each of the constituents, and also in a neighbouring control band in which none of the constituents is strongly absorbing. This can be done by rotating a suitable set of filters in front of the detector. The reflected component in any absorption band will decrease as the percentage of the corresponding constituent increases, and so by comparison with the signal in the "control" band the constituents can be measured. Equipments have been reported which sample in four wavelength bands to evaluate three constituents. They are claimed to be able to measure the cross machine profile of a paper web at up to an order faster than competing methods of moisture measurement.

Special transmission measurements can be carried out with the aid of double beam spectrometers. These have many uses, including the analysis of small samples, where the nondestructive nature of the test is often particularly important. Such analysis can be carried out on all phases—solid, liquid and gas. Where the constituents of samples are known, and it is only necessary to know their concentration, spot wavelength methods can be used, making use of the characteristic absorption bands of the constituents being tested. This allows the analysis to be carried out very rapidly, and there are numerous on-line applications to which this technique can be applied. For example, on-line gas analysis is used by anaesthetists in setting up rapidly the correct gas concentrations.

In materials which are opaque in the visible but transmitting in the infrared, defects and inclusions may be shown up if they give rise to an anomalous absorption or reflection effects. Investigations have been made on materials which become transparent in the near infrared using infrared image converters. Various types of defect have been observed in silicon and scattering inclusions have been seen in some samples of arsenic selenide infrared glass by this method.

In conclusion it should be emphasized that many of the applications mentioned are still being investigated and relatively few of the techniques have become widely established. There are, however, indications of a rapidly growing interest in the industrial possibilities offered by infrared nondestructive testing.

REFERENCES

Astheimer, R. W. and Schwarz, F. (1968). Thermal imaging using pyroelectric detectors. *Appl. Opt.* **7** (9), 1687.

Barnes, R. B. (1968). Diagnostic thermography. *Appl. Opt.* **7** (9), 1673.

Beardsley, W. and Sabey, J. W. (1968). An Absolute Radiometer for Infrared Radiance Measurements. Ministry of Technology, Unpublished Report.

Borg, S.-B. (1968). Thermal imaging with real time picture presentation. *Appl. Opt.* **7** (9), 1697.

Coleby, P. R. D. (1968). Fire-Alarm System using Infrared Flame Detector. Mullard Technical Communications, Vol. 10, pp. 93, 82.

Di Ward, R. and Wormser, E. M. (1959). Description and properties of thermal detectors. *Proc. Inst. Radio Engrs* **47**, 1508.

Dreyfus, M. G. (1963). Spectral variation of black body radiation. *Appl. Opt.* **2**, 1113.

Fergason, J. L. (1968). Liquid crystals in non-destructive testing. *Appl. Opt.* **7** (9), 1729.

Goodenough, J. G. (1959). Cooling techniques for infrared detectors. *Proc. Inst. Radio Engrs* **47**, 1514.

Jervis, M. H. (1968). Closed Circuit Infrared Television System. Mullard Technical Communications, Vol. 10, pp. 93, 108.

Kelsall, D. (1963). An automatic emissivity compensated radiation pyrometer. *J. scient. Instrum.* **40**, 1.

Martin, G. A. (1959). Infrared photoemission. *Proc. Inst. Radio Engrs* **47**, 1467.

Nichols, L. W. and Lamar, J. (1968). Conversion of infrared images to visible in colour. *Appl. Opt.* **7** (9), 1757.

Petritz, R. L. (1959). Fundamentals of infrared detectors. *Proc. Inst. Radio Engrs* **47**, 1458.

Sabey, J. W. (1966). An absolute radiometer for target and background measurements. *Infrared Phys.* **8**, 117.

Schultz, A. W. (1968). An infrared transient method for determining thermal inertia, conductivity and diffusivity of solids. *Appl. Opt.* **7** (9), 1845.

Shapiro, P. (1969). Infrared detector chart. *Electronics* **42**, 21.

Sundstrom, E. (1968). Wide-angle infrared camera for industry and medicine. *Appl. Opt.* **7** (9), 1763.

Wolfe, W. L. and Ballard, S. S. (1959). Optical materials, films and filters for infrared instrumentation. *Proc. Inst. Radio Engrs* **47**, 1540.

Yoder, J. R. (1968). Temperature measurement with an infrared microscope. *Appl. Opt.* **7** (9), 1791.

*General**

Hackforth, H. L. (1960). "Infrared Radiation." McGraw-Hill, New York.

Holter, M. R., Nudelman, S., Suits, G. H., Wolfe, W. L. and Zissif, G. J. (1962). "Fundamentals of Infrared Technology." Macmillan, New York.

Kruse, P. W., McGlauchlin, L. D. and McQuistan, R. B. (1962). "Elements of Infrared Technology." Wiley, New York.

* The references in this section are not mentioned in the text.

Smith, R. A., Chasmar, R. P. and Jones, F. E. (1957). "Detection and Measurement of Infrared Radiation." Oxford University Press, Oxford.

Wolfe, W. L. (ed.) (1965). "Handbook of Military Infrared Technology." Office of Naval Research, Washington.

Author Index

Applications Index

This index is provided as a guide to the applications of the techniques described and to the materials and components on which they may be used. It is not an index to the techniques themselves, since the various chapters are self-explanatory in this respect.

A

Active flaws, 2
Adaptive control techniques, 381
Adhesive
  bonds (*see* Bond testing)
  material, 275, 300
Aerial surveying, 474
Aircraft structures, 2, 167
Aluminium, 50, 80, 83, 84, 129, 147, 148, 175, 243, 245, 260, 274, 304, 388, 390, 392, 476
Amorphous materials, 53
Anaesthetics, 477
Arteries, 301
Art objects, 307
Axle boxes, 474

B

Babbitt, 388
Bearings, 474
Billet testing, 150
Biological
  organisms, 79
  processes, 301
Bismuth, 306
Blood vessels, 85
Board, 477 (*see also* Plaster board)
Bond testing, 79, 81, 167, 274, 300, 472
Bone, 85, 86
Bore inspection, 169, 301
Boron, 298, 299, 303
Brake
  blocks, 474
  linings, 475
Brass, 124, 125, 126, 127, 148, 162, 375, 387
Bridges, 1

Brittle fracture (*see* Fracture)
Bronze, 388
Burning rates, 431

C

Cadmium, 271, 280, 287, 295, 303
  alloys, 305
  plating, 304
Capsules, 1
Cardiovascular flow, 85
Casting defects, 263
Castings, 306, 307
Cavitation, 15, 25, 263
Ceramic
  coatings, 433
  ware, 475
Chemical processes, 301
Circuit boards, 284
Cladding inspection, 79, 399, 433
Coarse-grain materials, 150
Combustion monitoring, 474
Composite materials, 303, 304 (*see also* Laminates)
Conductivity measurement, 87
Copper, 50, 324
Corrosion monitoring, 2, 178, 304, 305, 433
Crack
  detection, 262, 393, 396
  formation, 1, 175
  growth, 6, 26 (*see also* Fatigue crack growth and Flaw growth)
  movement, 1, 28
Creep, 167, 174, 179
Crops, 474
Crucibles, 475
Cure state, 429